AF786746

Biochemistry and Physiology of Visual Pigments

Symposium Held at Institut für Tierphysiologie,
Ruhr-Universität Bochum/W. Germany,
August 27-30, 1972

For the Organizing Committee Edited by

Helmut Langer

With 202 Figures

Springer-Verlag
Berlin Heidelberg New York 1973

The Symposium was sponsored by

Ministerium für Wissenschaft und Forschung des Landes Nordrhein-Westfalen
Stiftung Volkswagenwerk
Stifterverband für die Deutsche Wissenschaft

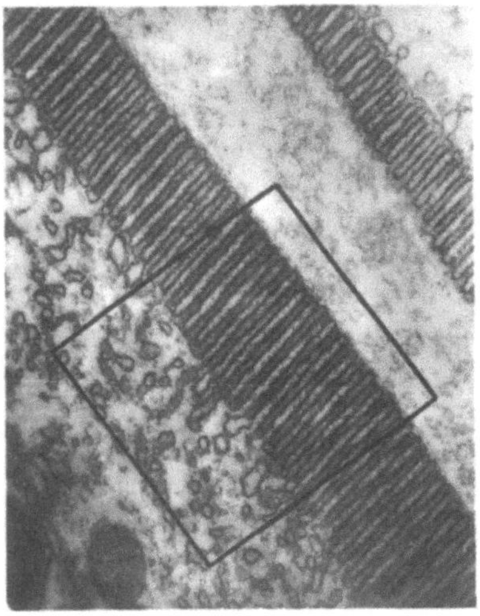

The cover shows a section of a rhabdomere and superimposed upon it the structural formula of 11-cis retinal, – the chromophore of the visual pigments of rhabdomeres as well as of rods and cones.
The background is an electron micrograph of a longitudinal section (parallel to the axes of the tubuli) through a peripheral rhabdomere in an ommatidium of the blowfly, *Calliphora erythrocephala* MEIGEN. To the right of the rhabdomere: interrhabdomeric extracellular space; to the left: intracellular subrhabdomeric vesicles. The outline in the figure on this page indicates the part of the micrograph used for the cover. Magnification on the cover 180,000 x. Authors: L. SCHNEIDER (Laboratory for Electron Microscopy, Zoologisches Institut I, Universität Würzburg, Würzburg/W. Germany) and H. LANGER (Tierphysiologisches Institut, Ruhr-Universität, Bochum/W. Germany).
The structural formula shows the conformation of 11-cis retinal1, as found in the crystalline state. The numbers represent distances between carbon atoms given in Å. Author: W. SPERLING (Institut für Neurobiologie, Kernforschungsanlage Jülich/W. Germany); see p. 21, Fig. 2.

This work is subject to copyright. All rights are reserved, whether the whole or part of the material is concerned, specifically those of translation, reprinting, re-use of illustrations, broadcasting, reproduction by photocoping machine or similar means, and storage in data banks.

Under § 54 of the German Copyright Law, where copies are made for other than private use, a fee is payable to the publisher, the amount of the fee to be determined by agreement with the publisher.

The use of registered names, trademarks, etc. in this publication does not imply, even in the absence of a specific statement, that such names are exempt from the relevant protective laws and regulations and therefore free for general use.

© by Springer-Verlag Berlin · Heidelberg 1973. Library of Congress Catalog Card Number 73-76922.
Softcovver reprint of the hardcover 1st edition 1973
ISBN 978-3-642-85771-3 ISBN 978-3-642-85769-0 (eBook)
DOI 10.1007/978-3-642-85769-0

Preface

This book is a report of a four-day Symposium on the Biochemistry and Physiology of Visual Pigments, which took place immediately after the VIth International Congress on Photobiology, held in Bochum, Federal Republic of Germany, in August 1972.

This meeting, which brought together about 50 investigators of various aspects of the visual process, was devoted to the visual cells of both vertebrates and invertebrates. Whereas the International Symposium on the Biochemistry of the Retina, held at Nijmegen, The Netherlands, in 1968, had concentrated on vertebrate photoreceptors, this Symposium dealt with invertebrate photoreceptors as well, so that workers in each field could become acquainted with recent progress in the other area.

The papers presented at the Symposium were divided into six main topics, to each of which a half-day session was devoted. The six parts of this book, following the introductory lecture, essentially correspond to these sessions. In addition to the invited contributions, the volume contains a number of short communications by other participants and two contributions by invited participants, who were unable to attend. The volume closes, as did the Symposium, with a General Discussion, prepared and moderated by S. L. Bonting, in which an attempt was made to integrate various new findings, and to reconcile certain points of disagreement.

While the usefulness of this type of meeting need hardly be stressed to the investigator, it is gratifying to note that governmental and private organizations for the advancement of science in the Federal Republic of Germany are also aware of this. The Symposium was made possible by their financial support, which is greatly appreciated.

The organizers are grateful to the staff of the Institut für Tierphysiologie, Ruhr-Universität Bochum, for technical assistance prior to and during the meeting. Special thanks are due to the secretary of the Symposium, Mrs. Elke Schürmann, and to Miss Claudia Retzlaff, who prepared the manuscript for publication. Finally, Springer-Verlag, Heidelberg, made every effort to publish the volume within the shortest possible time.

December 1972

Ch. Baumann S. L. Bonting
K. Hamdorf H. Langer

Contents

IV. Excitation and Adaptation of Photoreceptor Cells

V. Ionic Aspects of Excitation and Regeneration

VI. Enzymology and Molecular Architecture of the Light Sensitive Membrane

List of Participants

ABRAHAMSON, E. W., Case Western Reserve University, Department of Chemistry, Cleveland, OH 44106/USA

AKHTAR, M., University of Southampton, Department of Physiology and Biochemistry, Southampton 509 5NH/Great Britain

ARDEN, G., University of London, Institute of Ophthalmology, Judd Street, London WC1H 9QS/Great Britain

BAUMANN, CH., William G. Kerckhoff-Herzforschungs-Institut der Max-Planck-Gesellschaft, 635 Bad Nauheim/W. Germany

BITENSKY, M. W., Yale University, Department of Pathology, 310 Cedar Street, New Haven, CT 06510/USA

BONTING, S. L., University of Nijmegen, Department of Biochemistry, Kapittelweg 40, Nijmegen/The Netherlands

BOWNDS, D., University of Wisconsin, Laboratory of Molecular Biology, Madison, WI 53706/USA

BRIDGES, C. D. B., University of New York, Department of Ophthalmology, 550 First Avenue, New York, NY 10016/USA

COHEN, A. I., Washington University, Department of Ophthalmology, St. Louis, MO 63110/USA

DONNER, K. O., University of Helsinki, Department of Physiological Zoology, Arkadiankatu 7, Helsinki/Finland

DUNCAN, G., University of East Anglia, School of Biological Sciences, University Plain, Norwich NOR 88c/Great Britain

EBREY, T. G., Columbia University, Department of Biological Sciences, New York, NY 10027/USA

ERNST, W., William G. Kerckhoff-Herzforschungs-Institut der Max-Planck-Gesellschaft, 635 Bad Nauheim/W. Germany

FUTTERMAN, S., University of Washington, Department of Ophthalmology, Seattle, WA 98195/USA

GOLDSMITH, T. H., Yale University, Department of Biology, New Haven, CT 06520/USA

DE GRIP, W., University of Nijmegen, Department of Biochemistry, Kapittelweg 40, Nijmegen/The Netherlands

HALL, M. O., University of California, Jules Stein Eye Institute, Los Angeles, CA 90024/USA

HAMDORF, K., Ruhr-Universität Bochum, Institut für Tierphysiologie, 463 Bochum-Querenburg/W. Germany

HARA, R., Nara Medical University, Department of Biology, Kashihara, Nara/Japan

HARA, T., Nara Medical University, Department of Biology, Kashihara, Nara/Japan

HARRISON, E., University of Pennsylvania, Department of Biophysics and Physical Biochemistry, Philadelphia, PA 19104/USA

HELLER, J., University of California, Jules Stein Eye Institute, Los Angeles, CA 90024/USA

HENDRIKS, TH., University of Nijmegen, Department of Biochemistry, Kapittelweg 40, Nijmegen/The Netherlands

HÖGLUND, G., Karolinska Institutet, Fysiologiska Institutionen II, Solnavägen 1, 10401 Stockholm/Sweden

HORWITZ, J., University of California, Jules Stein Eye Institute, Los Angeles, CA 90024/USA

JAGGER, W., William G. Kerckhoff-Herzforschungs-Institut der Max-Planck-Gesellschaft, 635 Bad Nauheim/W. Germany

KARPLUS, M., Harvard University, Department of Chemistry, Cambridge, MA 02138/USA

KEMP, C. M., University of London, Institute of Ophthalmology, Judd Street, London WC1H 9QS/Great Britain

KLEINSCHMIDT, J., Harvard University, Biological Laboratories, Cambridge, MA 02138/USA

KNOWLES, A., University of Sussex. MRC Vision Unit, Falmer, Brighton BN 19QY/Great Britain

LANGER, H., Ruhr-Universität Bochum, Institut für Tierphysiologie, 463 Bochum-Querenburg/W. Germany

LIEBMAN, P. A., University of Pennsylvania, Department of Anatomy, Philadelphia, PA 19104/USA

MASON, W. T., Case Western Reserve University, Department of Chemistry, Cleveland, OH 44106/USA

PAK, W. L., Purdue University, Department of Biological Sciences, Lafayette, IN 47907/USA

PAULSEN, R., Ruhr-Universität Bochum, Institut für Tierphysiologie, 463 Bochum-Querenburg/W. Germany

REUTER, T., University of Helsinki, Department of Physiological Zoology, Arkadiankatu 7, Helsinki/Finland

ROSENKRANZ, J., Ruhr-Universität Bochum, Institut für Zellmorphologie, 463 Bochum-Querenburg/W. Germany

ROTMANS, J. P., University of Nijmegen, Department of Biochemistry, Kapittelweg 40, Nijmegen/The Netherlands

RÜPPEL, H., Technische Universität Berlin, Max-Volmer-Institut für Physikalische Chemie, 1 Berlin, Straße des 17. Juni 135/W. Germany

SCHENCK, G. O., Max-Planck-Institut für Kohlenforschung, Abteilung Strahlenchemie, 433 Mülheim/Ruhr, Stiftstraße 34-36/W. Germany

SCHWEMER, J., Ruhr-Universität Bochum, Institut für Tierphysiologie, 463 Bochum-Querenburg/W. Germany

SICKEL, W., Universität Köln, Institut für normale und pathologische Physiologie, 5 Köln, Robert-Koch-Str. 39/W. Germany

SPERLING, W., Kernforschungsanlage Jülich, Institut für Neurobiologie, 517 Jülich/W. Germany

Stavenga, D. G., Rijks-Universiteit, Natuurkundig Laboratorium, Westersingel 34, Groningen/Netherlands
Stieve, H., Kernforschungsanlage Jülich, Institut für Neurobiologie, 517 Jülich/ W. Germany
Täuber, U., Ruhr-Universität Bochum, Institut für Tierphysiologie, 463 Bochum-Querenburg/W. Germany
Wald, G., Harvard University, Biological Laboratories, Cambridge, MA 02138/USA
Weber, K. M., Ruhr-Universität Bochum, Institut für Tierphysiologie, 463 Bochum-Querenburg/W. Germany
Williams, T. P., Florida State University, Institute of Molecular Biophysics, Tallahassee, FL 32306/USA
Yoshikami, S., National Institute of Health, Laboratory of Physical Biology, Bethesda, MD 20014/USA
Yoshizawa, D., Kyoto University, Department of Biophysics, Faculty of Science, Kyoto/Japan
Zinkler, D., Ruhr-Universität Bochum, Institut für Tierphysiologie, 463 Bochum-Querenburg/W. Germany

Introductory Lecture

Visual Pigments and Photoreceptor Physiology

George Wald
Biological Laboratories, Harvard University, Cambridge, MA 02138/USA

We who work with the visual pigments are fortunate in our field. It offers unique opportunities to engage, and we hope perhaps to solve, some of the central problems of present-day biology. It projects also into some of the most interesting problems of present-day physical chemistry.

A central problem of modern biology is that of excitation. We do not yet understand the mechanism of any biological excitation, whether of receptors, nerve, muscle, or eggs - all of them loaded structures, that deliver a large response when triggered by a relatively miniscule stimulus. In vision, the stimulus of a dark-adapted rod has reached its ultimate physical limit - one photon. Presumably one photon excites a single visual pigment molecule; and our problem is how changes in a single molecule can trigger the enormously amplified response of the receptor cell.

Fortunately that molecule in this instance is part of a special cellular organelle, in vertebrate retinas the outer segment of a rod or cone. I think that an outer segment represents the highest degree of isolation of function to be found in the whole of physiology. Its only business is excitation. It is not distracted with vegetative functions as are whole cells; or even with conduction of excitation, as are nerve and muscle cells. The outer segments not only segregate excitation anatomically; fortunately for us they also break off readily from the rest of the retina, so that we can deal with them alone. It is hard to over-estimate the tactical advantage that this entails. It offers us reason to hope that this may be the first place for working out the mechanism of a biological excitation.

There is another regard in which I feel a little like that character in Molière's play who learned with delight that all his life he had been speaking prose. In the visual pigments we are working with membrane proteins, and membranes and their structure are of course now of central biological concern. The visual pigments are the first such proteins to be brought out of their membranes into aqueous solution. Willibald Kuehne, Professor of Physiology in Heidelberg, first showed how to do that in 1878, using the natural anionic detergent bile salts - a mixture of sodium glycocholate and taurocholate. A half century later Katherine Tansley (30) introduced digitonin to do this job, and this still remains the detergent of choice.

I think that our field offers the opportunity to explore another biological problem even more fundamental than those already mentioned: that of differentiation. Even those of us who confine our experiments to the visual pigments themselves should have this in our thoughts always.

The point is that in the retina we find an extraordinary degree of highly ordered differentiation laid out in a two-dimensional mosaic, in which each cell can report on its own. Problems of differentiation seem more approachable here than elsewhere in the organism. For example, how are cones segregated to the exclusion of rods in the primate fovea? How is it that in primates only the fovea and its immediately surrounding area deposit the carotenoid xanthophyll, $C_{40}H_{54}(OH)_2$, to form the distinctive yellow patch or macula lutea (32)? How is it that though all the cells contain the same genes, each primate cone appears to synthesize and deliver to its outer segment only one of the three opsins that with 11-cis retinal form the three color vision pigments (9)? And the rods still a fourth opsin that with 11-cis retinal makes rhodopsin?

Another instance of such differentiation offers a hint as to mechanism. The cones of the chicken retina contain brilliantly colored oil globules, one in each cone, placed at the juncture between the inner and outer segment where it can serve as an individual color filter. The globules are of three colors, light yellow, orange, and red. Some years ago Zussman and I (38) isolated from them three carotenoid pigments: a light yellow carotene, a golden xanthophyll, and the red astaxanthin, respectively a hydrocarbon, an alcohol and an acid. Those differences in their chemical nature made it very easy for us to separate the pigments by differential partition among various solvents, and by differential adsorption. It is tempting to think that the chicken takes advantage of the same properties to segregate these pigments among its cones.

The microspectrophotometry of single photoreceptors, particularly as it has come lately out of Paul Liebman's laboratory, is making great strides in defining this type of problem. Liebman (26) has recently brought most such measurements together in a fine and critical review.

With so much lying ahead in this field, it is good to have behind us a considerable distraction. The molecular weight of cattle rhodopsin is back again at about 40,000 grams per mole; the molar absorbance in digitonin at a little over 40,000; and the retinal chromophore is bound in rhodopsin to an ε-N of lysine. Bringing this information back to where it started took a lot of painstaking work in a number of laboratories. It has by no means been waste motion. We are much surer of this information now than we were; and in the course of this work have learned new and more sophisticated ways of handling rhodopsin than were available before - new solvents, adsorbents, procedures for purification, delipidation, and dephosphorylation. Opsin has been called at times a glycoprotein and sometimes a lipoprotein. I would rather think of it simply as a protein with interesting phospholipid associations. One can take away the phospholipid; and now ways are being found to put it back (45,19). Phospholipids even by themselves spin membranes remarkably like those found in cells (28); and putting them together with rhodopsin offers promising beginnings of a synthetic approach to outer segment and rhabdomere membrane structure.

Visual Excitation

The entire point of photosynthesis is to use light to do chemical work: to convert the light absorbed by chloroplast pigments as efficiently as possible into the internal energy of cellular metabolites. In vision, light plays an entirely different role; not to do work, but to excite - through the excitation of a visual pigment to trigger a nervous excitation. This is a fundamental distinction.

It emerges first in the little light required for vision. The absolute threshold for dark-adapted human rods lies at the ultimate physical limit: one photon. One photon is absorbed by one molecule of rhodopsin; and one must ask how so small an event can excite so large an effect. Clearly an enormous amplification is needed, of the order of at least one million.

Some years ago I suggested two mechanisms for achieving such amplification (33, 34,35). What I thought to be the less probable but the more interesting mechanism conceived of rhodopsin as a pro-enzyme, activated by light. The active enzyme could turn over a large number of substrate molecules, so constituting one stage of amplification. I pointed out that if the substrate were itself a proenzyme, activated by the first enzyme, that would make a second stage of amplification. One could add a third and fourth such proenzyme-enzyme couple, each a stage of amplification. In that way the outer segment of a rod or cone would become a photomultiplier.

Years later I was excited to learn that the system that governs human blood clotting is just such an enzyme cascade (15,27). I promptly got in touch with both Macfarlane and Ratnoff, asking whether, just as one activated molecule of rhodopsin may excite a rod, one molecule of Hagemann factor, the first proenzyme in the blood clotting mechanism, might activate a blood clot. I also laid out the very simple procedure, based on the same Poisson analysis that had proved so useful in photography and vision, that should provide the answer here (36). I have heard nothing further since. I still think that this would be an illuminating experiment in blood clotting; and that a cascade of proenzyme-enzyme couples remains an interesting possibility for achieving amplification in visual receptors.

The second suggestion was that the attack of light on a visual pigment might open a unimolecular hole in a photoreceptor membrane, through which a great many ions might diffuse. Opening such a channel in the plasma membrane of a rhabdomere might account for excitation in invertebrates, in which light depolarizes the photoreceptor membrane. But in vertebrate rods and cones, light hyperpolarizes the plasma membrane. The problem therefore is not to open holes, but to close them. It has lately been shown that a continuous current runs externally from the inner to the outer segment, carried by sodium ions. Light effectively shuts down the flow of sodium ions into the outer segment, so hyperpolarizing the cell.

Korenbrot and Cone (24) have lately studied this phenomenon in isolated rod outer segments of frogs and rats, using an osmotic shock procedure. They found that light increases the resistance to Na^+ influx without affecting other ions; and does so linearly with the number of rhodopsin molecules bleached. The excitation of one rhodopsin molecule reduces the Na^+ influx by about 1 per cent for about 1 sec, so preventing the entry of about 10^7 sodium ions. We have here a remarkably direct expression, all enclosed within the outer segment, of the way in which the absorption of one photon, by one rhodopsin molecule can trigger an enormously amplified effect. This may still be by opening a channel in a membrane. Hagins and his coworkers have found the effects of light on the sodium current to be mimicked by externally applied calcium ions (17). In rod outer segments, the rhodopsin is mainly in the disc membranes, some perhaps also in the plasma membrane (5). In cone outer segments, in which the transverse membranes are infoldings of the plasma membrane, the cone visual pigments are part of the continuous membrane that separates the intracellular from the extracellular space. Yoshikami and Hagins (41) and Hagins (17) suggest that the intradiscal spaces in rods and the extracellular medium in both rods and cones contain higher concentrations of Ca^{++} than the intracellular spaces. The action of light on the visual pigment increases the Ca^{++} permeability of the membranes, allowing Ca^{++} ions to flow out of the rod discs and in through the plasma membrane, so increasing the concentration of Ca^{++} in the intracellular space, with the net effect of closing down sodium channels through the plasma membrane. Presumably many Ca^{++} ions enter the intracellular space per visual pigment molecule that absorbs a photon; and each such Ca^{++} in turn shuts off the flow of many sodium ions, so accounting for the high gain.

Energetics

As already said, vision, unlike photosynthesis, is not concerned with getting light to do work. All that is wanted of light in vision is excitation, ultimately only a molecular excitation permitting molecular changes that lead to a nervous excitation. Said differently, in vision light promotes the kinetics of the system without necessarily contributing to its thermodynamics. Since almost all the transformations of visual pigments consequent upon the absorption of light are thermally irreversible, as is their regeneration, what we can measure readily are only kinetic parameters - the energies and entropies of activation. They tell us how the molecules get ready to react, not the energy changes of the completed reactions. The latter would constitute the thermodynamics of the system; and of that we know as yet almost nothing.

In itself that is nothing to complain about. It represents another interesting problem to tackle and eventually solve. The present trouble is that lacking the thermodynamics, there is a strong tendency to treat the kinetics as though it were thermodynamics. When one is told that a given process has a high positive entropy of activation, for example, it is with the strong suggestion that the process itself results in a large increase in entropy. Actually of course it tells us nothing of the final result. A protein for example could in its excited state have opened up configurationally - i.e. gained in entropy - precisely as a precondition for rearranging into a much more highly organized structure - i.e. with a large decrease in entropy. The activation parameters tell us only the conditions for surmounting potential energy barriers to reaction, nothing about the beginning and final energy states.

What we can guess of the real thermodynamics of visual pigment transformations at present looks peculiar. If the action of light in visual excitation is limited, as now seems likely, to activating the isomerization of the 11-cis retinal chromophore of a visual pigment to all-trans, that in itself is energetically a downhill reaction, for the all-trans configuration has the lowest ground state of all the retinal isomers. The first product of this reaction, a prelumi- or batho-pigment, to be discussed shortly, is highly unstable, and goes spontaneously in the dark through a series of further transformations, the intermediates of bleaching. All these highly spontaneous reactions, from the prelumi-pigment to free all-trans retinal and opsin, involve - by the definition of spontaneous reactions - decreases in free energy. Not only has the retinal gone into its most probable and stable configuration - all-trans - but the opsin has opened up its structure (c.f. 23,29). One has no thermodynamic data for these essentially irreversible processes, but certainly the whole appearance is of continuously decreasing free energy and increasing entropy.

Then what? To go back to the visual pigment, one has first to re-isomerize the all-trans retinal to 11-cis. Hubbard (22) has measured this reaction in n-heptane, the one reaction in which, since one can measure equilibria, we have the thermodynamic parameters. For the isomerization from 11-cis to all-trans retinal, ΔF is -1.1 to -1.4 kcal per mole, ΔH is negligible (she estimates it at +150 cal per mole), and ΔS is also very small, + 4.4 e.u. (cal/degree/mole). So in isomerizing all-trans retinal to 11-cis one has at last an uphill reaction, yet very small indeed, with an input of free energy of a little over 1 kcal per mole.

Where that energy comes from when this reaction occurs thermally as it surely does, we do not know. But once it has occured, the 11-cis retinal combines in the dark with opsin to regenerate a visual pigment as a highly spontaneous - i.e. free energy-yielding, exergonic - reaction, capable of doing work. In the vertebrate visual cycle, the regeneration of visual pigment does a very important job of work -

it pulls the equilibrium between retinol and retinal in the oxidative direction by trapping retinal, a rare example of a physiological trapping reaction (34).

So the energetics of vision seems strange. Light, by producing a high-energy product, a prelumi- or batho-pigment, activates a series of downhill reactions that end with a net loss of free energy and gain of entropy. Just the contrary of storing the light energy, as in photosynthesis, apparently visual excitation degrades energy. Then there occurs a single uphill reaction - the isomerization of all-trans retinal to 11-cis - with its very small energy input; and then a further downhill process, albeit with an apparent decrease in entropy, in the regeneration of visual pigment. Is this the true picture? It suggests that the processes that trigger visual excitation are remarkably economical - that in this regard it costs organisms virtually nothing to go on seeing.

The Colors of Visual Pigments: Chromophore-opsin Interactions

We still are trying to understand the colors of visual pigments. The essential point is how, by combining the chromophore, 11-cis retinal, with a visual protein, an opsin, the brilliant colors of many of the visual pigments are achieved. 11-cis retinal itself, with λ_{max} about 380 nm, is very light yellow. Anchoring it to opsin by a Schiff base linkage with the ε-N of a lysine residue, in itself makes the situation worse, since such Schiff bases of retinal have λ_{max} about 365 nm. Protonating the Schiff base brings the λ_{max} up to about 440 nm. That leaves the further shift of spectrum to explain, to about 500 nm in most rhodopsins, or 562 nm in an iodopsin.

To explain these further shifts of spectrum, Hubbard (21) and Kropf and Hubbard (25) introduced the idea of side-chain interactions between the retinal and contiguous portions of the opsin. It was thought that such interactions might demand closeness of fit, so explaining the necessity for a specific geometry of retinal, the 11-cis isomer. So also, the 9-cis isomer, which most resembles 11-cis in geometry, sufficiently satisfies the requirement of closeness of fit as to form the parallel series of iso-pigments. It was pointed out also that the presence of a negative group on opsin some distance up along the side-chain might help to pull the positive charge on the protonated Schiff base up into the conjugated system of the chromophore, promoting its resonance form, and so exalting the spectrum (21,25). Dartnall and Lythgoe (14) have introduced further considerations in the nature of these interactions; and Abrahamson and Wiesenfeld (1) have provided a provocative quantum-mechanical discussion.

The Batho-pigment Problem

These views raise knotty problems in accounting for the spectra of the pre-lumi pigments, which Yoshizawa has lately suggested be given the much preferable name, batho-pigments (42). We now know four batho-pigments - the bathorhodopsins of cattle and squid, chicken bathoiodopsin, and carp bathoporphyropsin. All of them are more intense pigments than the visual pigments from which they are derived - both in that their spectra are displaced very considerably toward the red, and their specific absorbances considerably increased.

If the colors of the visual pigments depend upon closeness of fit with opsin, going with the specific geometry of the chromophore, 11-cis retinal, and if the effect of light is to isomerize the chromophore to all-trans, so producing a batho-pigment, how can that change of geometry, which would be expected to destroy the closeness of fit, yield a heightened pigmentation?

To deal with this problem, Yoshizawa and I (43) introduced a further consideration: that the interaction between the side-chain of the chromophore and the opsin involves a degree of attachment; the retinal and opsin hold on to each other and so mutually constrain each other geometrically. As a result, when a visual pigment absorbs a photon, though the chromophore is isomerized to all-trans, its interaction with opsin does not permit it to assume the relaxed all-trans geometry. We conceive of the batho-pigments as in a high state of strain, and that this accounts for their enormous instability. They can be held only in rigid solvents at very low temperatures, $-150^{\circ}C$ in the case of bathorhodopsin, about $-170^{\circ}C$ in the case of bathoiodopsin. We think that they represent virtually excited states, of high potential energy. Hence relatively little additional energy is needed - a relatively small photon - to raise a batho-pigment to its first electronically excited state, so accounting for the large bathochromic shifts.

This way of thinking of the batho-pigments proved useful when we encountered a further problem. Bathorhodopsin, warmed to $-150^{\circ}C$, goes spontaneously in the dark to lumirhodopsin, the next step in the bleaching sequence; but bathoiodopsin, on warming to $-170^{\circ}C$ goes mainly back to iodopsin (44). To explain this, we suggested that when a batho-pigment in its highly strained and highly unstable state, is allowed to react, as in this case by warming, something has to give. In the case of rhodopsin, it is the opsin that gives, changing its configuration to yield lumirhodopsin. In the case of bathoiodopsin, it is the chromophore that gives, going back to 11-cis.

Recently a further consideration has come into this area. Honig and Karplus (20) on the basis of quantum-mechanical considerations, and Gilardi et al. (16) on the basis of crystallography, have pointed out that the most stable geometry of 11-cis retinal at ordinary temperatures is 11-cis, 12-s-cis. As one can readily see on a space-filling model of this molecule, rotating the chain at the 12-13 single bond into a cis configuration considerably relieves the overcrowding - the steric hindrance - brought about by the 11-cis linkage. There is evidence that such an s-cis linkage should shift the spectrum considerably toward the red. The presence of the 12-s-cis linkage in free 11-cis retinal would help greatly to explain why its spectrum is so little degraded by the sterically hindered 11-cis linkage. If - as is by no means sure - the retinal chromophores of the visual pigments are not only 11-cis but 12-s-cis, that would help to explain why their spectra lie at such long wavelengths, so putting less burden upon the hypothesis of side-chain interactions with opsin. Furthermore, Honig and Karplus suggest that the batho-pigments represent isomerizations from 11-cis to 11-trans, while still retaining the 12-s-cis linkage. That would account for a large further bathochromic shift. In the visual pigments the bathochromic effect of the s-cis linkage is pitted against the hypsochromic effect of the 11-cis linkage. Relieved of the latter, the 12-s-cis linkage could display its full bathochromic effect. These authors suggest that the next step is to all-trans retinal, the condition in lumirhodopsin.

This very interesting and provocative development now needs careful appraisal. To what degree does it take the place of, to what degree add to, previous ways of regarding chromophore-opsin relationships? They are by no means mutually exclusive. The reasons why 11-cis retinal should assume the 12-s-cis configuration are not clearly applicable to the chromophores of the visual pigments. Where the free molecule tends to avoid steric hindrance, the chromophores of the visual pigments seem to seek it out. If the visual pigments were anxious to avoid steric hindrance, why do they not use 9-cis retinal as chromophore? In choosing the greatly hindered 11-cis linkage, they picked an intrinsically unstable chromophore, putting the molecule so to speak on hair trigger. What I am saying is that though free retinal seeks a geometry that represents minimum potential energy, a visual pigment on the contrary may be seeking something else, e.g. the highest possible quantum efficiency for isomerization. This may be what led visual pigments to the 11-cis linkage in the first place.

I think further that we still have important uses for the concepts of side-chain interaction between the retinal chromophore and opsin, and the mutual geometrical constraints occasioned by it, including the highly strained condition of the batho-pigments. So, interested as I am in the 12-s-cis linkage, I think it may add to, rather than supersede the previous considerations.

Hypsorhodopsin

That brings us to a new problem, that of hypsorhodopsin (Yoshizawa and Horiuchi, cited in 42). These authors have found that when cattle rhodopsin is irradiated at liquid helium temperature (-269°C or 4°K) with wavelengths longer than 540 nm, a new intermediate, hypsorhodopsin, is formed, with λ_{max} about 430 nm and a specific absorbance about 0.9 that of rhodopsin. Hypsorhodopsin can be converted with short wavelength light to bathorhodopsin. What is much more interesting, however, is that on warming slightly, to about -250°C, this enormously unstable pigment goes in the dark to bathorhodopsin.

That suggested that hypsorhodopsin might be the first photoproduct; but there are difficulties in so regarding it. As Yoshizawa (42) says, "All that can be said with certainty is that at the temperature of liquid helium any of the chromo-proteins - rhodopsin, isorhodopsin, hypsorhodopsin, and bathorhodopsin - can be converted into any other either directly or indirectly, by suitable irradiation".

What then is hypsorhodopsin? I am reluctant to intrude into this extraordinary area that has just been opened by Yoshizawa's experiments; but in the nature of a dialogue with him should like to offer the following thoughts.

It seems to me highly significant that when rhodopsin is irradiated at liquid helium temperature with light at 579 nm, which is well absorbed by rhodopsin but hardly absorbed by isorhodopsin, that yields isorhodopsin with very little hypsorhodopsin. On the other hand, when rhodopsin is irradiated at 546 nm, a wavelength absorbed strongly also by isorhodopsin, much hypsorhodopsin is formed (42). To me, this suggests that hypsorhodopsin is not formed directly from rhodopsin, and that isorhodopsin may be its immediate precursor. That is, I would suggest that to go from rhodopsin to hypsorhodopsin takes at least 3 photons; whereas perhaps one photon may convert isorhodopsin to hypsorhodopsin.

It seems to me significant furthermore that the spectrum of hypsorhodopsin lies close to that of a simple protonated Schiff base of retinal, calling for neither special effects owing to side-chain interaction with opsin, or exaltation owing to an s-cis linkage. It seems to me possible that hypsorhodopsin represents a highly unstable form in which the retinal chain has been twisted out of all interacting contact with opsin. This state is perhaps more readily achieved from isorhodopsin than from rhodopsin itself; and is so highly strained that, given the opportunity, i.e. slight warming, it relaxes in the dark to bathorhodopsin.

A little earlier, however, I invoked the concept of molecular strain to explain the long-wavelength spectra of the batho-pigments. Now I am associating still higher strain and consequent instability with a large hypsochromic shift. Actually, hypsorhodopsin lies somewhat to the short wavelength side of an ordinary protonated Schiff base of retinal - at about 430 nm rather than 440 nm. Perhaps one should assume that hypsorhodopsin represents an unprotonated Schiff base - which intrinsically might lie at about 365 nm - in which strain has shifted the spectrum up to 430 nm. That suggests the speculation that the subtraction or addition of a proton may be the fundamental difference between hypso- and bathorhodopsin.

All of this obviously is little more than rhetoric at present. I feel confident that Yoshizawa's experimentation will soon tell us something more.

The First Photoproduct

A new investigation definitely identifies bathorhodopsin as the first product of the action of light on rhodopsin. Rentzepis and co-workers at the Bell Telephone Laboratories (10) have worked out an extraordinary procedure of picosecond photometry, and have now applied it to the bleaching of cattle rhodopsin. The method is simple in principle, though surprising at first contact. One ordinarily thinks of light as traveling so rapidly - $3 \cdot 10^{10}$ cm per sec - that it is hard to imagine that the time taken to travel 1 cm could have any significance. Actually of course it takes light $33 \cdot 10^{-12}$ sec - 33 picosec - to travel 1 cm. In essence, Rentzepis exposes a rhodopsin solution to a single laser pulse of average width 6 picosec and 530 nm, and simultaneously to interrogating pulses at 561 nm, extracted from the Raman spectrum of the 530 nm pulse. The interrogating pulses are reflected from an echelon mirror in which the successive spaces are 6 mm apart; hence the reflected interrogating pulses pass through the rhodopsin solution and are recorded on a photographic plate at 20 picosec intervals. In later experiments, a finer echelon mirror was used that reflected these pulses at 2 picosec intervals.

It turns out that when a rhodopsin solution was irradiated in this way, a new species absorbing strongly at 561 nm, apparently bathorhodopsin, was recorded within the pulse width of the exciting flash, 6 picosec. The bathorhodopsin decayed with a lifetime at room temperature of about 30 nanosec. The measured decay rates were: $2.7 \pm 0.2 \cdot 10^7$ per sec at 17.5°; $3.7 \pm 0.2 \cdot 10^7$ at 22.5°; and $4.1 \pm 0.5 \cdot 10^7$ per sec at 29.3°C, not very different from those previously measured in retinas by Cone (12).

We already had an apparently latency-free manifestation of the absorption of light by rhodopsin in retinas in the first wave (R_1) of the early receptor potential (11). That however involved a time resolution of about 0.5 microsec. These new experiments show with a time resolution almost six orders of magnitude greater that bathorhodopsin is the first product of the irradiation of rhodopsin in solution.

As Rentzepis and co-workers indicate, the time for this response is embarrassingly short, even for the rotation that should accompany the isomerization from 11-cis to 11-trans. It seems to me that here again the concept of mutual geometrical constraint of chromophore and opsin is helpful. According to this concept, rhodopsin having absorbed a photon, goes only into the highly strained state of bathorhodopsin in which isomerization around the 11-12 bond has not yet been completed. To induce that state of torsion might take less time than to complete a rotation.

Orientations and Motions

In spite of their great disparity in mass and volume, the retinal chromophore occupies an astonishingly long space on the rhodopsin molecule. The latter if spherical is about 40 Å in diameter; while 11-cis retinal is about 15 Å long. Anchored to opsin by its Schiff base linkage to the ε-N of a lysine residue, it lies in a bed that seems to consist almost exclusively of the lipophilic side-chains of neutral amino acids (7). By recent estimates the rod disc membranes are composed some 35 per cent of mixed phospholipids, and 80-90 per cent of their protein is rhodopsin.

Rhodopsin has a quite ordinary distribution of amino acids in its structure, with enough acidic and basic residues to make it potentially soluble. That it is insoluble in aqueous solutions, an almost necessary condition for a membrane protein, is probably owing to the fact that many of the polar, hydrophilic residues in its structure are folded inside, and not accessible to water. A first indication of this was Albrecht's failure to find either a C-terminal or an N-terminal amino acid in cattle rhodopsin (2). If the whole molecule does not present an essentially lipophilic surface, that surely seems to be true of the portion that holds the chromophore. We can expect this to lie in close association with the fatty acid chains of the phospholipid.

Many years ago, W.J. Schmidt showed that in frog rod outer segments the chromophores are mainly oriented in the planes of the transverse membranes, whereas the fatty acid chains of the phospholipids lie perpendicular to the membranes, i.e. parallel to the rod axis. The transverse orientation of the rhodopsin chromophores has since been measured spectrophotometrically by Denton, Liebman, and in our laboratory. In frog rod outer segments we found the ratio of rhodopsin absorbances in light polarized perpendicular to as compared with parallel to the rod axis (the dichroic ratio) to be as high as 4.5. That argues that the rhodopsin chromophores lie almost wholly in the planes of the disc membranes. We found evidence also that the vitamin A released by bleaching rhodopsin in the rods keeps the same orientation. On the other hand, when rhodopsin is bleached in the presence of hydroxylamine, the retinal oxime formed takes up a more highly axial orientation: the dichroic ratio reverses so as to be about 2:1 in the axial plane. That is, the retinal oxime lies partly perpendicularly to the disc membranes, whether each molecule at an angle, or as is more likely some molecules in the plane of the membranes and others of them perpendicular to it, as are the fatty acid chains of the phospholipids (37).*

The high degree of orientation of rhodopsin in the rod disc membranes emerges strikingly in another connection. Cone and Brown (13), Cone measuring the early receptor potentials (ERP) and Brown doing the microspectrophotometry on parallel pairs of eyes from the same rats, found that when such pairs of eyes were heated for 10 minutes at temperatures up to 48 °C, neither the ERP nor the orientation of the rhodopsin was affected. Between 48 °C and 58 °C, however, the ERP dropped to zero; and in exact parallel with its decay, the rhodopsin, though not bleached by the heat treatment, lost its orientation, becoming as randomly oriented as in solution.

We have lately made the strange observation that rhodopsin, cast in a dry gelatine film, exhibits behaviour strikingly like that in the disc membranes (39). The rhodopsin chromophores are oriented primarily in the plane of the film. In edge-on measurements with plane-polarized light we found the dichroic ratio, parallel/perpendicular, to be 2 to 3. The retinal oxime resulting from bleaching in the presence of hydroxylamine is oriented more perpendicularly to the plane of the film.

*It should be realized that the axial component of orientation is twice as effective in absorbing light as transverse orientation; for of the latter only the crosswise component absorbs, not the end-on component.

What is more remarkable, shearing a wet paste of rhodopsin-digitonin micelles between glass slides orients the rhodopsin chromophores in the direction of shear. On bleaching in the presence of hydroxylamine, the retinal oxime swings around so as to lie more perpendicularly to the direction of shear.

These observations, particularly the shear orientation, imply a large asymmetry in the rhodopsin-digitonin micelles, in which the chromophore lies parallel with the long axis of the micelle. In the absence of other indications that rhodopsin is highly asymmetric, we have suggested the possibility that under the circumstances of these experiments rhodopsin may form linear polymers, as do many other proteins in special circumstances.

As for the possible asymmetry of rhodopsin itself, Wu and Stryer (40) have labelled cattle rhodopsin at 3 specific sites with fluorescent dyes, then irradiated each dye and measured the efficiency of transfer of excitation to the rhodopsin chromophore. Using Förster's theory for such radiationless energy transfer, they calculated that the furthest transmission of excitation was about 75 Å. Since a spherical rhodopsin having molecular weight about 40,000 gm per mole would have a diameter of only about 40 Å, this measurement implies a considerable asymmetry. I think this is an interesting yet not compelling argument. Much as I would like to accept it, I would have more confidence in the direct visualization of the rhodopsin micelle in the electron microscope.

In the rod disc membranes, rhodopsin seems to float in an oily bed composed apparently of the fatty acid chains of the phospholipids (6). The three major classes of phospholipid are phosphatidyl ethanolamine, choline, and serine in the approximate ratio 42:46:10. The content of polyunsaturated fatty acids in these lipids, located mainly in the α-position, is among the highest yet reported in membranes (3). Hence the lipid portions of these membranes are unusually fluid (cf. also 4).

One consequence of this condition is the recent demonstration that rhodopsin rotates in the disc membrane (8). By partly bleaching rhodopsin in the retina with plane-polarized light, it should be possible to induce dichroism. The chromophores most aligned with the plane of polarization of the irradiation should be bleached selectively, leaving the remaining chromophores oriented in the perpendicular plane. Hagins and Jennings (18), having failed to photoinduce dichroism in this way, suggested among other explanations that most probably rhodopsin undergoes Brownian rotation in the membrane.

Brown has now shown that if a frog retina is fixed with glutaraldehyde, then partly bleached with plane-polarized light, it becomes permanently highly dichroic. Apparently the bifunctional glutaraldehyde forms crosslinks that hold the rhodopsin from rotating. They also make it inextractable with digitonin. Such a monofunctional fixative as formaldehyde does neither.

Cone (12) has since flash-bleached rhodopsin in fresh frog retinas with plane-polarized light, inducing a transient photodichroism that is dissipated within about 80 microsec at 6°C. That is presumably the length of time it takes for rhodopsin molecules in the membranes, by Brownian rotation, to randomize their orientations. The half-relaxation time is 3.0 ± 1.5 microsec at 20°C, with a temperature coefficient, Q_{10} of about 3.0. It should be noted that throughout such rotations the chromophores remain in the planes of the membranes. According to Cone's measurements, rhodopsin in the disc membrane is as though floating in a light oil, about as viscous as olive oil.

The disc membranes seem to present both hydrophilic and hydrophobic phases, as does rhodopsin itself; and one can suppose that rhodopsin orients itself in, indeed forms part of the interfaces between such phases: "The photopigments may be oriented in the protein layers of the outer limb so that their carotenoid groups project into the lipoidal interstices. In this event the multiple interfaces between protein and lipid are actually composed largely of photopigment. If, as seems likely, surface structure plays as large a part in the excitation and recovery of rods and cones as it does in nerve, this consideration provides a ready explanation of the role of the photopigments in the visual process." (Wald, 1944) (31). It is this orientation that apparently keeps the rhodopsin chromophore in the plane of the membrane.

References

1. ABRAHAMSON, E.W., J.R. WIESENFELD: The structure, spectra and reactivity of visual pigments. In: Handbook of Sensory Physiology, vol. VII/1, ed. H.J.A. DARTNALL, pp. 69-121. Springer, Heidelberg (1972).
2. ALBRECHT, G.: Terminal amino acids of rhodopsin. J.Biol.Chem. 229, 477-487 (1957).
3. ANDERSON, R.E., L. SPERLING: Positional distribution of the fatty acids in the phospholipids of bovine retina rod outer segments. Arch.Biochem.Biophys. 144, 673-677 (1971).
4. BLASIE, J.K.: The location of photopigment molecules in the cross-section of frog retinal receptor disk membranes. Biophys.J. 12, 191-204 (1972).
5. BLASIE, J.K., C.R. WORTHINGTON, M.M. DEWEY: Molecular localization of frog retinal receptor photopigment by electron microscopy and low-angle X-ray diffraction. J.Mol.Biol. 39, 407-416 (1969).
6. BLASIE, J.K., C.R. WORTHINGTON: Planar liquid-like arrangement of photopigment molecules in frog retinal receptor disk membranes. J.Mol.Biol. 39, 417-439 (1969).
7. BOWNDS, D.: Site of attachment of retinal in rhodopsin. Nature (Lond.) 216, 1178-1181 (1967).
8. BROWN, P.K.: Rhodopsin rotates in the visual receptor membrane. Nature New Biol. 236, 35-38 (1972).
9. BROWN, P.K., G. WALD: Visual pigments in single rods and cones of the human retina. Science 144, 45-52 (1964).
10. BUSCH, G.E., M.L. APPLEBURY, A.A. LAMOLA, P.M. RENTZEPIS: Fortion and decay of prelumirhodopsin at room temperatures. Proc.Nat.Acad. Sci. U.S. 69, 2802-2806 (1972).
11. CONE, R.A.: Early receptor potential: photoreversible charge displacement in rhodopsin. Science 155, 1128-1131 (1967).
12. CONE, R.A.: Rotational diffusion of rhodopsin in the visual receptor membrane. Nature New Biol. 236, 39-43 (1972).
13. CONE, R.A., P.K. BROWN: Dependence of the early receptor potential on the orientation of rhodopsin. Science 156, 536 (1967).
14. DARTNALL, H.J.A., J.N. LYTHGOE: The spectral clustering of visual pigments. Vision Res. 5, 81-100 (1965).
15. DAVIE, E.W., O.D. RATNOFF: Waterfall sequence for intrinsic blood clotting. Science 145, 1310-1312 (1964).
16. GILARDI, R., I.L. KARLE, J. KARLE, W. SPERLING: Crystal structure of the visual chromophores, 11-cis and all-trans retinal. Nature 232, 187-189 (1971).
17. HAGINS, W.A.: The visual process: excitatory mechanisms in the primary rereceptor cells. Ann.Rev.Biophys.Bioeng. 1, 131-158 (1972).

18. HAGINS, W.A., W.H. JENNINGS: Radiationless migration of electronic excitation in retinal rods. In: Energy Transfer with Special Reference to Biological Systems, Symp. Faraday Soc. 27, 180-190 (1960).
19. HONG, K., W.L. HUBBELL: Preparation and properties of phospholipid bilayers containing rhodopsin. Proc. Nat. Acad. Sci. U.S. 69, 2617-2621 (1972).
20. HONIG, B., M. KARPLUS: Implications of torsional potential of retinal isomers for visual excitation. Nature (Lond.) 229, 558-560 (1971).
21. HUBBARD, R.: The chromophores of the visual pigments. In: Visual Problems of Colour, ed. W.S. STILES, pp. 151-169. H.M. Stationery Office (1958).
22. HUBBARD, R.: The stereoisomerization of 11-cis retinal. J. Biol. Chem. 241, 1814-1818 (1966).
23. HUBBARD, R., D. BOWNDS, T. YOSHIZAWA: The chemistry of visual photoreception. Cold Spring Harbor Symp. 30, 301-315 (1965).
24. KORENBROT, J.I., R.A. CONE: Dark ionic flux and the effects of light in isolated rod outer segments. J. Gen. Physiol. 60, 20-45 (1972).
25. KROPF, A., R. HUBBARD: The mechanism of bleaching rhodopsin. Ann. N.Y. Acad. Sci. 74, 266-280 (1958).
26. LIEBMAN, P.A.: Microspectrophotometry of photoreceptors. In: Handbook of Sensory Physiology, vol. VII/1, ed. H.J.A. DARTNALL, pp. 481-528. Springer, Heidelberg (1972).
27. MACFARLANE, R.G.: An enzyme cascade in the blood clotting mechanism and its function as a biochemical amplifier. Nature (Lond.) 202, 498-499 (1964).
28. REVEL, J.P., S. ITO, D.W. FAWCETT: Electron micrographs of myelin figures of phospholipids simulating intracellular membranes. J. Biophys. Biochem. Cytol. 4, 495-498 (1958).
29. SHICHI, H., M.S. LEWIS, F. IRREVERRE, A.L. STONE: Purification and properties of bovine rhodopsin. J. Biol. Chem. 244, 529-536 (1969).
30. TANSLEY, K.: The regeneration of visual purple: its relation to dark adaptation and night blindness. J. Physiol. 71, 442-458 (1931).
31. WALD, G.: The molecular organization of visual processes. In: Colloid Chemistry, vol. 5, ed. J. ALEXANDER, pp. 753-762. Reinhold Publishing Co., New York (1944).
32. WALD, G.: Human vision and the spectrum. Science 101, 653-658 (1945).
33. WALD, G.: In: Nerve Impulse, Trans. 4th Conference, ed. D. NACHMANSOHN, pp. 11-57. Josiah Macy Foundation, New York (1954).
34. WALD, G.: The biochemistry of visual excitation. In: Enzymes: Units of Biological Structure and Function, ed. T.P. SINGER, pp. 355-367. Academic Press (1956).
35. WALD, G.: The molecular organization of visual systems. In: Light and Life, ed. W.D. MC ELROY and B. GLASS, pp. 724-750. Johns Hopkins Press, Baltimore (1961).
36. WALD, G.: Visual excitation and blood clotting. Science 150, 1028-1030 (1965).
37. WALD, G., P.K. BROWN, I.R. GIBBONS: The problem of visual excitation. J. Opt. Soc. Amer. 53, 20-35 (1963).
38. WALD, G., H. ZUSSMAN: Carotenoids of the chicken retina. J. Biol. Chem. 122, 449-460 (1938).
39. WRIGHT, W.E., P.K. BROWN, G. WALD: The orientation of rhodopsin and other pigments in dry films. J. Gen. Physiol. 59, 201-212 (1972).
40. WU, C.W., L. STRYER: Proximity relationships in rhodopsin. Proc. Nat. Acad. Sci. U.S. 69, 1104-1108 (1972).
41. YOSHIKAMI, S., W.A. HAGINS: Light, calcium and the photocurrent of rods and cones. Biophys. J. Abstracts 11, 47a
42. YOSHIZAWA, T.: The behaviour of visual pigments at low temperatures. In: Handbook of Sensory Physiology, vol. VII/1, ed. H.J.A. DARTNALL, pp. 146-179. Springer, Heidelberg (1972).

43. YOSHIZAWA, T., G. WALD: Pre-lumirhodopsin and the bleaching of visual pigments. Nature (Lond.) 197, 1279-1286 (1963).
44. YOSHIZAWA, T., G. WALD: Photochemistry of iodopsin. Nature (Lond.) 214, 566-571 (1967).
45. ZORN, M., S. FUTTERMAN: Properties of rhodopsin dependent on associated phospholipid. J.Biol.Chem. 246, 881-886 (1971).

I. Pigment Structure and Chemical Properties

Theory of Retinal and Related Molecules

M. Karplus
Department of Chemistry, Harvard University, Cambridge, MA 02138/USA

A theoretical investigation has been made of retinal isomers and related molecules (e.g., polyenes, retinal Schiff bases) by means of semiempirical quantum-mechanical techniques that provide reliable potential surfaces for conjugated molecules. Ground-state geometries and vibrational frequences, excitation energies, and Franck-Condon factors for electronic transitions have been calculated. In contrast to previous treatments, double-excited states are included in the configuration-interaction formulation; these significantly change the ordering of the π-electron excited states (e.g., an excited ^1A_g state appears in the spectrum of all-trans polyenes below the ^1B_u state that corresponds to the strongly allowed transition).

The method has been used to predict the ground-state geometry of 11-cis retinal. It is best described as distorted 11-cis, 12-s-cis, in contrast to the assumption made by all previous workers that the molecule has an 11-cis, 12-s-trans conformation; the theoretically predicted structure has been confirmed by x-ray analysis (see the following paper by W. Sperling). Calculations on retinal Schiff bases suggest that the protonated forms have their strong absorption maximum near 450 nm, in agreement with solution spectra, so that the position of the observed rhodopsin peak (~ 500 nm) requires environmental perturbations; calculations indicate that distortion of the Schiff base by the protein and/or the presence of charged groups (either positive or negative) can yield shifts of the correct order, while polarizability effects are unlikely to be large enough.

The results of the above calculations make possible some suggestions concerning the behavior of the chromophore during the initial stages of visual excitation. The observed spectral shifts and the interconversion properties of the intermediates

$$\text{rhodopsin} \xrightarrow{h\nu} \text{prelumi} \xrightarrow{\text{dark}} \text{lumi}$$

can be understood if they have chromophore conformations

$$\text{11-cis,12-s-cis} \xrightarrow{h\nu} \text{11-trans,12-s-cis} \xrightarrow{\text{dark}} \text{11-trans,12-s-trans.}$$

More details concerning the method and its applications can be found in the published papers listed under references.

Acknowledgement. Grateful acknowledgement is made to B. Honig, B. Hudson, D. Kliger, K. Schulten, B. Sykes, and A. Warshel, who collaborated on various aspects of the work summarized above.

References

1. HONIG, B., M. KARPLUS: Implications of torsional potential of retinal isomers for visual excitation. Nature (London) 229, 558-560 (1971); erratum: Nature (London) 231, 67 (1971).

2. HONIG, B., B. HUDSON, B.D. SYKES, M. KARPLUS: Ring orientation in β-ionone and retinal. Proc. Nat. Acad. Sci. U.S. 68, 1289-1293 (1971).
3. SCHULTEN, K., M. KARPLUS: On the origin of a low-lying forbidden transition in polyenes and related molecules. Chem. Phys. Letters 14, 305-309 (1972).
4. WARSHEL, A., M. KARPLUS: Calculations of ground and excited state potential surfaces and conjugated molecules. I. Formation and parametrization. J. Am. Chem. Soc. 94, 5612-5625 (1972).
5. WARSHEL, A., M. KARPLUS: Vibrational structure of electronic transitions in conjugated molecules. Chem. Phys. Letters 17, 7 (1972).

Discussion

T.P. Williams: What is the relative concentration of the s-conformers which can exist at carbon atoms 12 and 13 at ordinary temperature?

M. Karplus: From the potential function, the ratio of 11-cis, 12-s-cis to 11-cis, 12-s-trans at room temperature is calculated to be about 8 to 1. However, this ratio is highly approximate since the method of calculation can have significant errors in the relative energies of the two conformers.

G. Wald: How explain the lack of fine structure in spectra of vitamin A and retinals even in EPA glasses at liquid nitrogen temperatures? Incidentally, astaxanthin (α-dihydroxy, diketo β-carotene), though it has no vibrational structure at room temperature, develops full fine structure in EPA at $-180^{o}C$.

M. Karplus: The lack of fine structure in retinal and related molecules (e.g. β-ionone) comes primarily from the difference in the ring orientation for the ground and excited state and the resulting sensitivity of the spectrum to slight changes in the effective ground-state orientation. Since the ground-state potential is very flat in the neighborhood of the minimum, low frequency vibrational modes and environmental pertubations can both contribute to having a range of orientations that is sufficient to wipe out the "usual" fine structure.

T.G. Ebrey: Could the unexpectedly long fluorescence life-time to the forbiddeness of the lowest excited state predicted by your calculations?

M. Karplus: The long life-time of the ^1A_g state is predicted by the calculation in the sense that the $^1A_g \rightarrow {}^1A_g$ (ground-state) is forbidden in a symmetric polyene without vibronic or environmental pertubations. In retinal, the corresponding state is calculated to be weakly allowed due to the presence of the oxygen and the non-planarity of the ring.

T.G. Ebrey: What effect would the addition of a double bond to the ring have on the allowedness of the lowest excited state? Both vitamin A_2 and porphyropsin (at $77^{o}K$) seem to have fine structure.

M. Karplus: An additional double-bond in the ring would make the state slightly more allowed.

Conformations of 11-cis Retinal

W. Sperling
Institut für Neurobiologie, Kernforschungsanlage Jülich, Jülich/W. Germany

Isoprenic polyenes have two different kinds of double bonds depending on the nature of the substituents, X and X':

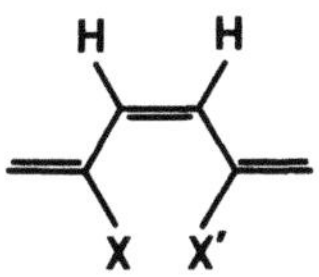

1) X = X' = H

2) X = H
X'= CH_3

The question if compounds of type 2 could form cis bonds about the central double bond has a long and controversial history. Linus Pauling (1) stated in 1939 that bonds of this type "must have the trans configuration" because of steric hindrance between the hydrogen and the methyl group. In the meantime compounds of type 2 were not only synthesized in the laboratory, but a representative of this class, 11-cis retinal, was found to be the chromophore of the visual pigments (2-8). Photoisomerization of this chromophore from 11-cis to all-trans retinal is believed to be the trigger of the visual process. According to this model, 11-cis retinal is bound to the lipoprotein of the visual pigment, and the configurational change associated with the cis-trans isomerization causes simultaneous and subsequent conformational changes in the protein. These conformational changes by means of changing some physical state of the membrane lead to the further events in the chain of the visual process.

When rhodopsin is exhaustively irradiated with visible light, it is bleached and finally separated into the isomerization product all-trans retinal and opsin. By adding 11-cis retinal to opsin in vitro, it is possible to reconstitute rhodopsin, as defined by its absorption maximum. Only a very few aldehydes, for example 9-cis retinal and some 11-cis retinal derivatives, combine with opsin. This specificity establishes great importance of the stereochemistry, both of the retinal and the lipoprotein complex.

Polyenes can undergo a variety of conformational and configurational changes. The energetically preferred form is normally the one with "single" and "double" bonds in a planar conformation. The alternating system of single and double bonds gives a certain double bond character to single bonds, but not nearly enough to overcome the still alternating carbon-carbon bond distances in the system, even for infinite length. In principle, because of the coplanarity of the system, both kinds of bonds should produce cis-trans isomers. But usually, in an unhindered conjugated system, the trans form of single bonds (s-trans) is energetically favored over the cis-form (s-cis)*. The energy barrier for rotation around "single" bonds is low enough to make isolation of s-cis isomers impossible under normal conditions.

---

*By definition a single bond is called s-trans if the angle α between the two twisted planes of the conjugated system is $0^\circ \leqq \alpha < 90^\circ$, and s-cis, if the angle is $90^\circ < \alpha \leqq 180^\circ$. $\alpha = 0^\circ$ corresponds to the planar s-trans conformation.

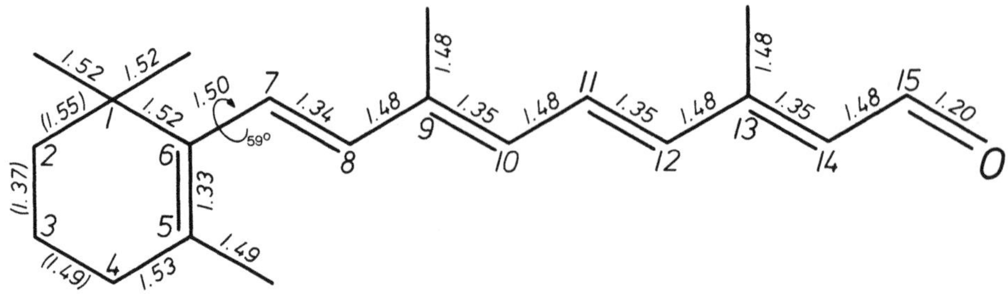

Fig. 1. 6s-cis retinal

Crystal Structure of 11-cis and all-trans retinal

The knowledge of the crystal structure might be of great help for an elucidation of the conformation of 11-cis retinal in solution and especially in rhodopsin. The results of the x-ray analysis (9) of all-trans and 11-cis retinal are shown in Fig. 1 and 2.

In all-trans retinal the situation is as follows:
 The side chain 6-16 is planar.
 The 6-7 bond is s-cis, and the segment 5-6-7-8 is twisted 59° about the 6-7 bond.
In the case of 11-cis retinal the following features are remarkable:
 The chain segment 6-13 is planar.
 The chain segment 12-16 is planar.
 Both segments are twisted about the 12-13 single bond.
 The torsion is 39° from the planar s-cis conformation, or 141° from the planar s-trans conformation.
 The 6-7 bond is s-cis, the segment 5-6-7-8 is twisted 40° about the 6-7 bond.
This means that 11-cis retinal in the crystal structure has three cis bonds, and the complete name should be 6s-cis, 11-cis, 12s-cis retinal (correspondingly the all-trans retinal would be called 6s-cis retinal).

The torsions about the 6-7 and the 12-13 bond, respectively, give rise to molecular asymmetries with the faculty to form different atropisomers. Two atropisomers should exist for all-trans retinal with its one asymmetric center, four should exist in the case of 11-cis retinal with its two asymmetric centers. All these atropisomers are optically active. In the strongly asymmetric protein environment of the membrane only one of these isomers would be expected to occur. Depending on the (not yet known) energy barrier between these different atropisomers, it may be possible to isolate them, for example at low temperature, in a rigid medium or, as already done for two of them, in the crystal state.

Spectral Characteristics of Retinals

All sterically hindered cis-polyenes of the above-mentioned type 2 have features which their non-hindered cis- and trans-isomers do not possess. Fig. 5, for example, shows the solution spectrum of 11-cis retinal (heavy line) at room temperature. The long-wavelength band is much smaller than the corresponding bands of 9-cis, 13-cis, and all-trans retinal, whereas the absorption at shorter wavelengths is big-

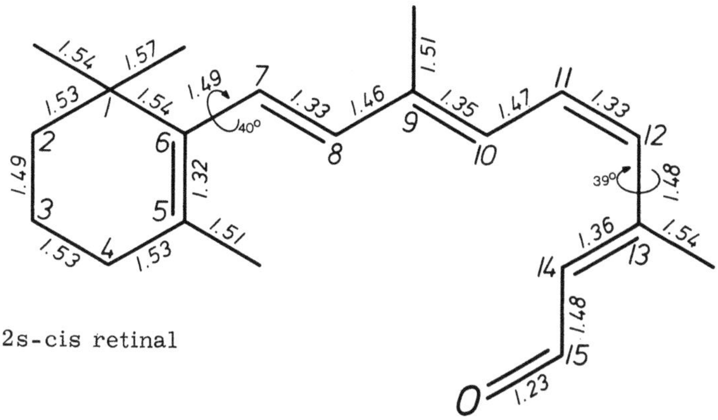

Fig. 2. 6s-cis, 11-cis, 12s-cis retinal

ger (heavy lines in Figs. 3,4,6). Wald and coworkers (11) measured the absorption spectra of three different sterically hindered carotenoids, including 11-cis retinal, at liquid nitrogen temperature, and found that the size (oscillator strength) of the long-wavelength band increased markedly on cooling. They did not measure the interesting wavelength region below about 320 nm. These investigators attributed the increase of the band at lower temperature to relief of steric hindrance, "probably by decreasing the van der Waals radii of the overlapping substituent groups (7,11). To obtain comparable data the four isomers all-trans, 13-cis, 11-cis, and 9-cis retinal were measured (8; Sperling and Rafferty, unpublished data) at two different temperatures, 77° and 295°K, and in two different solvent systems, EPA (= diethylether: 2-methylbutane: ethanol = 5:5:2, v/v) and PMh (= 2-methylbutane: methylcyclohexane = 5:1, v/v). The pathlength of the cuvettes was 1 mm and the concentrations of the different retinals between 2.0 and $2.7 \cdot 10^{-4}$ mole/liter (measured with an accuracy of 0.01 mole/liter). The spectra are shown in Figs. 3-10. Table I summarizes the absorption maxima and molar absorption coefficients of all spectra.

Table I. Absorption maxima and molar absorption coefficients of the spectra shown in Figs. 3-10

	λ_{max} [nm] (295°K)	(77°K)	$\Delta\lambda_{max}$	ϵ_{max} [liter $mole^{-1}cm^{-1}$] (295°K)	(77°K)
	A) Retinal in EPA				
all-trans	374.5	387.2	12.7	46100	52650
13-cis	369.0	381.9	12.9	38300	44200
11-cis	370.0	387.5	17.5	26400	46350
9-cis	366.2	378.7	12.5	37100	43050
	B) Retinal in PMh				
all-trans	368.3	386.5	18.2	48900	44450
13-cis	363.4	382.0	18.6	39550	36950
11-cis	362.5	389.0	26.5	26900	34300
9-cis	360.8	382.5	21.7	39100	35250

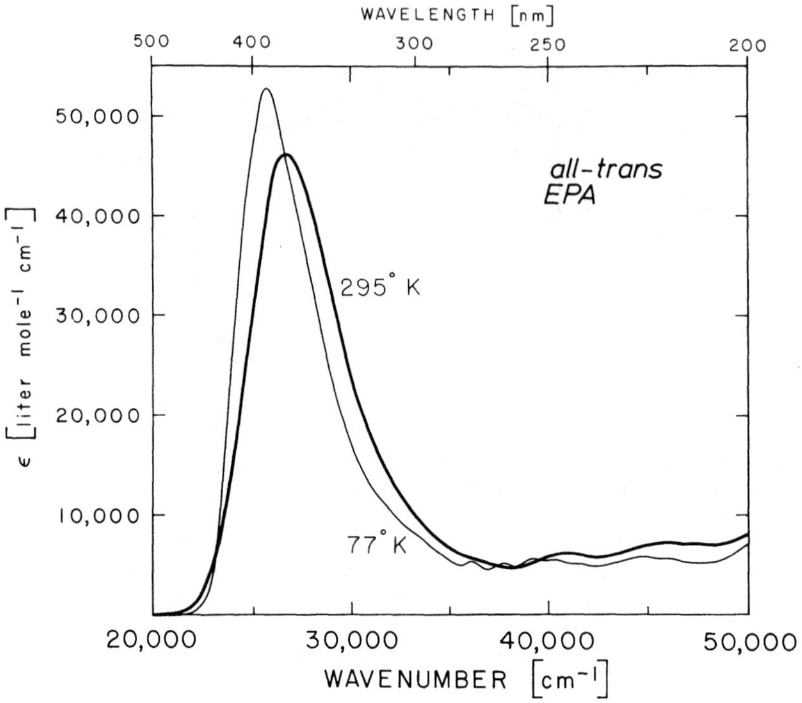

Fig. 3

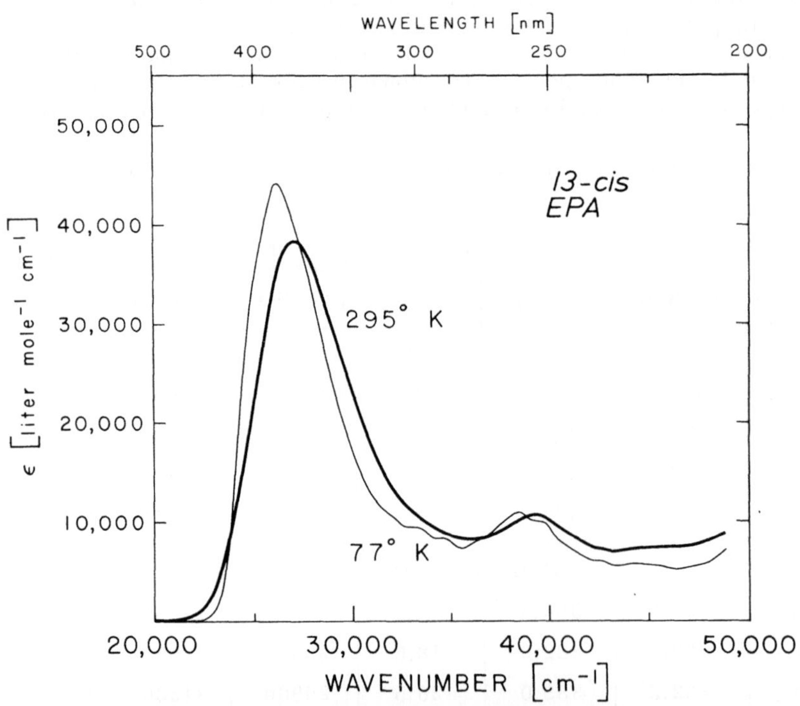

Fig. 4

Figs. 3-10. Absorption spectra of all-trans, 13-cis, 11-cis, and 9-cis retinal in EPA (Figs. 3-6) and PMh (Figs. 7-10). All spectra are measured at room temperature (heavy lines) and liquid nitrogen temperature (light lines)

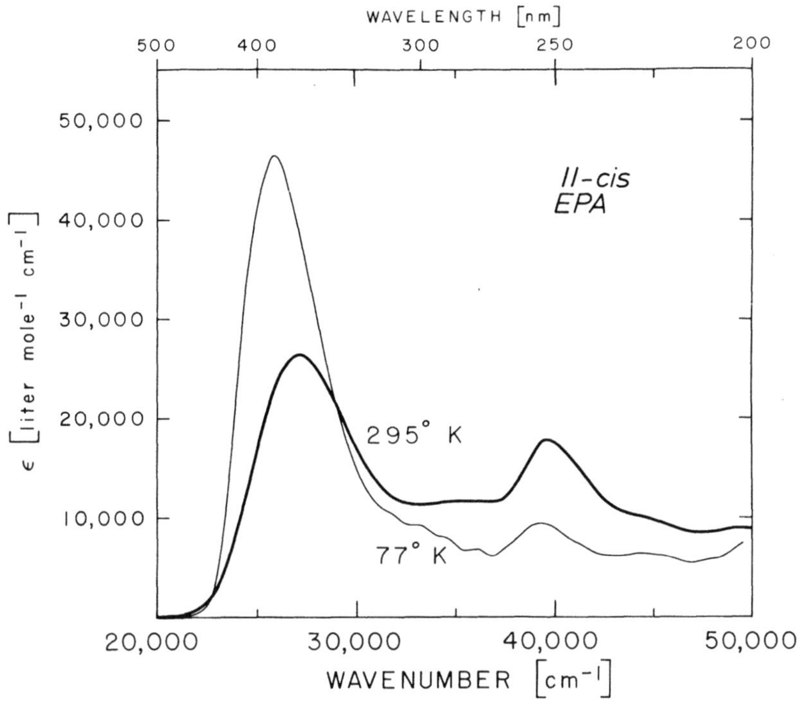
WAVELENGTH [nm]
500
400
300
250
200
50,000
40,000
30,000
20,000
10,000
ϵ [liter mole⁻¹ cm⁻¹]
11-cis
EPA
295° K
77° K
20,000
30,000
40,000
50,000
WAVENUMBER [cm⁻¹]

Fig. 5

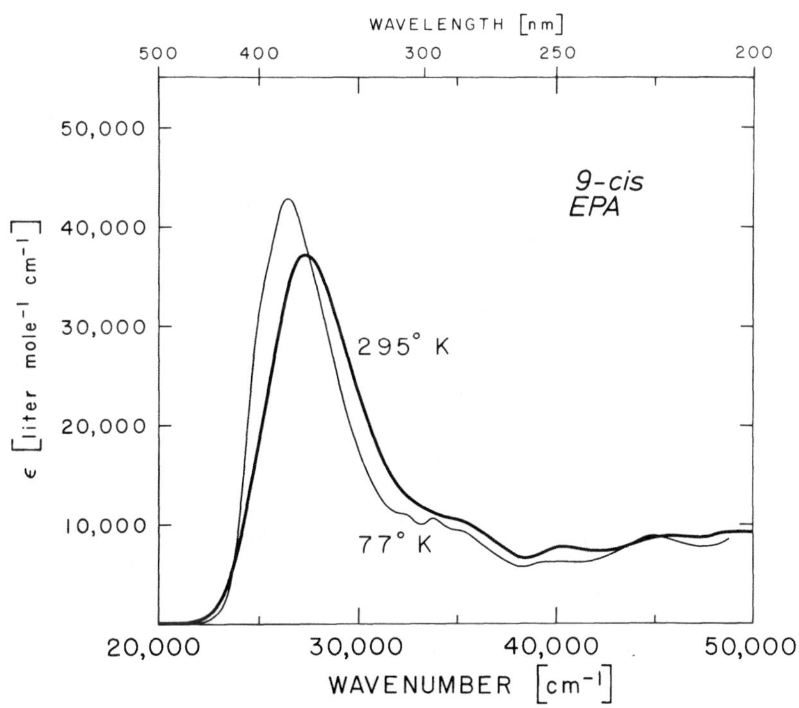
WAVELENGTH [nm]
500
400
300
250
200
50,000
40,000
30,000
20,000
10,000
ϵ [liter mole⁻¹ cm⁻¹]
9-cis
EPA
295° K
77° K
20,000
30,000
40,000
50,000
WAVENUMBER [cm⁻¹]

Fig. 6

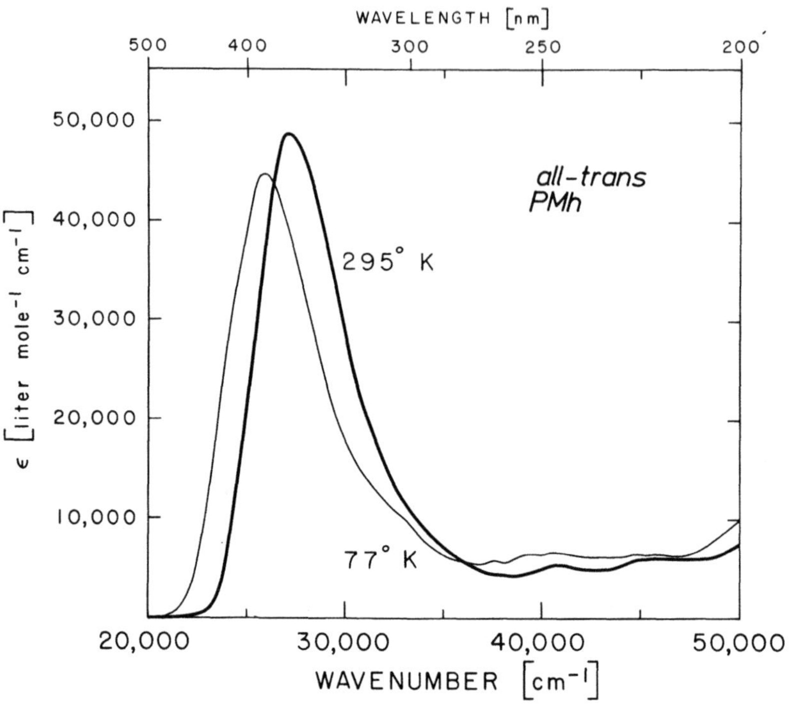
WAVELENGTH [nm]
500
400
300
250
200
50,000
40,000
30,000
20,000
10,000
ε [liter mole⁻¹ cm⁻¹]
all-trans
PMh
295° K
77° K
20,000
30,000
40,000
50,000
WAVENUMBER [cm⁻¹]

Fig. 7

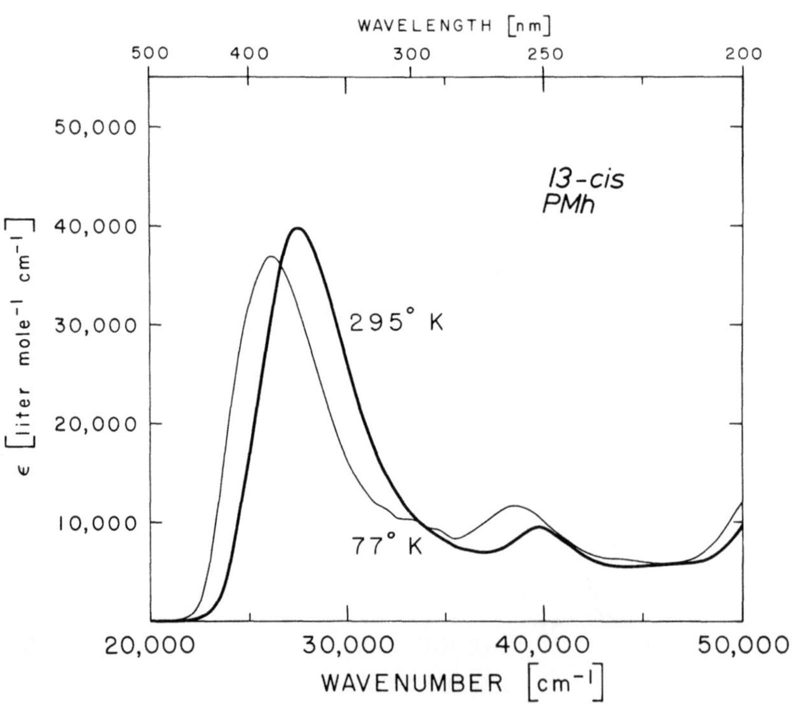
WAVELENGTH [nm]
500
400
300
250
200
50,000
40,000
30,000
20,000
10,000
ε [liter mole⁻¹ cm⁻¹]
13-cis
PMh
295° K
77° K
20,000
30,000
40,000
50,000
WAVENUMBER [cm⁻¹]

Fig. 8

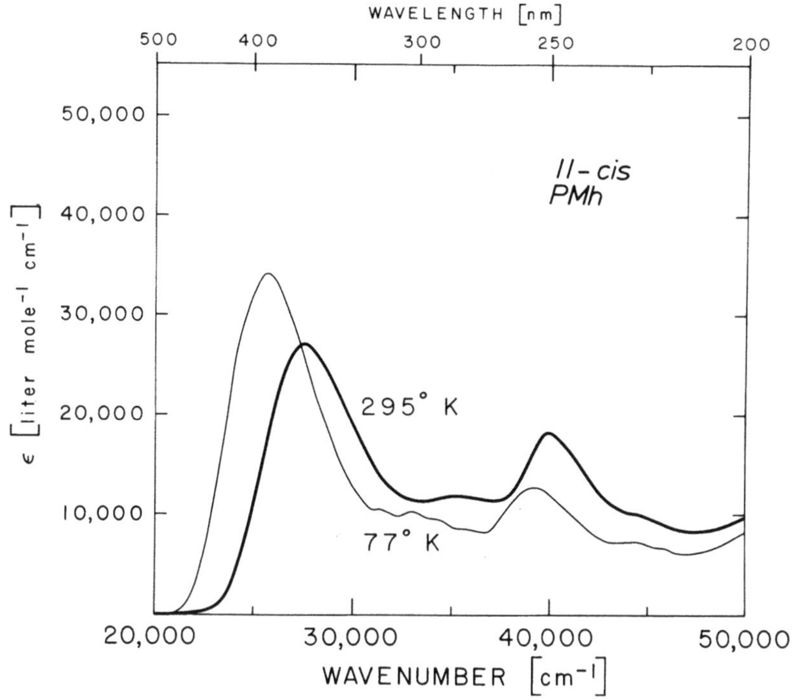
WAVELENGTH [nm]
500
400
300
250
200
50,000
40,000
30,000
20,000
10,000
ε [liter mole⁻¹ cm⁻¹]
11-cis
PMh
295° K
77° K
20,000
30,000
40,000
50,000
WAVENUMBER [cm⁻¹]

Fig. 9

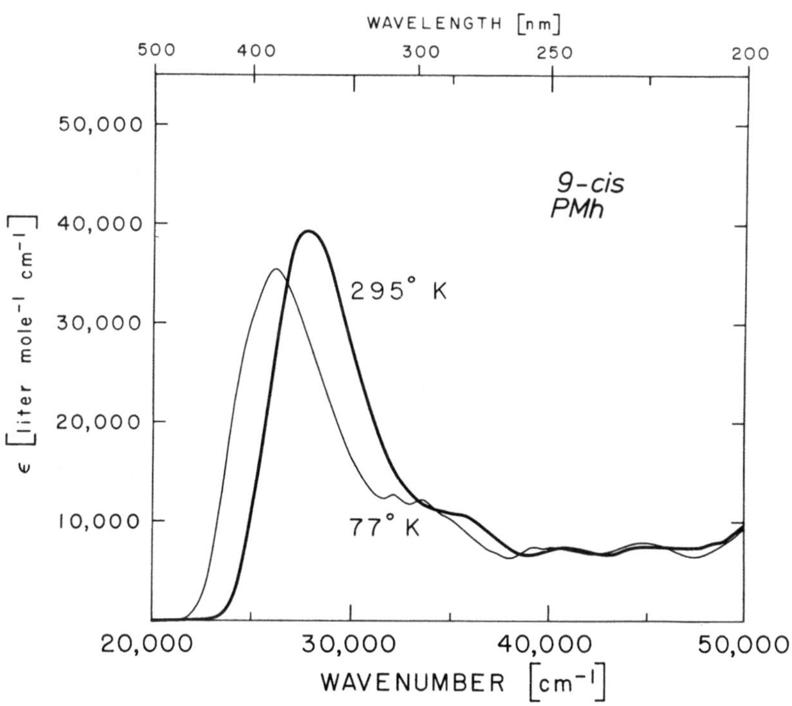
WAVELENGTH [nm]
500
400
300
250
200
50,000
40,000
30,000
20,000
10,000
ε [liter mole⁻¹ cm⁻¹]
9-cis
PMh
295° K
77° K
20,000
30,000
40,000
50,000
WAVENUMBER [cm⁻¹]

Fig. 10

Table II. Relative oscillator strengths

	EPA		PMh	
	295°K	77°K	295°K	77°K
	A) Whole spectrum from 20,000 to 49,000 cm^{-1}			
all-trans	2098	2012	2040	2087
13-cis	2016	1935	1926	2000
11-cis	2032	2020	1965	1951
9-cis	1992	1969	1964	1947
	B) Main band (half band width in cm^{-1})			
all-trans	1546 (5680)	1425 (4830)	1538 (5430)	1352 (5600)
13-cis	1316 (5970)	1250 (4980)	1299 (5700)	1114 (5750)
11-cis	877 (6630)	1228 (4840)	870 (6330)	1000 (5620)
9-cis	1234 (5870)	1190 (4950)	1246 (5630)	1082 (5900)
	C) Half band			
all-trans	586	527	555	551
13-cis	515	437	497	465
11-cis	371	468	350	437
9-cis	492	456	465	453

Table II gives the relative oscillator strengths (f-values). Section A lists the overall f-values of all bands in the measured region between 20,000 and 49,000 cm^{-1}. Section B gives the f-values of the main absorption bands. For these bands the shorter wavelength slopes were estimated and are probably accurate within 10%. Behind these values (in parentheses) are the half band widths in cm^{-1}. Section C gives the f-value of the half band, i.e. the area measured from 20,000 cm^{-1} to the absorption maximum.

The most striking feature of this series of spectra is the enormous increase of the main absorption band of 11-cis retinal at lower temperatures, accompanied simultaneously with a large decrease in absorption in the 40,000 cm^{-1} region. On the other hand, the spectra of all-trans, 13-cis, and 9-cis retinal change only slightly with temperature. The absorption maximum shifts to the red with lower temperature for all isomers and in both solvents. The spectra measured in PMh differ in one major characteristic from those in EPA. In EPA the main absorption band becomes sharper, and has a higher molar absorption coefficient (ε_{max}) at lower temperature, whereas in PMh, with the exception of 11-cis retinal, ε_{max} becomes smaller, and the band broadens. All eight liquid nitrogen-temperature spectra show fine structure adjacent to and at the high wavenumber side of the main band, with an interband distance of about 1450 cm^{-1}. The main band shows no fine structure in any case. The spectra are probably fairly accurate up to 48,000 to 49,000 cm^{-1}, whereas solvent absorption above 49,000 cm^{-1}, especially for the low-temperature spectra, may cause a certain inaccuracy. It should be noted that the position of the wavelength maximum depends very much on the solvent and temperature. At liquid nitrogen temperature, for example, 11-cis retinal has the largest λ_{max}.

The spectra of 11-cis retinal (Fig. 5 and 9) suggest that dramatic changes in the π-electron system of the molecule occur with change of temperature. These changes are certainly accompanied by a change in the degree of planarity of the conjugated system.

A conformational change in the carbon 10-13 region of 11-cis retinal is also supported by NMR data (together with Stanley Bernstein, Yellow Springs, Ohio; unpublished). Whereas the chemical shifts of the hydrogens of the methyl groups 1, 5, and 9 do not change appreciably with lowering the temperature, there is a distinct shift of the CH_3-group at carbon atom 13 when going from +30°C to -80°C. By comparison in all-trans retinal the chemical shifts of all five methyl groups in position 1, 5, 9, and 13 do not change with temperature under equal conditions.

It is tempting, of course, to speculate on these conformational changes. The simplest assumption is a temperature-dependent equilibrium between two conformers. One conformer is probably identical with the structure realized in the crystal state, i.e. the molecule would be s-cis concerning the 12-13 bond with a twist of about 40°. For the second conformer I would favor an s-trans conformation (9,10). At any temperature there would be an equilibrium between these two according to the Boltzmann contribution:

$$\text{12s-cis} \rightleftharpoons \text{12s-trans}$$

If this is true it should be possible to freeze the high-temperature equilibrium and keep it at lower temperatures. By using polyethylene as a solvent this can be realized: If 11-cis retinal is dissolved in polyethylene at room temperature, the spectrum looks similar to that in other saturated hydrocarbon solvents. By lowering the temperature the viscosity of the polyethylene increases rapidly and the spectrum stops changing (except for those small changes common to all isomers in a hydrocarbon solvent).

It is surprising that the main band of the low-temperature conformer of 11-cis retinal (see Table II, section B and C) has about the same oscillator strength as the main bands of the other two cis isomers. Usually the oscillator strength of the long-wavelength band of aromatic and conjugated systems is extremely sensitive to changes in the coplanarity of the system. Small deviations from planarity cause a decrease of this band. On the other hand, there can be no doubt that the side chain of 9-cis and 13-cis retinal is coplanar.

The sum of the f-values of all absorption bands of a system with q π-electrons should be a constant: $\Sigma f = q$ (12). Table II, section A, shows that the overall oscillator strength between 20,000 and 49,000 cm^{-1} is about the same for all four isomers, independent of solvent and temperature. This is the case in spite of the fact that the absorption spectra are artificially cut off at 49,000 cm^{-1}, and the far UV-region not taken into account.

The degree to which the 5-6 double bond participates in the π-electron system of the chain depends on the twisting angle about the 6-7 bond. By gradually twisting the molecule about this bond small changes in λ_{max} and ε_{max} can be achieved. It should be noted that there is a marked difference of this angle between all-trans and 11-cis retinal.

This paper could only touch on the interesting problems of the stereochemistry of 11-cis retinal and the consequences for the visual process. We have still a long way to go for a complete understanding of this molecule and its function in the membrane of the photoreceptor.

References

1. PAULING, L.: Recent work on the configuration and electronic structure of molecules; with some applications to natural products. Fortschr. Chem. org. Naturstoffe 3, 203-235 (1939).
2. KARRER, P., R. SCHWYZER, A. NEUWIRTH: Oxydation von 4-Methyl-o-benzochinon zu cis-cis-β-methylmuconsäureanhydrid. Helv. Chim. Acta 31, 1210-1214 (1948).
3. PAULING, L.: Zur cis-trans-Isomerierung von Carotinoiden. Helv. Chim. Acta 32, 2241-2246 (1949).
4. DIETERLE, J.M., CH. D. ROBESON: Crystalline neoretinene b. Science 120, 219-220 (1954).
5. OROSHNIK, W.: The synthesis and configuration of neo-b-vitamin A and neoretinene b. J. Amer. Chem. Soc. 78, 2651-2652 (1956).
6. ZECHMEISTER, L.: Cis-trans isomeric carotenoids, vitamin A, and acrylpolyenes. Springer-Verlag, Wien (1962).
7. HUBBARD, R., G. WALD: Pauling and carotenoid streochemistry. In: Structural Chemistry and Molecular Biology, ed. A. RICH, N. DAVIDSON. Freeman, San Francisco and London (1968).
8. SPERLING, W., CH. N. RAFFERTY: Relationship between absorption spectrum and molecular conformations of 11-cis retinal. Nature (Lond.) 224, 591-594 (1969).
9. GILARDI, R., I.L. KARLE, J. KARLE, W. SPERLING: Crystal structure of the visual chromophores, 11-cis and all-trans retinal. Nature (Lond.) 232, 187-189 (1971).
10. HONIG, B., M. KARPLUS: Implications of torsional potential of retinal isomers for visual excitation. Nature (Lond.) 229, 558-560 (1971).
11. JURKOWITZ, L., J.N. LOEB, P.K. BROWN, G. WALD: Photochemical and stereochemical properties of carotenoids at low temperatures. Nature (Lond.) 184, 614-624 (1959).
12. KUHN, W.: Über die Gesamtstärke der von einem Zustande ausgehenden Absorptionslinien. Zeitschr. f. Physik 33, 408-412 (1925).

Discussion

T.P. Williams: I would like to ask the same I have already asked Dr. Karplus: What is the relative concentration of the s-conformers which exist at carbon atoms 12 and 13 at ordinary temperature?

W. Sperling: The energy difference between the two conformers is small. One can calculate a ratio of about 35/65 for the "high-energy" (12-s-cis?) and the "low-energy" (12-s-trans?) conformer (for room temperature).

The Binding Site of Retinaldehyde in Native Rhodopsin

W. J. de Grip, S. L. Bonting, and F. J. M. Daemen
Department of Biochemistry, University of Nijmegen, Nijmegen/The Netherlands

It has been shown conclusively that the chromophore in metarhodopsin II (MII) is bound to an ε-amino group of a lysine residue of opsin (1,2) and that in native rhodopsin the chromophore is not bound to a phospholipid amino group (3,4,5).

In the course of studies concerning the influence of aminogroup modification on the properties of rhodopsin, we succeeded in obtaining almost complete modification without much denaturation of the pigment. The first purpose of our study then became the identification of the binding site in rhodopsin. During this study we developed a procedure for the preparation of rod photoreceptor membranes with reproducible and maximal rhodopsin content, which we call the "enrichment procedure".

Enrichment Procedure

Vertebrate rod photoreceptor membranes are usually isolated by separating the outer segments, which are easily broken off from the inner segment during mild homogenization, from other cells or cellular organelles by means of flotation on a sucrose cushion or sucrose gradient density centrifugation. In our laboratory we use the gradient technique to isolate cattle outer segments, since it yields better results than flotation (6). We noticed however that the rhodopsin content of our preparations, stored as lyophilisates, varied considerably during the year, being highest in the winter and lowest in summer (Fig. 1; open bars). Addition of 11-cis retinaldehyde to the preparations in all cases increased the rhodopsin content. Hence, we concluded that the preparations contain opsin and that the excised eyes are not able to regenerate all opsin present during the two hours of dark adaptation

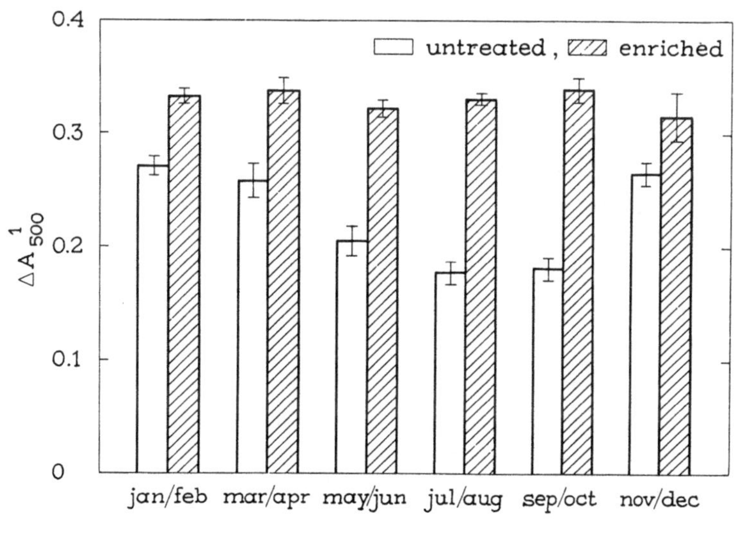

Fig. 1. Seasonal fluctuation in rhodopsin content of rod outer segment membrane lyophilisates. The rhodopsin content is expressed as the ΔA^1_{500}, which is the ΔA_{500} of 1 mg lyophilisate in 1 ml detergent solution

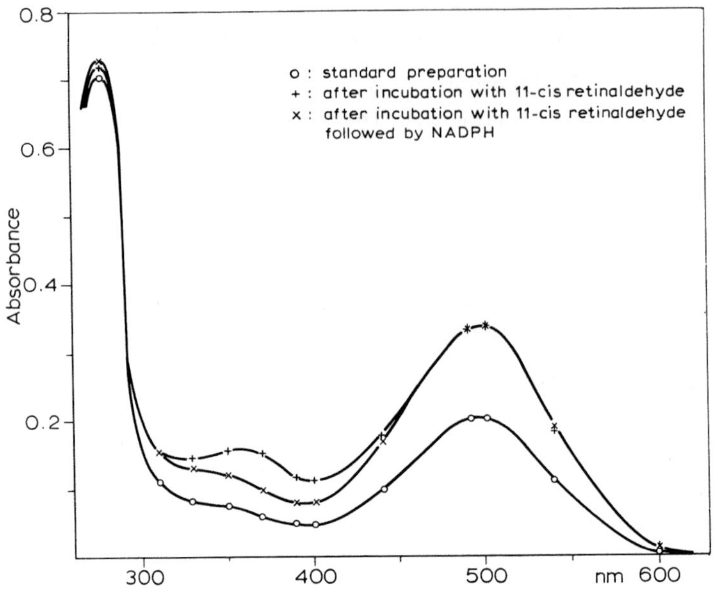

Fig. 2. Spectra of normal membrane preparation and of preparations, during the isolation of which retinaldehyde or retinaldehyde and NADPH were added. Solutions (1% digitonin in 1/15 M phosphate buffer pH 6.5) contained 1 mg lyophilisate per ml

at room temperature prior to pigment isolation. We are able to convert all opsin to rhodopsin by addition of excess 11-cis retinaldehyde to the outer segment suspension isolated from the gradient. This increases the rhodopsin content throughout the year to a constant level of 8 nMoles per mg of lyophilisate (Fig. 1, hatched bars). The absorbance spectrum of these enriched preparations, however, show an extra band at 380 nm due to excess retinaldehyde (Fig. 2), which cannot be removed by washing. This problem can be overcome by adding NADPH after the enrichment procedure, which leads to the reduction of the excess retinaldehyde to retinol by the retinol-dehydrogenase present in the outer segment membranes. The retinol can be nearly completely washed out by a second gradient centrifugation, followed by two washings with buffer solution (Fig. 2). Complete removal can be accomplished by the addition of a small amount of retinol binding protein to these washing solutions. After these treatments, which take about 3 hours at room temperature, electronmicrographs show that the outer segments are still morphologically intact.

Analysis of the resulting preparations reveals a constant composition (Table I): about 36% of the dry weight of the membrane lyophilisates consists of protein, 87% of which is accounted for by rhodopsin-protein (7); about 50% is due to lipids and the rest comes mainly from residual water and carbohydrates. Thus the procedure outlined above, yields, without requiring the use of any detergent, reproducible preparations of maximal rhodopsin content, the spectral characteristics of which are almost as good as those of rhodopsin, chromatographically purified in detergent solution (6).

Significance of Primary Amino Groups

Chemical modification of functional side chains is a valuable tool for the protein chemist in obtaining information on the structural or functional roles of these side chains. In view of the significance of primary amino groups (binding of the chro-

Table I. Chemical composition of rod sac membrane (lyophilized)

Protein		36%	
	rhodopsin protein		87%
Lipids		50%*	
	phosphatidyl ethanolymine		29%
	" choline		28%
	" serine		9%
	other phospholipids		20%
	cholesterol		8%
	other lipids		6%
Carbohydrates		4%	
Water		6%	
Unknown		4%	

* Fatty acids: unsaturated 52%, incl. 34% C22:6

mophore, electrostatical interactions, catalytic functions), we decided to investigate the influence of amino group modification on photoreceptor membrane properties. The binding site of the chromophoric group in rhodopsin has been subject of much investigation. Bownds et al. (1) and Akhtar et al. (2) have shown that the labile aldimine bond, linking the retinaldehyde to an amino group of opsin, can be stabilized by reduction with sodium borohydride to a secondary amine during illumination. Subsequent hydrolysis yields an ε-retinyl-amino-lysine-residue. Since the reduction only occurs during simultaneous illumination, this proves the binding site in MII, but not necessarily in native rhodopsin. Model studies in our laboratory (7), namely, have shown that retinaldehyde very easily transiminizes from one amino group to another and that this can take place upon denaturation or illumination of the rhodopsin complex. This study and a recent one of Fager et al. (8) involving reduction of rhodopsin with cyanoborohydride are indicative, however, for the involvement of an ε-amino-lysine group in rhodopsin too.

Determination of Primary Amino Groups

We first had to know which primary amino groups are present in the photoreceptor membrane. In determining these primary amino groups three independent methods have been used: amino acid analysis, dansylation and trinitrophenylation. The dansylation procedure involves labeling of the primary amino groups with dansylchloride, yielding fluorescent dansylderivatives, which withstand acid hydrolysis and thereupon can be identified and determined by quantitative thin layer chromatography. The dansylation is carried out in 50% acetone, which denatures the membrane and liberates the chromophore so that the original binding site is also labeled. The reagent, dansylchloride, is not completely specific for primary amino-groups, but

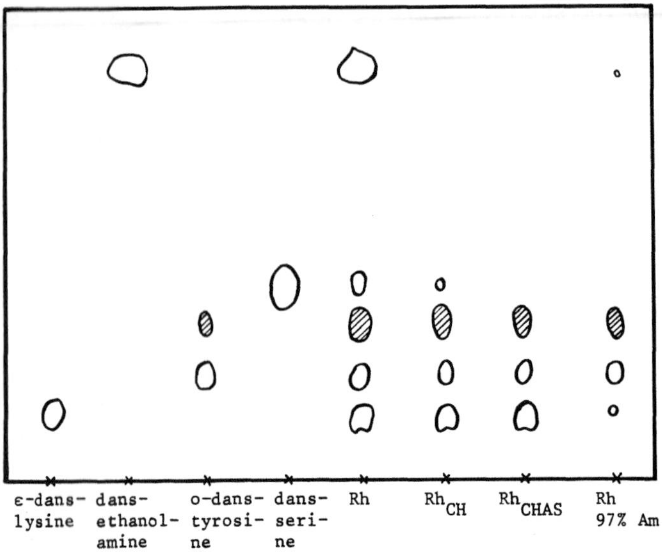

Fig. 3. Drawing of a thin layer chromatogram (silica gel) obtained after dansylation and subsequent hydrolysis of normal (Rh) and enzymatically delipidated (Rh_{CH}, Rh_{CHAS}; cf. 5) rod outer segment membrane preparations. Solvent system Chl-MeOH-HAC 15:4:1 (v/v). Excitation wavelength: 350 nm. The open spots represent yellow fluorescing substances, while the hatched spot comes from the blue fluorescing dansylic acid

only the amino derivatives and the tyrosinehydroxy derivative are stable towards acid hydrolysis. The trinitrophenylation involves reacting the intact preparation with trinitrobenzenesulfonic acid (TNBS), whereafter the total amount of primary amino groups present is determined from the 340 nm absorbance of the yellow TNP-amino derivatives.

An example of a thin layer chromatogram of a hydrolysate of a dansylated normal membrane preparation is shown in Fig. 3. The dansylderivatives of ε-amino-lysine groups, arising from membrane proteins, and of ethanolamine and serine, arising from phospholipids, can be distinguished in addition to the blue fluorescing dansylic acid, a hydrolysis artefact, and o-dansyl-tyrosine. Glucosamine, detected by Heller (9) in rhodopsin, is presumably completely acetylated, since its dansylderivative, which moves just ahead of serine in this solvent system, is not observed. No α-amino derivatives other than serine are present. The serine derives completely from PS, since it disappears upon enzymatic delipidation of the membrane (4,5) prior to dansylation (Fig. 3, R_{CH} and R_{CHAS}). This implies that the N-terminal amino acid of rhodopsin is blocked, in agreement with earlier observations of Albrecht (10). These conclusions are further supported by the quantitative determinations (Table II), showing good agreement between individual and total amounts of primary amino groups. Finally, amino acid analysis of dansylated membranes demonstrates that only the amount of lysine, ethanolamine, serine, and tyrosine has decreased upon dansylation (Table III). Our conclusion is therefore that the primary amino groups in the membrane are only provided by ε-amino-lysine, ethanolamine, and serine residues. Per mole of rhodopsin we detect about 16 moles of lysine, 28 moles of ethanolamine, and 8 moles of serine.

Chemical Modification of Amino Groups

Next we sought a method for the chemical modification of amino groups, which would influence the protein structure as little as possible to avoid affecting the chromophore binding. A suitable reagent was found in methylacetimidate (Fig. 4), which has several advantages: it is a small molecule, it preserves the charge of the ori-

Table II. Determination of free amino groups in rod outer segment membranes (nMol/mg lyophilisate)

	(n = 2) AAA	DNS-C1	TNBS	(n = 5) AAA	TNBS
Lys	122	127			
Eth-NH_2	216	227			
Ser	70	60			
Total	408	414	418	414	423
	±8	±20	±9	±7	±6

ginal amino group, it is active at physiological pH and room temperature and it is strictly specific for primary amino groups. Since it rapidly hydrolyses in water, we add a 20 fold excess to a membrane suspension in phosphate buffer of pH 7.5 and remove the hydrolysis products after about half an hour at room temperature by centrifugation. The sediment is resuspended in the same buffer and a second cycle is started by another addition of a reagent, and so on. The extent of modification is determined by the reaction of the remaining amino groups with TNBS.

The effect of modification is shown in Fig. 5. After five cycles nearly complete amidination is acclomplished, while about 70% of the rhodopsin is left intact. Under these circumstances 1.9 ± 0.4 primary amino group is left per molecule of rhodopsin, including the chromophore binding group. The regeneration capacity of the mo-

Table III. Amino acid analysis* of ROS-membranes before and after dansylation (per mole rhodopsin; n = 3)

	Before	After	Δ		Before	After	Δ
Gly	31.1	31.9	+ 2%	Phe	30.5	30.9	+ 1%
Asp	31.4	31.3	± 0%	His	7.4	7.3	- 1%
Thr	31.0	32.0	+ 4%	Arg	10.3	10.3	± 0%
Glu	38.9	40.4	+ 4%	Gluc-NH_2	3.4	3.8	+ 12%
Pro	25.8	25.3	- 2%	Tyr	17.8	6.0**	- 66%
Ala	36.5	37.1	+ 2%	Lys	15.6	0.3	- 98%
Val	28.8	28.8	± 0%	Ethanolam	27.2	0.8	- 97%
Met	11.2	14.4	+ 19%				
Ileu	18.4	18.4	± 0%	Ser(protein)	22.8	23.0	+ 1%
Leu	35.7	36.8	+ 3%	Ser(lipid)	9.1	0.1	- 99%

*Cys and Try (partially) destroyed during hydrolysis.
**Not corrected for de-DANSylation during hydrolysis.

AMIDINATION

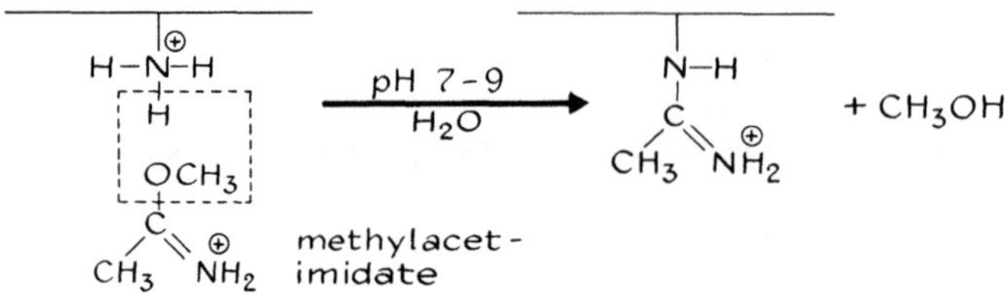

+ Reaction in H_2O at pH 7-9
+ small reagent
+ same charge as primary NH_2-group } no disturbance of tertiary structure
+ specific for primary aminogroups
- not stable during acid hydrolysis

Fig. 4. Reaction of methylacetimidate with primary amino groups

dified rhodopsin (relative amount of rhodopsin regenerated after illumination and subsequent incubation with excess 11-cis retinaldehyde in the dark), has also only slightly decreased to about 70%. Thus, the membrane structure is not severely affected by the chemical modification. Yet, when the membranes are illuminated before amidination, a different picture is obtained. As compared to "dark-adapted" membranes, illuminated membranes show a rapid decrease in regeneration capacity after about 60-70% modification. This is further proof for the involvement of an amino group in the chromophore binding. In rhodopsin this group is protected against amidination by the bound chromophore, but upon illumination the group is exposed and can be amidinated, after which it is no longer capable of binding retinaldehyde.

Table IV. ROS-membrane primary amino groups* before and after five amidination cycles

	Before	After		
		darkness	illumination	illumination + NADPH
(I) ε-amino lysine	16 + 1	1.4 + 0.3		
(lipid)ethanolamine	28 + 1	0.5 + 0.2		
(lipid)serine	8 + 1	0.1		
(II) ε-amino lysine	16 + 1	1.0 + 0.2	0.3 + 0.2	0.2
(lipid)ethanolamine	28 + 1	0.4 + 0.2	0.4 + 0.2	0.2
(lipid)serine	8 + 1	0.1	0.5 + 0.2	0.1

* including aldimine link

(I) : relative to the amount of rhodopsin present after amidination

(II): relative to the amount of rhodopsin present before amidination

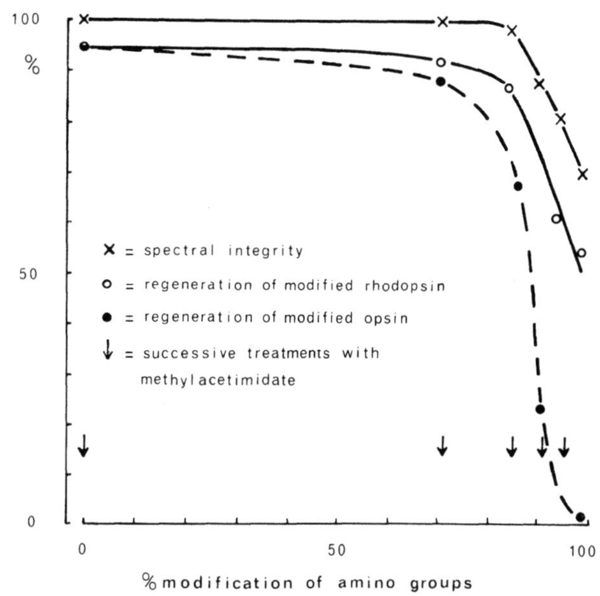

Fig. 5. Effect of amidination on the spectral integrity of rhodopsin and on the regeneration capacity of rhodopsin and opsin. The extent of modification was determined with TNBS

The fact that after extensive amidination less than one primary amino group is left in addition to the binding site, affords the opportunity to identify the binding site. This has been accomplished by treating the amidinated preparation with dansylchloride. After hydrolysis the thin layer chromatogram shows that all dansyl-amino-derivatives diminish with increasing extent of amidination. Quantitative analysis (Table IV) demonstrates that in extensively amidinated rhodopsin only the ε-amino-lysine group is present in sufficient amount (1.4 mole per mole of rhodopsin) to ac-

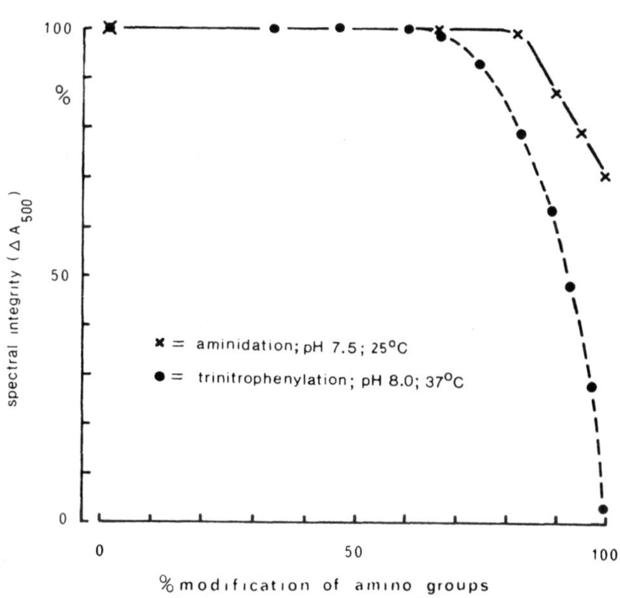

Fig. 6. Effect of amidination and trinitrophenylation on spectral integrity of rhodopsin

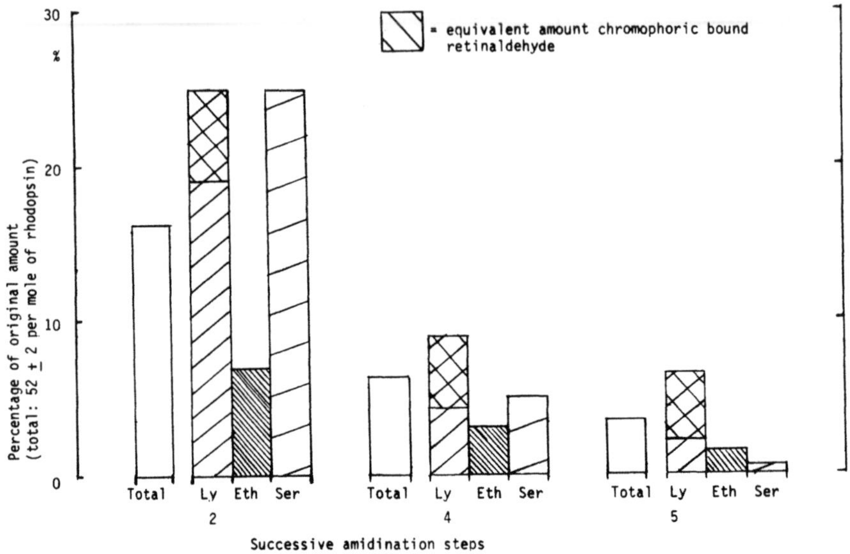

Fig. 7. Relative amounts of various primary amino groups left after various extent of amidination

---

count for the binding of retinaldehyde. This is further supported by the fact that amidination after illumination yields a 70% decrease in the amount of ε-amino-lysine that can be dansylated. This decrease is about equivalent to the amount of rhodopsin left after five amidination cycles. The "liberated" retinaldehyde appears to remain linked partly to membrane amino groups, since practically complete modification is only obtained after reduction by NADPH-addition during the illumination prior to amidination (Table IV).

These findings present in our opinion incontrovertible proof for the involvement of an ε-amino-lysine group in the binding of the chromophoric group in native rhodopsin and MII. Whether this involves the same lysine residue in rhodopsin and MII, is discussed in the paper by Bonting et al. (11).

In view of the different properties of TNBS as compared to those of methylacetimidate (MAI) (large group, no preservation of charge), we have also used this reagent for modification. Fig. 6 shows the difference between the two reagents. While MAI influences the membrane properties only to a small extent, trinitrophenylation at 37 °C causes disruption of the protein structure and ultimately abolishes the spectral integrity and the regeneration capacity of rhodopsin. A striking fact is that in both cases these effects manifest themselves only after the greater part of the amino groups has been modified. This is further emphasized by the fact that trinitrophenylation at room temperature virtually comes to a halt after about 70% modification without any effect on the 500 nm absorbance band of rhodopsin. Similar results were obtained by Dratz et al. (12) with another negative charged reagent: fluorescein-isothiocyanate. We wondered therefore whether the initial 70% modification might involve only amino groups of phospholipids, which account for 70% of the total number of free amino groups (ethanolamine 55%, serine 15%). Therefore, we analysed membranes amidinated to various extent by dansylation. The results show (Fig. 7) that phosphatidylethanolamine (PE) is very rapidly amidinated, while phosphatidylserine (PS) is more resistant and a small part of the ε-amino-lysine groups are

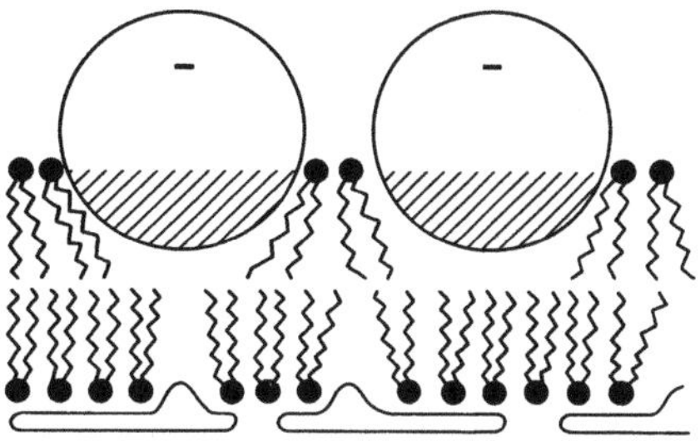

Fig. 8. Structure of the rod disc membrane, as proposed by Blasie (13)

also not easily accessible. Preliminary analysis of trinitrophenylated preparations gives similar results: At first PE-amino groups appear to be trinitrophenylated almost exclusively.

These results are understandable in the light of the membrane structure shown in Fig. 8, which is derived from the studies of Blasie (13). Most PE and PC groups presumably are not tightly associated with rhodopsin protein, which is also suggested by the ease with which they are hydrolised by phospholipase C (4). PS seems partly more intimately associated with the membrane proteins. The net negative charge on the rhodopsin protein could indicate a more interior location of part of the lysine residues. The negatively charged groups will at first tend to bind the positively charged MAI molecules, thus shielding the protein amino groups, so that amidination will first take place mainly at the phospholipid level. The negatively charged TNBS molecules will be repelled by the negative charge on the rhodopsin molecule, explaining why trinitrophenylation at first occurs almost exclusively at the phospholipids and why it does not proceed further at room temperature. Thus, the relatively slow modification of protein amino groups by certain reagents does not require postulation of a new concept of membrane structure, as proposed by Dratz et al. (12), in which the proteins are completely embedded in the lipid phase.

Conclusion

An enrichment procedure for rod photoreceptor membrane preparation has been worked out, which yields a pure membrane preparation with maximal and reproducible rhodopsin content without the use of detergents. Chemical analysis of the free amino groups of this preparation shows that they consist entirely of ε-amino groups of lysine residues in the protein and of serine and ethanolamine of the phospholipids phosphatidylserine and phosphatidylethanolamine. Treatment with methylacetimidate, followed by dansylation, hydrolysis, and quantitative thin layer chromatography prove that in native rhodopsin the chromophore is bound to an ε-amino group of lysine.

References

1. BOWNDS, D., G. WALD: Reaction of the rhodopsin chromophore with sodium borohydride. Nature (Lond.) 205, 254-257 (1965).
2. AKHTAR, M., P.T. BLOSSE, P.B. DEWHURST: The reduction of a rhodopsin derivative. Life Sci. 4, 1221-1226 (1965).
3. DAEMEN, F.J.M., P.A.A. JANSEN, S.L. BONTING: Biochemical aspects of the visual process XIV. The binding site of retinaldehyde in rhodopsin studied with model aldimines. Arch.Biochem.Biophys. 145, 300-309 (1971).

4. BORGGREVEN, J.M.P.M., J.P. ROTMANS, S.L. BONTING, F.J.M. DAEMEN: Biochemical aspects of the visual process XIII. The role of phospholipids in cattle rhodopsin studied with phospholipase C. Arch.Biochem.Biophys. 145, 290-299 (1971).
5. BORGGREVEN, J.M.P.M., F.J.M. DAEMEN, S.L. BONTING: Biochemical aspects of the visual process XVII. Removal of amino group containing phospholipids from rhodopsin. Arch.Biochem.Biophys. 151, 1-7 (1972).
6. DE GRIP, W.J., F.J.M. DAEMEN, S.L.BONTING: Biochemical aspects of the visual process. XVIII. Enrichment of rhodopsin in rod outer segment membrane preparations. Vision Res., in press.
7. DAEMEN, F.J.M., W.J. DE GRIP, P.A.A. JANSEN: Biochemical aspects of the visual process XX. The molecular weight of rhodopsin. Biochim.Biophys. Acta 271, 419-428 (1972).
8. FAGER, R.S., PH. SEGNOWSKI, E.W. ABRAHAMSON: Aqueous cyanohydrideborate reduction of the rhodopsin chromophore. Biochem.Biophys.Res. Commun. 47, 1244-1247 (1972).
9. HELLER, J.: Structure of visual pigments I. Purification, molecular weight and composition of Bovine visual pigment$_{500}$. Biochemistry 8, 2906-2913 (1968).
10. ALBRECHT, G.: Terminal amino acids of rhodopsin. J.Biol.Chem. 229, 477-487 (1957).
11. BONTING, S.L., J.P. ROTMANS, F.J.M. DAEMEN: Chromophore migration after illumination of rhodopsin. This volume pp. 39-44.
12. DRATZ, E.A., J.E. GAW, S. SCHWARTZ, W.H. CHING: Molecular organization of photoreceptor membranes of rod outer segments. Nature New Biology 237, 99-102 (1972).
13. BLASIE, J.K.: Net electric charge on photopigment molecules and frog retinal receptor disk membrane structure. Biophys.J. 12, 205-213 (1972).

Chromophore Migration after Illumination of Rhodopsin

S. L. Bonting, J. P. Rotmans, and F. J. M. Daemen
Department of Biochemistry, University of Nijmegen, Nijmegen/The Netherlands

The main facts known about the fate of the chromophore of rhodopsin after illumination can be summarized as follows. The chromophore in native rhodopsin, 11-cis retinaldehyde, is bound to an amino group by means of an aldimine bond (-C = N-). It is not bound to a phospholipid amino group (1,2,3). In metarhodopsin II (M II) the isomerized chromophore all-trans retinaldehyde is bound to the ε-amino group of a lysine residue of opsin. The decay of M II, possibly through an intermediate metarhodopsin III (M III) or pararhodopsin, leads <u>in vivo</u> and <u>in vitro</u> in the presence of NADPH to release of retinol, and <u>in vitro</u> in the absence of NADPH to release of all-trans retinaldehyde. The NADPH-dependent release of retinol is due to the action of a retinoldehydrogenase present in the rod sac membrane, which is capable of reduction of free retinaldehyde as well as of reductive hydrolysis of retinylidene aldimines (4,5).

In the previous paper by de Grip <u>et al.</u> (6) it is shown by means of chemical modification of all free amino groups of rhodopsin and displacement of the chromophore by a dansyl group that in native rhodopsin retinaldehyde is also attached to the ε-amino group of lysine.

Since transiminization of retinaldehyde easily occurs (1) and there are some 16 lysine groups per rhodopsin molecule, the question arises whether the chromophore remains attached to the same lysine group during the conversion of rhodopsin to M II. Another question is what happens to the chromophore after the decay of the latter intermediate: it is released completely or not, and how is the interaction with the retinoldehydrogenase system? These questions are considered in this report.

The approach used in this study was to illuminate the rhodopsin in a suspension of rod outer segment (ROS) membranes and then to fix the chromophore to its binding site by treatment with $NaBH_4$, which reduces the labile aldimine bond -C = N- to the stable amine bond -C - N-. The original chromophore binding site was then probed by treatment with 11-cis retinaldehyde: if the site is vacated by migration of the chromophore, a new light-sensitive pigment should result (Fig. 1).

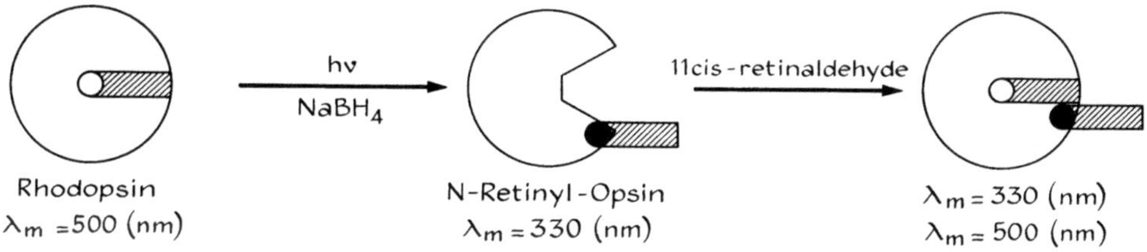

Fig. 1. Formation of retinylrhodopsin upon incubation of illuminated and reduced rhodopsin with 11-cis retinaldehyde (o aldimine bond at original binding site, ● reduced aldimine bond). This reaction can only occur, if the chromophore migrates to a new binding site prior to reduction

Scheme I: Reduction by $NaBH_4$ during and after illumination and reaction with 11-cis retinaldehyde

25 nmol rhodopsin (ROS) suspended in

1 ml 0.25 M phosphate buffer (pH 7.0; 20°C)

2 mg $NaBH_4$ is added to suspension: a) kept in dark
b) immediately illuminated
c) 1-60 min after illumination

10 min illumination (orange + infrared filter).

$NaBH_4$ washed out of ROS preparation

Resuspended in:

1 ml 0.067 M phosphate buffer (pH 6.3; 20°C)

5-fold molar amount 11-cis retinaldehyde (in 15 μl acetone)

3 hr incubation at 20°C.

ΔA_{500} determined (50 mM NH_2OH, Triton X-100).

Binding Site in Metarhodopsin II

The experimental details are outlined in Scheme I. In all these experiments aqueous suspensions of an enriched ROS preparation (7) were used without application of detergents in order to approach physiological conditions as much as possible.

When $NaBH_4$ was added immediately prior to illumination, we found a considerable amount of photopigment formed after treatment with 11-cis retinaldehyde when the reduction (visually observed as "bleaching") was slow (~1 min); however, when reduction was fast (< 5 sec), no photopigment was formed (Table I). Addition of $NaBH_4$ at various time intervals after illumination yielded gradually increasing amounts of the photopigment (Fig. 2). The resulting photopigment has a 500 nm and a 330 nm absorption peak, indicating the presence of a retinyl group as well as a retinylidene group, and is called retinylrhodopsin. The lack of retinylrhodopsin formation when reduction is complete within 5 sec of illumination indicates that in MII the chromophore is bound to the same lysine residue as in rhodopsin, which is also suggested by induced circular dichroism studies (8). The increasing amount of retinylrhodopsin formation with an increasing time interval between illumination and reduction (Fig. 2) indicates that upon decay of MII the chromophore migrates away from the original binding site.

Table I. Effect of reduction rate on photopigment formation

Rhodopsin	$\xrightarrow[NaBH_4]{h\nu}$	92% reduction ~1 min	$\xrightarrow[\text{retinaldehyde}]{\text{11-cis}}$	72% photopigment
Rhodopsin	$\xrightarrow[NaBH_4]{h\nu}$	100% reduction < 5 sec	$\xrightarrow[\text{retinaldehyde}]{\text{11-cis}}$	0% photopigment

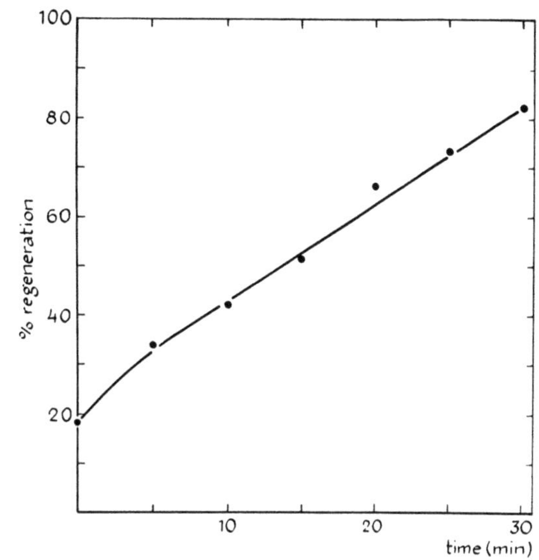

Fig. 2. Amount of retinylrhodopsin formed by reaction of illuminated and reduced rhodopsin with 11-cis retinaldehyde as a function of the time interval between illumination and addition of $NaBH_4$

State of Chromophore after Metarhodopsin II Decay

The binding state of the chromophore during MII decay was determined by two methods: protonation and $NaBH_4$ reduction. The former method involves stabilization of the aldimine bond by protonation through acidification, and calculation of the percentage of bound retinaldehyde from measurements of the absorbances at 380 and 440 nm (4). The latter method employs reduction by $NaBH_4$, followed by extraction and determination of retinol, which results from free retinaldehyde only. The results of the two methods, which are in good agreement, are shown in Fig. 3. Liberation of the chromophore is a slow process, which reaches a maximum of only 40% after 30 min at 20°C. About 60% of the chromophore remains bound to sites other than the original site. Extraction with chloroform-methanol, followed by thin layer chromatography, indicates that 30 min after illumination part of the chromophore is bound to phospholipid amino groups and part to protein amino groups.

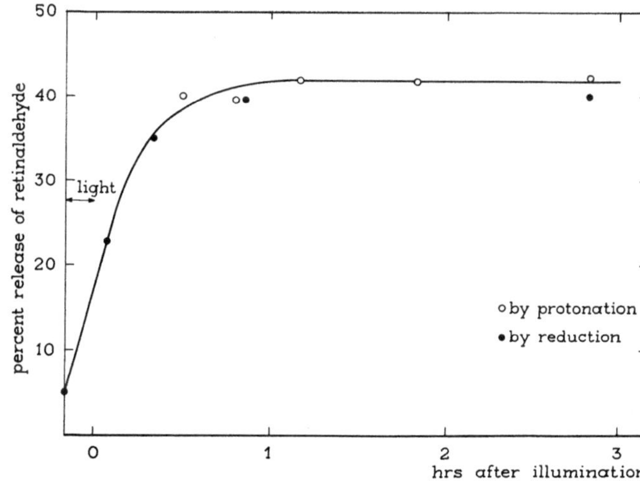

Fig. 3. Retinaldehyde release from rhodopsin after illumination as a function of the time elapsing after illumination

Scheme II. Retinoldehydrogenase assay

25 nmol rhodopsin (ROS: untreated, $NaBH_4$ treated, or $NaBH_4$-light treated) suspended in
200 µl 0.067 M phosphate buffer (pH 6.0)
30 nmol all-trans retinaldehyde
150 nmol NADPH
10 min incubation at 37°C
25 µl aliquots for retinaldehyde determination with thiobarbituric acid reagent.

Enzymatic Reduction of Chromophore

Since in the presence of NADPH during this period enzymatic reduction of free and aldimine linked retinaldehyde to retinol takes place, we assume the occurence of an intermediary linking of retinaldehyde to an amino group of the active site of the retinoldehydrogenase present in the photoreceptor membrane. Evidence for such an intermediate has been obtained in two ways. In the first approach we measured the retinoldehydrogenase activity as a function of the time elapsing between illumination and addition of $NaBH_4$. Experimental details are outlined in Scheme II. Presence of retinaldehyde on the active site during $NaBH_4$ addition will yield a stable amine bond, thus making the enzyme inactive. The finding of a minimum activity about 15 min after illumination (Fig. 4) suggests that at that time retinaldehyde is bound to a considerable extent to retinoldehydrogenase, while thereafter it appears to become bound to amino groups of phospholipids and proteins.

Confirmation of this finding has been obtained in an entirely different way by means of the amidination experiments of de Grip et al. (6). When the first four treatments with methylacetimidate (MAI) are carried out in the dark, 40% of the retinoldehydrogenase activity is left, while 92% of the amino groups are blocked (Table II). A fifth treatment with MAI completes amidination, leaving 15% retinoldehydrogenase activity if carried out in the light and no activity if carried out in the dark. In the former case the retinaldehyde can migrate to the enzyme protecting the active site

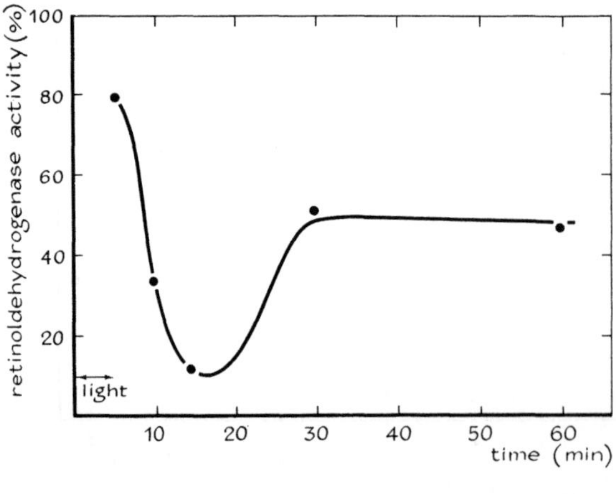

Fig. 4. Retinoldehydrogenase activity of rod sac membranes as a function of the time interval between illumination and addition of $NaBH_4$

Table II. Effect of NH_2-blocking on retinoldehydrogenase activity

treatment	NH_2 blocked %	rhodopsin left %	regeneration capacity %	retinol DH activity %
MAI*, 4 x in dark	92	80	75	40
MAI, 5th time, light	97	2	0	15
MAI, 5th time, dark	97	70	70	0

* MAI = methylacetimidate

against amidination, while in the latter case no migration of the chromophore occurs and the active site remains unprotected.

The migrational sequence, indicated by these experiments, is schematically represented in Fig. 5. The rhodopsin molecules ("rhod") are floating in the phospholipid layer of one face of the rod sac membrane. A molecule of retinoldehydrogenase ("ret. DH") is thought to be interspersed between the rhodopsin molecules. After the photolytic formation and decay of M II (no. 1) the chromophore wanders to the enzyme (no. 2), and if not reduced it proceeds to the amino groups of phospholipids (no. 3a) and proteins (no. 3b, incl. rhodopsin). The times indicated are those obtained under our experimental conditions ($20^{\circ}C$), and could be considerably shorter in vivo. In the presence of an adequate supply of NADPH the reduction to retinol could be nearly complete and hence, the binding to phospholipids and proteins slight.

Conclusion

Upon illumination of a detergent-free photoreceptor membrane suspension (pH 7.0; $20^{\circ}C$) the chromophoric group remains bound to the same lysine residue during photolytic conversion of rhodopsin to M II. During decay of the latter substance it wanders to the active site of retinoldehydrogenase, and if not reduced to retinol it migrates on to amino groups of phospholipids and proteins present in the rod sac membrane, while only 40% of the non-reduced chromophore is released under these conditions.

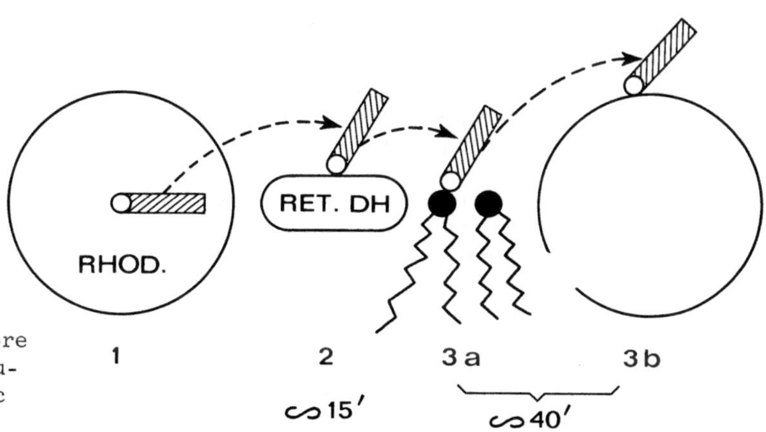

Fig. 5. Chromophore migration after illumination of rod sac membranes

References

1. DAEMEN, F.J.M., P.A.A. JANSEN, S.L. BONTING: Biochemical aspects of the visual process. XIV. The binding site of retinaldehyde in rhodopsin studied with model aldimines. Arch.Biochem.Biophys. 145, 300-309 (1971).
2. BORGGREVEN, J.M.P.M., J.P. ROTMANS, S.L. BONTING, F.J.M. DAEMEN: Biochemical aspects of the visual process. XIII. The role of phospholipids in cattle rhodopsin studied with phospholipase C. Arch.Biochem.Biophys. 145, 290-299 (1971).
3. BORGGREVEN, J.M.P.M., F.J.M. DAEMEN, S.L. BONTING: Biochemical aspects of the visual process. XVII. Removal of amino group containing phospholipids from rhodopsin. Arch.Biochem.Biophys. 151, 1-7 (1972).
4. DE PONT, J.J.H.H.M., F.J.M. DAEMEN, S.L. BONTING: Biochemical aspects of the visual process. VII. Equilibrium conditions in the formation of retinylidene imines. Arch.Biochem.Biophys. 140, 267-274 (1970).
5. DE PONT, J.J.H.H.M., F.J.M. DAEMEN, S.L. BONTING: Biochemical aspects of the visual process. VIII. Enzymatic conversion of retinylidene imines by retinoldehydrogenase from rod outer segments. Arch.Biochem.Biophys. 140, 275-285 (1970).
6. DE GRIP, W.J., F.J.M. DAEMEN, S.L. BONTING: The binding site of retinaldehyde in native rhodopsin. This volume pp. 29-38 (1972).
7. DE GRIP, W.J., F.J.M. DAEMEN, S.L. BONTING: Biochemical aspects of the visual process. XVIII. Enrichment of rhodopsin in rod outer segment membrane preparations. Vision Research 12, 1697-1707 (1972).
8. WAGGONER, A.S., L. STRYER: Induced optical activity of the metarhodopsins. Biochemistry 10, 3250-3254 (1971).

Discussion

Ch. Baumann: According to your reaction scheme, the chromophore migrates back to the opsin if the dehydrogenase is not active. Have you any information about the absorptive properties of this newly formed chromoprotein?

S.L. Bonting: The chromophore need not go only to retinoldehydrogenase or opsin; it might also go to amino groups of the 13% non-rhodopsin protein present in cattle photoreceptor membrane. Although we have not made a specific study of the newly formed chromoprotein(s), we have no evidence for a MIII (λ_{max} 470 nm) in the protonation procedure, only for a non-protonated aldimine link (λ_{max} 360 nm).

D. Bownds: You propose that the retinyl group in rhodopsin might migrate to a new site in MII (Fig. 1). One might then add 11-cis retinal to N-retinyl-opsin made from MII and regenerate rhodopsin's red color. This can be demonstrated only if one quantitatively forms and reduces MII and then finds that addition of 11-cis retinal causes the appearance of 500 nm absorption. Has this experiment been done?

S.L. Bonting: We propose that if the retinylidene group upon illumination of rhodopsin migrates to a new site and is then fixed by $NaBH_4$ reduction, the original binding site should be available for reaction with 11-cis retinaldehyde, resulting in a retinylrhodopsin (with maxima at 500 and 330 nm). Our finding is that upon complete reduction within 5 sec of illumination, no retinylrhodopsin is formed by subsequent incubation with 11-cis retinaldehyde. However, when reduction takes 1 min or more, this artificial pigment is formed in amounts increasing with time. Our conclusion is that at the moment that MII is formed, the chromophore is still attached to the original binding site, but that during the decay of metarhodopsin it migrates to other binding sites. In our experiments the amount of retinylrhodopsin formed approximates quantitatively the amount of rhodopsin bleached to MII.

II. Photolysis and Intermediates of the Pigments

The Kinetics of Early Intermediate Processes in the Photolysis of Visual Pigments

Edwin W. Abrahamson
Department of Chemistry, Case Western Reserve University, Cleveland, OH 44106/USA

The sensory process of vision is unique in that the molecules actually receiving the sensory stimulus have characteristic visible absorption spectra. This property not only permits their ready identification but also affords a very sensitive means of monitoring molecular events that follow light absorption. As these spectral changes can be measured precisely on a time scale of the order of milliseconds and less, concomitant with early electrophysiological events, the possibility of relating the two becomes iminent.

One of the methods of pointing up key molecular features of biological processes is through the study of the comparative biochemistry of related systems. We have, accordingly, followed this path in trying to gain insight into the key molecular properties and events in the photolysis of vertebrate (bovine) and invertebrate (squid) visual pigments.

From a structural point of view photoreceptor cells are of two types, the ciliary cell of vertebrates and the rhabdomeric cell of invertebrates. The arrangement and disposition of the membranes is distinctly different in the two systems. In vertebrates the visual pigment is fixed in an ordered pattern in the surface of unitary disclike sac membranes which form a stack of about 500 in number surrounded, in most cases, by an outer plasma membrane (1). In the case of the more numerous rod cells the outer segments are clearly imbedded in a lipid pigment epithelium layer which appears to mediate the transport of vitamin A between the outer segment and the vascular system. In the rhabdomeric outer segment of the invertebrates the visual pigment is contained in the walls of tubular microvilli which project in a honeycomb pattern laterally (perpendicular) to the principal axis of the receptor cell (1). There appears to be no direct exchange of vitamin A with the vascular system.

Visual pigments are chromolipoproteins whose chromophore is the same in both vertebrates and intervertebrates, being derived from 11-cis retinal or its 3-dehydroderivative (Fig. 1). This chromophore, in the native state, is linked to a protein, opsin (2), which the weight of evidence suggests is a protonated Schiff base linkage (3). The protein component, however, is distinctly different in the two systems. In vertebrates the protein has a molecular weight about one-half that of the invertebrate squid (1) and the amino acid compositions, though highly hydrophobic in both cases, are distinctly different on a mole fraction basis (1).

Fig. 1. Structure of 11-cis retinal (left) and 11-cis-3-dehydro retinal (right)

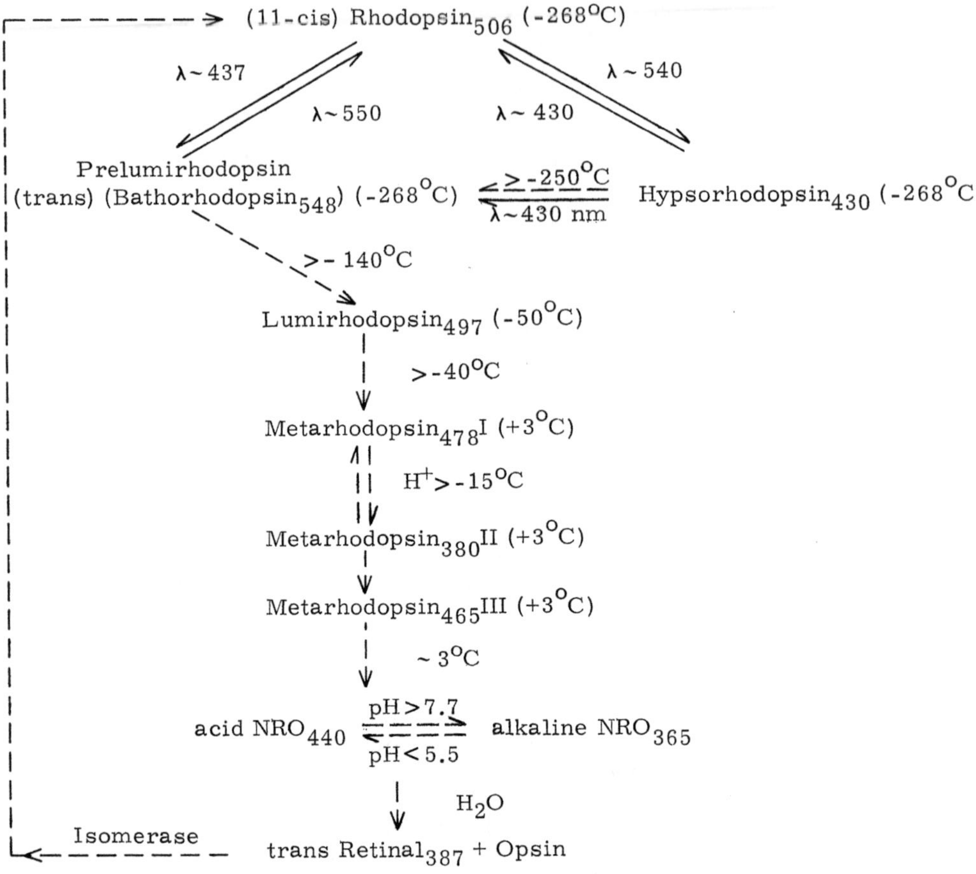

Fig. 2. Intermediate sequence in the photolysis of vertebrate (bovine) rhodopsin. Arrows with solid lines denote photoreactions while those with dotted lines denote thermal, dark reactions. Temperatures given in parentheses denote temperatures at which spectra of intermediates were taken

---

The lipid composition of vertebrate and invertebrate visual pigment membranes is also quite different. Although both show a predominance of phospholipids the vertebrate has a much higher degree of unsaturation in its fatty acid chains and this is concentrated more in the larger chain fatty acids than in the invertebrate squid (3). This tends to bear out the x-ray diffraction evidence suggesting that the vertebrate membrane has a distinct fluidity (4). The apparent function of the squid microvilli in detecting polarized light (5) suggests that the chromophore is fixed in a rigid membrane structure. This is consistent with the observed deformation properties of the membrane (6).

Although the lipid is apparently not involved in a direct linkage to the chromophore as we have earlier suggested (7) it is essential to the formation of the native visual pigment and the maintenance of its native conformation (8,9). Zorn and Futterman (8) as well as Shichi (9) have found that the incubation of 11-cis retinal with bovine opsin will not result in pigment formation unless phospholipid is present, although the specific lipid requirement for formation has not yet been determined. Shichi

Table I. Thermal decay of prelumirhodopsin$_{548}$

I) Grellman <u>et al.</u> (34)

Rhodopsin solutions	Rate constants (sec^{-1}) -50°C	-67°C	t (38°C)[a] sec	$\Delta H^{\ddagger}$ kcal/mole	log A	$\Delta S^{\ddagger b}$ e.u.
k_1	3960	1270	$1.5 \cdot 10^{-7}$	10±2	13.7	5
k_2	715	219	$8.6 \cdot 10^{-7}$	10±2	12.9	0.9
k_3	251	53	$3.8 \cdot 10^{-5}$	10±2	12.2	-2

II) Pratt <u>et al.</u> (12)

Rod particles	-50°C	-65°C				
k_1		2580	$1.3 \cdot 10^{-8}$	12.5±3	16.5	18
k_2	5180	417	$7.8 \cdot 10^{-8}$	12.5±3	15.7	14
k_3	1026	118	$2.8 \cdot 10^{-7}$	12.5±3	15.2	12

III) Busch <u>et al.</u> (11)

Rhodopsin solution	17.5°C	29.3°C				
k	$2.7 \cdot 10^{7}$	$4.1 \cdot 10^{7}$	$1.0 \cdot 10^{-8}$	6.0		-5

[a] Calculated from Arrhenius equation.
[b] Log A was reported. $\Delta S^{\ddagger}$ calculated for T = -60°C.

has also found that the circular dichroism (CD) spectrum of rhodopsin which normally undergoes little change on photolysis changes significantly when delipidated previous to photolysis (10).

The intermediate sequence in the photolysis of vertebrate rhodopsin is shown in Fig. 2. From the point of view of the generation of a neural signal only the first four processes through metarhodopsin II (MII) are important (1,3). The processes beyond this are much too slow to be involved in this function although they may play a role in the regeneration of the native visual pigment.

The primary photochemical process is the photoconversion of rhodopsin to prelumirhodopsin, and this appears to be a simple photoisomerization process about the 11-12 double bond of the chromophore occuring in times less than 20 picoseconds (11). The subsequent conversion of prelumirhodopsin to lumirhodopsin has been studied at low temperatures by Pratt et al. (12) using the conventional flash photolysis technique (Table I). More recent data using a picosecond laser technique has been reported at room temperature which yields essentially the same activation parameters (11). At physiological temperatures (~37°C) the first-order half-life of the reaction appears to be about 10 nsec and the activation parameters are consistent with the notion that it is a simple thermal isomerization about a single double bond, - perhaps the 12-13.

Table II. The thermal decay of lumirhodopsin$_{497}$ treated as simultaneous first-order processes

	Process	Temp °C	k sec^{-1}	$\Delta F^{\ddagger}$ kcal/mole	$\Delta H^{\ddagger}$ kcal/mole	$\Delta S^{\ddagger}$ cal/degree mole
Rapp (15)	k_2	-13.8	0.0635	16.5	17.6	4.0
66% glycerol	k_3		0.0046	17.9	18.4	1.9
in aqueous digitonin	k_2	-20.0	0.0224	16.6		
(pH = 7)	k_3		0.0020	17.9		
	k_2	-30.7	0.0061	16.6		
	k_3		0.0038	17.9		
Erhardt _et al._ (16)	k_1	-21.4	0.03	16.9	16.3	-2.3
50% glycerol in	k_2		0.0097	17.0	3.60	-53.0
aqueous digitonin	k_3		0.0015	17.9	5.29	-52.0
Matthews and Wald as quoted by Hubbard _et al._ (17) treated as simple first-order process	k	-20.0	0.00045	19.0	60	+160

One point which is common to the early intermediate processes is that they appear to involve more than one form of a particular intermediate (1, 3). This is particularly true for certain detergent preparations of the visual pigment such as digitonin although the group at Bell Telephone Laboratories (13) now find a single first-order process for the decay of prelumirhodopsin and for metarhodopsin$_{478}$I (MI) in the detergent lauryl diamine oxide (LDAO). Von Sengbusch observes only a single first-order process in the decay of MI when digitonin solutions are stored at liquid nitrogen temperatures (14). The reason for these apparent multiforms of the intermediates is not known but in all probability it is related to the configuration of the lipoprotein moiety, possibly a function of the lipid composition of the micelle. Single first-order processes have been observed in our laboratory for rod outer segment suspensions in the case of MI→MII process (15) but multiform processes have been observed by others for the earlier processes.

The thermal conversion of lumirhodopsin→ MI is a process somewhat difficult to interpret in molecular terms because of the inconsistencies in the experimental data (Table II). The more recent flash photolysis data from our own laboratory on digitonin micelles when interpreted in terms of multiform first-order processes indicates little in the way of configurational change of the lipoprotein moiety which is consistent with our earlier findings that the process involves no pH changes or exposure to thiol groups (16). However, the Harvard laboratory reports only a single first-order process with parameters indicating substantial configurational changes in the lipoprotein moiety (17). This data was, however, apparently taken after steady illumination and may represent only the latter stages of the process.

Table III. Rate data for bovine metarhodopsin$_{478}$I → metarhodopsin$_{380}$II treated as simultaneous first-order processes

		Rate constants sec^{-1}		$\Delta G^{\ddagger}$ $\frac{kcal}{mole}$	$\Delta H^{\ddagger}$ $\frac{kcal}{mole}$	$\Delta S^{\ddagger}$ $\frac{cal}{degree\ mole}$
Wulff et al. (35)		at 28°C	at 36.7°C			
aqueous digitonin	k_1	3,300	22,000	12.5	36.2	78
pH = 7	k_2	120	2,700	14.5	50.2	119
	k_3	21.0	200	15.8	43.0	90
Abrahamson et al. (19)		at 6.7°C	at 15.8°C			
aqueous digitonin	k_1	577				
pH = 7	k_2	45.5	303.0	15.0	37.1	76.5
	k_3	0.88	9.9	15.6	43.1	95.2
Ostroy et al. (21)		at -9.2°C	at -13.7°C			
33% glycerol-aqueous digitonin pH = 5.85	k_2	0.0353	0.0051		59.9	162
pH = 4.5	k_2	0.0836	0.0121		58.7	159
Pratt et al. (12)		at 5°C	at 25°C			
33% glycerol-aqeuous digitonin	k_2	0.45	126		43	97
pH = 7	k_3	0.09	17		41	83
Matthews and Wald after Hubbard et al. (17)		Rate constants not reported		19	60	160
Von Sengbusch (14)		at 37°C				
digitonin centrifugated 4000 g	k	1,150 ± 110		E_A = 33.0 ± 1.5 kcal/mole		
Pratt et al. (12)		at 10°C	at 20°C			
bovine ROS	k_1	6.0	50	35	28	70
	k_2	1.3	12	37	29	74
Rapp (15)		at 18°C	at 39.4°C			
Sonicated bovine ROS	k	41.4	1480	14-15	30.7	54.8
Hagins (32)		at 12°C	at 26°C			
excised rabbit eye	k	30.0	600	13.0	37.6	80.0
Ebrey (33)		at 60°C	at 13.0°C			
rat retina measured as decay of ERP	k	~3.5	~28	~15	41 ± 8	91+20
Von Sengbusch (14)		at 37°C				
bovine ROS suspension	k	1,180 ± 90		E_A = 33.7 ± 1.0 kcal/mole		

The decay of MI was the intermediate process for which the kinetics were first studied in solution (18,19). As pointed out previously the data indicated that several different spectrally identical forms of the intermediate were apparent each with different first-order decay constants. This is the first process in the intermediate sequence to show pronounced environmental effects (20,21) in that water and hydrogen ion are required for the reaction. In Table III are compiled the data on this reaction from various laboratories including studies in aqueous digitonin micelles, in whole and sonicated rod outer segments (ROS), and in the retina.

The pH effect on the rate is quite evident from the low temperature data of Ostroy et al. (21) and the more recent room temperature data from our laboratory (22).

Falk and Fatt (23) using conductivity and Erhardt et al. (16) and Emrich (24) using dye indicators show an uptake of one proton in the process and this follows the kinetics of the 478 to 380 nm change in spectral absorbance maxima. Oddly enough the spectral change is indicative of the loss of a proton at the binding site so that the overall process must reflect changes in the periphery of the lipoprotein, a point borne out by the activation parameters which are consistent with substantial configurational changes in the lipoprotein moiety. This point, however, is not supported by the ORD and CD data (25,26).

The decay of MI in rod outer segments (ROS), whether whole or sonicated, appears to be much simpler than in micellar suspensions. Aside from the data of Pratt et al. (12), only a single first-order decay process is apparent which probably reflects the stabilizing character of the membrane milieu on rhodopsin molecules. We have, however, noted a small deviation in the very initial stages of the decay which might indicate the presence of a small fraction of a faster decaying species but, quite likely, this is an artifact.

It is interesting to compare the multiform rate data of micellar rhodopsin with the single decay rates of MI in ROS and in retina. There is no great difference in the two as shown in the decay of lumirhodopsin$_{497}$ but, in general, the ROS decay rates tend more toward the slower of the multiform processes of micellar rhodopsin. It is also of interest to note how well the data on sonicated bovine ROS agree with the much less precise data on the intact excised eyes of the rabbit and rat.

As the MI→MII process is the earliest one in the sequence to show chemical involvement of the environment, i.e. pH effects, and significant macromolecular configurational changes according to its kinetics, it appears to be the most likely canditate reaction of the early intermediate processes for initiating of neural activity, particularly since it appears to correlate temporally with the early receptor potential (ERP) of the mammalian electroretinogram (27).

The intermediate picture in squid rhodopsin is decidedly different than in vertebrates although there are striking similarities in the spectra of the native pigment and its earliest (prelumirhodopsin) intermediate. It is generally assumed that the sequence established for squid by the Harvard laboratory (28,29) and shown in Fig. 3 is common to all invertebrates although it is a point yet to be proved.

The primary photochemical process in the photolysis of squid rhodopsin appears to be identical to the vertebrate case. No substance analogous to hypsorhodopsin, however, has yet been identified. Beyond the prelumirhodopsin stage events clearly differ from the vertebrate case. Nothing is known, to date, as to the character of the prelumirhodopsin → lumirhodopsin conversion but the conversion of lumirho-

Fig. 3. Intermediate sequence in the photolysis of squid rhodopsin

(11-cis) Rhodopsin (25°C)
↓ hν, -195°C
(trans) Bathorhodopsin$_{550}$ (-195°C)
↓ -170°C
Lumirhodopsin$_{530}$ (-50°C)
↓ H_3O^+, -20°C
acid Metarhodopsin$_{500}$ ⇌ alkaline Metarhodopsin$_{380}$

dopsin → metarhodopsin has a character more in keeping with the MI → MII process of vertebrate rhodopsin. Erhardt and Abrahamson (30) have found that the lumirhodopsin → metarhodopsin process is acid catalyzed and thiol groups become titratable. The terminal intermediate, metarhodopsin, is an acid base indicator like N-retinylidene opsin although its acid form very closely resembles the spectra of native squid (Loligo pealii) rhodopsin save for its enhanced molar absorbance.

So far only the kinetics of lumirhodopsin to metarhodopsin have been studied. Our early studies (30) showed the process to proceed at classically measurable rates at low temperatures exhibiting a single first-order decay rate with activation parameters consistent with the notion that substantial lipoprotein configurational changes were occuring. This has been borne out by the recent studies in our laboratory (31)

Table IV. Rate data for the lumirhodopsin to metarhodopsin conversion in aqueous digitonin micellar preparation of squid rhodopsin

process	rate constants sec^{-1}			$\Delta H^{\ddagger}$ kcal/mole	$\Delta S^{\ddagger}$ cal/degree mole
Abrahamson and Erhardt (30), pH = 6.6 in 2:1 glycerol: H_2O after one minute illumination				40	90
Brauner (unpublished), flash photolysis of aqueous digitonin suspensions pH = 9.7	at 3.9°C 8.8	at 20.3°C 86.9	at 31.8°C 332.4	20.5±1	20±1
pH = 7.5	at 0°C 2.89	at 10.8°C 7.66	at 16.5°C 59.0	26.5±1	
pH = 6.65	at 0°C 135.9	at 12.3°C 2303	at 19.3°C 8366	34.5	78

using flash photolysis. These later studies show more clearly the pronounced effect of pH on the reaction (Table IV). Conversion to acid metarhodopsin (pH = 6.65) is faster by two orders of magnitude near room temperature than conversion to alkaline metarhodopsin (pH = 9.7) but the $\Delta H^{\ddagger}$ for the latter process is almost 15 kcal lower. This would imply a considerable configurational change in the peripheral lipoprotein in the conversion.

If one compares the vertebrate and invertebrate systems so far studied two common features emerge: 1) The primary photochemical process the conversion of rhodopsin to prelumirhodopsin (bathorhodopsin), i.e. the photoisomerization of the chromophore from the 11-cis to trans form and, 2) a near temporal equivalence of the early intermediate processes metarhodopsin$_{478}$I $\rightarrow$ metarhodopsin$_{380}$II in vertebrates and the lumi$\rightarrow$ metarhodopsin in squid vertebrate, both occuring under physiological conditions in a fraction of a millisecond and involving pronounced environmental (pH) sensitivity and apparently significant configurational changes in their lipoprotein moieties. One might logically conclude, therefore, that the common primary photochemical process gives rise to these key thermal processes which, in their respective system, initiates the chain of electrophysiological events.

Acknowledgement. The work reported herein was supported by grants No. EY00209 and No. EY00471 from the Eye Institute of the National Institutes of Health.

References

1. ABRAHAMSON, E.W., J.R. WIESENFELD: The structure, spectra and reactivity of visual pigments. In: Handbook of Sensory Physiology, Vol VII/1, ed. H.J.A. DARTNALL, pp. 67-121. Springer, Berlin-Heidelberg-New York (1972).
2. FAGER, R.S., P. SEJNOWSKI, E.W. ABRAHAMSON: Aqueous cyanoborohydride reduction of the rhodopsin cohromophore. Biochem.Biophys.Res. Comm. 47, 1244-1247 (1972).
3. ABRAHAMSON, E.W., R.S. FAGER: The chemistry of vertebrate and invertebrate visual photoreceptors. In: Current Topics in Bioenergetics, ed. R.SANADI. Academic Press, N.Y. (1973).
4. BLASIE, J.K., C.R. WORTHINGTON: Planar liquid-like arrangement of photopigment molecules in frog retinal receptor disk membranes. J.Mol.Biol. 39, 417-439 (1969).
5. MOODY, M.F., J.R. PARRIS: The discrimination of polarized light by octopus: Behavioral and morphological study. Z.Vergl.Physiol. 44, 268-291 (1961).
6. MASON, W.T.: unpublished observations (1972).
7. POINCELOT, R.P., P.G. MILLAR, R.L. KIMBEL, E.W. ABRAHAMSON: Lipid to protein chromophore transfer in the photolysis of visual pigments. Nature (Lond.) 221, 256-257 (1969).
8. ZORN, M., S. FUTTERMAN: Properties of rhodopsin dependent on associated phospholipid. J.Biol.Chem. 246, 881-886 (1971).
9. SHICHI, H.: II. Phospholipid requirement and opsin conformation for regeneration of bovine rhodopsin. J.Biol.Chem. 246, 6178-6182 (1971).
10. SHICHI, H.: Circular dichroism of bovine rhodopsin. Photochem.Photobiol. 13, 449-502 (1971).
11. BUSCH, G.E., M.L. APPLEBURY, A.A. LAMOLA, P. RENTZEPIS: The kinetics of prelumirhodopsin at physiological temperatures. Proc.Nat.Acad. Sci. U.S., in press (1972).

12. PRATT, O., R. LIVINGSTON, K.H. GRELLMAN: Flash photolysis of rod particle suspensions. Photochem.Photobiol. 3, 121-127 (1964).
13. LAMOLA, A.A.: private communication (1972).
14. SENGBUSCH, G.VON, H. STIEVE: Flash photolysis of rhodopsin. I. Measurements on bovine rod outer segments. Zeitschr.Naturforsch. 26b, 488-489 (1971).
15. RAPP, J.: Ph.D.Thesis: Case Western Reserve University, Cleveland, Ohio (1971).
16. ERHARDT, F., S. OSTROY, E.W. ABRAHAMSON: Protein configuration changes in the photolysis of rhodopsin. I. The thermal decay of cattle lumirhodopsin in vitro. Biochem.Biophys.Acta 112, 256-264 (1966).
17. HUBBARD, R., D. BOWNDS, T. YOSHIZAWA: The chemistry of visual photoreception. Cold Springs Harbor Symp.Quant.Biol. 30, 301-315 (1965).
18. LINSCHITZ, H., V. WULFF, R. ADAMS, E.W. ABRAHAMSON: Light initiated changes of rhodopsin in solution. Arch.Biochem. 68, 233-236 (1957).
19. ABRAHAMSON, E.W., J. MARQUISEE, P. GAVUZZI, J. ROUBIE: Flash photolysis of visual pigments. Z.Elektrochem. 64, 177-180 (1960).
20. MATTHEWS, R., R. HUBBARD, P. BROWN, G. WALD: Tautomeric forms of metarhodopsin. J.Gen.Physiol. 47, 215-240 (1963).
21. OSTROY, S., F. ERHARDT, E.W. ABRAHAMSON: Protein configuration changes in the photolysis of rhodopsin. II. The sequence of intermediates in thermal decay of cattle metrahodopsin in vitro. Biochem.Biophys.Acta 112, 265-277 (1966).
22. KEAR, K., W. ALDRED, A. BRAUNER: unpublished data (1972).
23. FALK, G., P. FATT: Rapid hydrogen uptake of rod outer segments and rhodopsin solutions on illumination. J.Physiol. 183, 211-224 (1966).
24. EMRICH, H.M.: Optical measurements of the rapid pH-change in the visual process during the metarhodopsin I-II reaction. Zeitschr.Naturforsch. 26b, 352-356 (1971).
25. CASSIM, J., C.N. RAFFERTY, D. MC CONNELL: Ultraviolet circular dichroic studies on retinal photoreceptor outer segment membranes. Biophys. Soc.Abstr. 16 Ann.Meeting, Toronto 205a (1972).
26. WAGGONER, A.S., L. STRYER: Induced optical activity of the metarhodopsins. Biochemistry 10, 3250-3254 (1971).
27. CONE, R.A., W.H. COBBS: Rhodopsin cycle in the living eye of the rat. Nature (Lond.) 221, 820-822 (1969).
28. HUBBARD, R., R.C.C. ST.GEORGE: The rhodopsin system of the squid. J.Gen.Physiol. 41, 501-528 (1958).
29. YOSHIZAWA, T., G. WALD: Transformations of squid rhodopsin at low temperatures. Nature (Lond.) 201, 340-345 (1964).
30. ABRAHAMSON, E.W., F. ERHARDT: Protein configuration changes in the photobleaching of rhodopsin. Fed.Proc. 23, 384 (1964).
31. BRAUNER, A.: unpublished data (1972).
32. HAGINS, W.A.: Ph. D. Thesis: University of Cambridge (1957).
33. EBREY, T.: Ph. D. Thesis: University of Chicago (1967).
34. GRELLMAN, K H., R. LIVINGSTON, D. PRATT: A flash photolytic investigation of rhodopsin at low temperatures. Nature (Lond.) 193, 1258-1260 (1962).
35. WULFF, V.J., R.G. ADAMS, H. LINSCHITZ, E.W. ABRAHAMSON: Effect of flash illumination on rhodopsin in solution. Ann.N.Y.Acad.Sci. 74, 281-290 (1958).
36. ROBINSON, W.E., A. GORDON-WALKER, D. BOWNDS: Molecular weight of frog rhodopsin. Nature (Lond.) 235, 112-114 (1972).
37. DAEMEN, F.J.M., W.J. DE GRIP, P.A.A. JANSEN: Biochemical aspects of the visual process. XX. The molecular weight of rhodopsin. Biochem.Biophys. Acta 271, 419-428 (1972).

Discussion

S.L. Bonting: With regard to the discrepancy between the molecular weights of rhodopsin obtained from amino acid analysis and from sodium dodecylsulfate gel-electrophoresis, I should like to mention that agreement between the values from these two methods has recently been obtained for frog rhodopsin by Bownds and co-workers (36) and for cattle rhodopsin by our group (37). The phospholipid content does not cause any discrepancy, because we have shown that delipidated rhodopsin gives the same values with SDS-gelelectrophoresis as non-delipidated rhodopsin. The detergent apparently removes the phospholipids.

E.W. Abrahamson: By acrylamide electrophoresis our laboratory obtains a molecular weight of 35-37,000 for bovine rhodopsin both by SDS methods and in emulphogene micelles of the native form using the method of Smith and Hedrick. It seems likely that the earlier estimates by AAA may have been too low.

Photoselection and Linear Dichroism of Retinal Isomers and Visual Pigments

Joram Heller and Joseph Horwitz
Jules Stein Eye Institute, UCLA School of Medicine, Los Angeles, CA 90024/USA

It has been known for about ten years that when frozen solutions (glasses) of visual pigments at $77^{o}K$ are illuminated with visible light, there are changes in the absorption spectrum of the pigments (1,2,3). These changes in the absorption spectrum, which in general are fairly small (less than 20 nm shifts in the absorption maxima), are dependent on the wavelength of the illuminating light and show the typical behavior of the photochromic systems.

We have exploited the method of photoselection to measure small changes in the absorption spectrum of illuminated samples which are embedded in rigid media. Isotropic samples of frozen solutions, in which the photosensitive molecules are trapped in their original random orientation, were illuminated with monochromatic light, which was linearly polarized in the x-axis (Fig.1). Light was absorbed only by those molecules whose transition moment was favorably oriented with respect to the illuminating light, since the absorption of light depends on the square of the scalar product of the transition moment vector of the chromophore and the vector of the electric field of the propagating light. The absorption of light by these photosensitive molecules produced stable photoproducts that had different absorption spectra from that of the molecules that did not absorb. The photoproducts were all in a particular orientation since they were produced by linearly polarized light

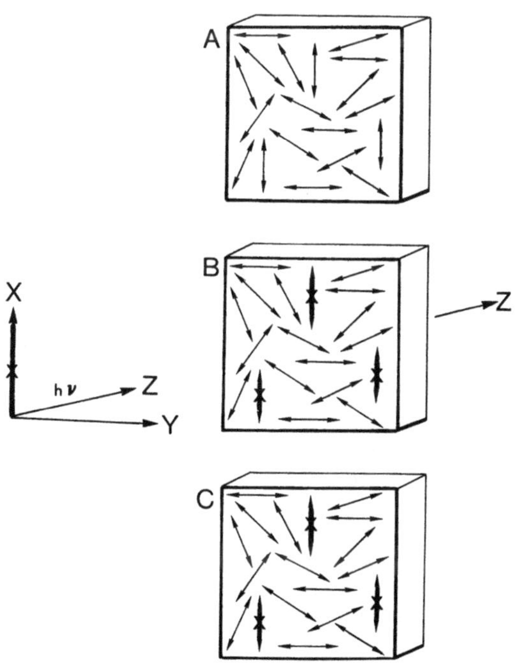

Fig. 1. Principle of photoselection. By freezing the sample the molecules are initially trapped in a rigid medium at their random orientation in solution and the sample as a whole is isotropic (a). Illumination with light that is linearly polarized along the x-axis is absorbed only by molecules that have their transition moments favorably oriented with respect to the light (b). The stable photoproducts which are thus produced, and which show a different absorption spectrum, are oriented and the sample as a whole exhibits linear dichroism

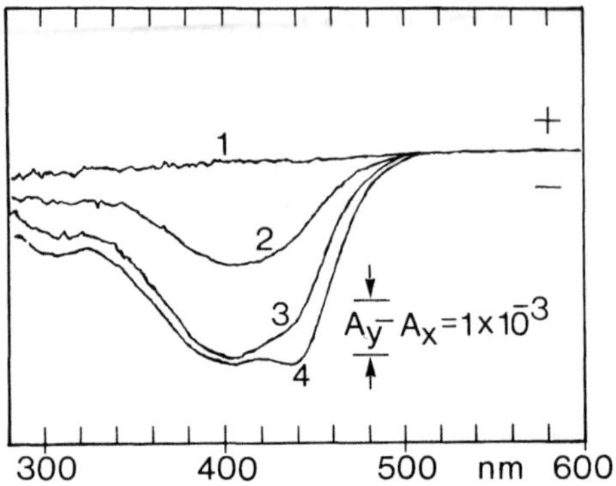

Fig. 2. Instrument traces of the linear dichroism spectra of 0.73 mM 11-cis retinal dissolved in glycerol:ethanol (1:1 v/v). Path length 0.10 mm, scanning speed 200 nm/min, time constant 100 msec. Curve 1: Linear dichroism spectrum of the sample at 77° K before illumination. Curve 2: After 15 min illumination at 360 nm with a spectral half-intensity bandwidth of 5 nm. Curve 3: After a total of 45 min illumination. Curve 4: After total of 150 min illumination

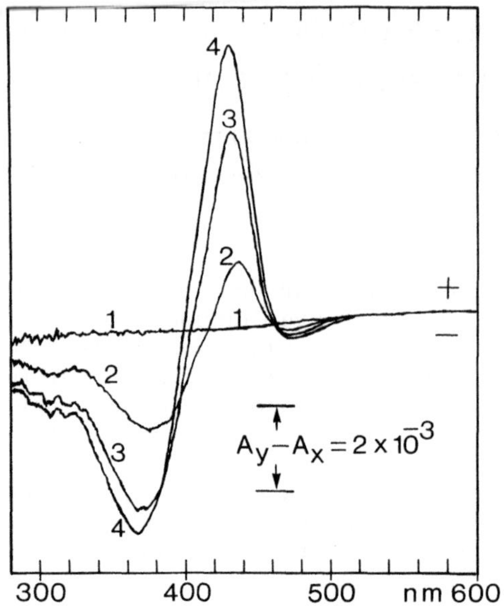

Fig. 3. Instrument traces of the linear dichroism spectra of 0.66 mM 11-cis retinal dissolved in glycerol:ethanol (1:1 v/v). Path length 0.10 mm, scanning speed 200 nm/min, time constant 100 msec. Curve 1: Linear dichroism spectrum of the sample at 77°K before illumination. Curve 2: After 15 min illumination at 440 nm with a spectral half-intensity bandwidth of 10 nm. Curve 3: After a total of 60 min illumination. Curve 4: After a total of 120 min illumination

(photoselection). The molecules that did not absorb light were oriented too, since they were arranged in such a fashion as not to have interacted with the linearly polarized light. Thus the sample as a whole became anisotropic and exhibited linear dichroism (Fig. 1). The linear dichroism (LD) spectra were recorded as described in detail elsewhere (Horwitz and Heller, submitted for publication, 1972). The LD spectrophotometer measures A_y-A_x: the absorbance in the y-axis minus the absorbance in the x-axis. The instrument is capable of measuring an A_y-A_x of 10^{-5} absorbance units.

When samples of all-trans, 9-cis, 11-cis, or 13-cis retinal were dissolved in ethanol:glycerol (1:1) and frozen to 77°K, the samples were isotropic and did not exhibit any linear dichroism as shown by the completely flat baseline (Figs. 2,3,4). If these samples were then illuminated for various times with monochromatic, linearly polarized light at wavelength corresponding to the main absorption peak of retinal, namely between about 345 and 450 nm, they became anisotropic, i.e., they exhibited a linear dichroism signal (Fig. 2 and 3). The increments of the linear dichroism signal became smaller at longer times of illumination, until finally there

Fig. 4. Linear dichroism spectra obtained from 0.75 mM 9-cis, 0.52 mM 11-cis, 0.36 mM 13-cis, and 0.57 mM all-trans retinal dissolved in glycerol: ethanol (1:1 v/v) after 150 min of irradiation at 360 nm, spectral half-intensity bandwidth 5 nm. The horizontal traces are the linear dichroism spectra of the samples at 77°K before illumination. Path length 0.10 mm, scanning speed 200 nm/min, time constant 100 msec

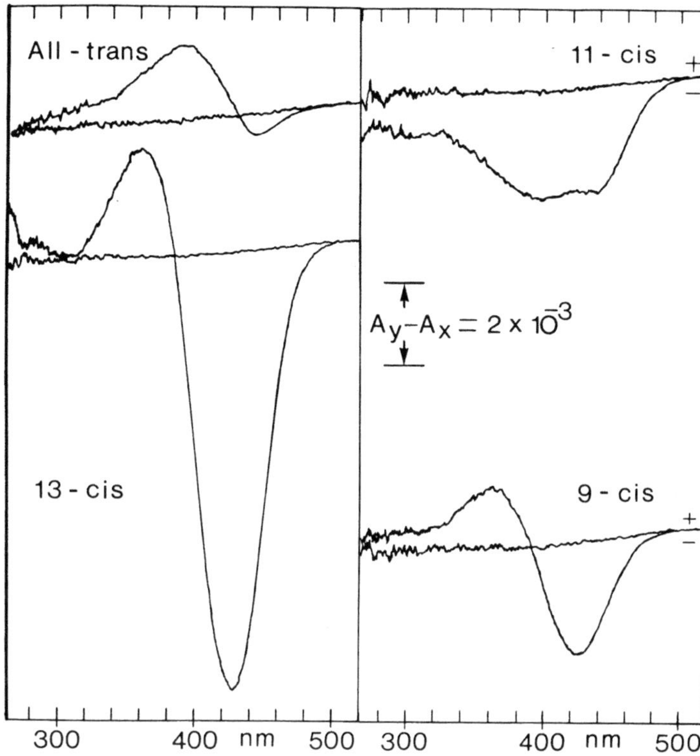

were only minimal changes with further illumination (Fig. 2). Under our conditions of illumination this took about 90 minutes for an initial A_{380} of less than 0.3 at a 0.1 mm light path.

The shape of the linear dichroism signal was dependent on two factors: (1) the wavelength of illumination, and (2) the nature of the isomer. When, for instance, the 11-cis isomer was illuminated at 360 nm, a particular linear dichroism was produced with broad negative peaks at about 430, 400, and 300 nm (Fig. 2). When, on the other hand, the same isomer was illuminated at 440 nm, different linear dichroism spectra were produced with a sharp positive peak at about 430 nm and a negative peak at about 365 nm (Fig. 3). At intermediate wavelengths between 360 and 440 nm, illumination of the 11-cis isomer produced an apparent mixture of the photoproducts which characterize the 440 and 360 nm illumination.

When the four retinal isomers were illuminated with linearly polarized light at 360 nm (5 nm half-intensity bandwidth) for a long enough time (about 150 min for A_{max} of less than 0.3), each isomer showed a characteristic linear dichroism spectrum as shown in Fig. 4. Since the spectra were so different, it is possible to identify uniquely each isomer from its linear dichroism spectrum. In other words, the linear dichroism spectra could serve as finger prints in identifying isomers. Moreover, in the concentration range ($A_{max} \leqq 0.3$ at 0.1 mm light path) the magnitude of the linear dichroism spectrum of each isomer was linearly proportional to concentration within ± 3%.

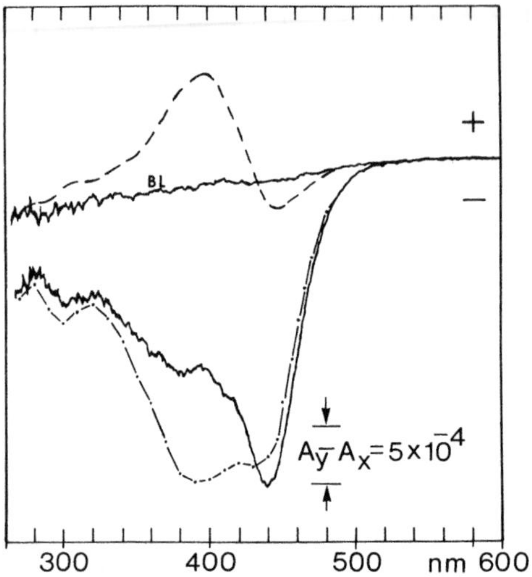

Fig. 5. Linear dichroism spectra obtained from a mixture of 0.35 mM all-trans retinal and 0.60 mM 11-cis retinal. — — — the linear dichroism spectra obtained from the all-trans part of the mixture. —·—·— the linear dichroism spectra obtained from the 11-cis part of the mixture. ——— (solid curve) the obtained spectra of the mixture. Illumination conditions as in Fig. 4

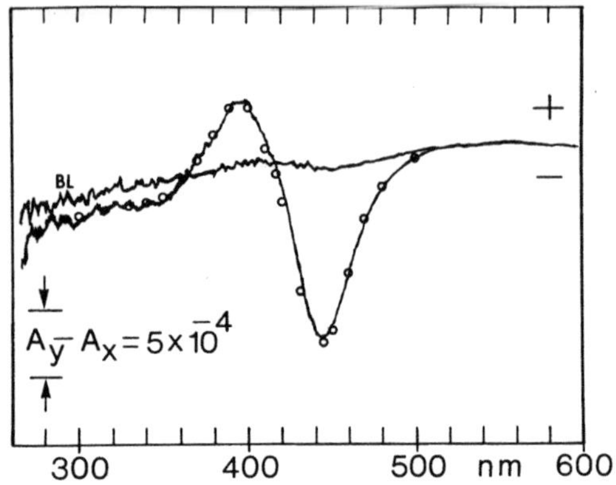

Fig. 6. Linear dichroism spectrum of a mixture of 0.22 mM 11-cis and 0.54 mM all-trans retinal. The solid line is the experimental curve, o o o is the calculated spectrum obtained by a linear addition of the spectra of pure 11-cis and all-trans retinal. Other conditions as in Fig. 4

For analyzing mixtures we used either 360 or 440 nm as the standard wavelength. When a mixture of retinal isomers was illuminated in the same manner, a different linear dichroism spectrum was obtained. The new linear dichroism spectrum was always the linear summation of the individual linear dichroism spectra of the components that made up the particular mixture (Fig. 5). Thus, whenever a mixture of 11-cis and all-trans retinal in known proportion was illuminated, the resulting linear dichroism spectrum could be precisely duplicated by the linear combination of the linear dichroism spectra of the 11-cis and all-trans isomers at these particular proportions. This observation was tested for mixtures of other isomers and for other proportions and was found to be always true. In other words, when a mixture of retinal isomers was illuminated, each isomer behaved as if it were the only component in the system (no interaction between the different isomers) and the resulting spectra simply reflected the independent summation of the various components of the mixture. Therefore, the linear dichroism spectra could be used to identify the components of an unknown mixture of isomers and to determine quantitatively the amount of each isomer in the mixture. Such an analysis was performed both graphically, as shown in Fig. 6, and by a computer program.

The accuracy of either the graphical method or the computer program was dependent on the kind of isomers present in the mixture because of the large variation in the signal size for the various isomers. For example, since the 13-cis gave a large signal, it was possible to determine accurately 2 to 3% of 13-cis in a mixture of other isomers. Conversely, to measure accurately all-trans, which gives a fairly small signal, in a mixture containing large amounts of 13-cis it had to be present at levels of more than 10%. The average accuracy of our computer program for mixtures of retinals which did not include large amounts of the 13-cis isomer was less than 5%.

When the retinal obtained from native visual pigment was investigated by the method of photoselection (linear dichroism), about 95% of the material was found to be in the 11-cis conformation. These results were confirmed by thin layer chromatography. Another 5% seemed to be an oxidation product of retinal, most probably retinoic acid. After illumination of rhodopsin retinals (95% of the total) which were obtained contained about 91% all-trans, 9% 13-cis, and some 5% of the same oxidation product. These results were again confirmed by thin layer chromatography. The products obtained after illumination with a 440 nm or 570 nm cut-off filter were identical. Because of the very large signal of the 13-cis isomer, it was possible to establish unambiguously that this isomer was not obtained from native visual pigment but was seen only after illumination.

When the same method of photoselection and linear dichroism was applied to the various geometrical isomers of retinol we found to our surprise that none of them gave a detectable signal under any condition of illumination. All-trans retinoic acid, on the other hand, gave a relatively large LD signal. These observations indicated to us that the production of detectable LD signals is dependent on various aspects of the chemical structure of the compound at hand. We have investigated a number of model compounds in an attempt to delineate some of the necessary structural parameters.

1) The potential for cis-trans isomerization is not a structural prerequisite to obtain LD signal by the method of photoselection at 77 °K. The compound β-phenylcinnamaldehyde (I) or pulegon (II) which do not have cis-trans isomers shows large LD signals upon illumination.

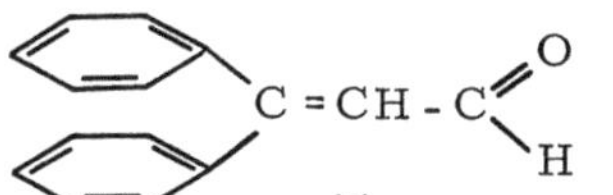
(I)

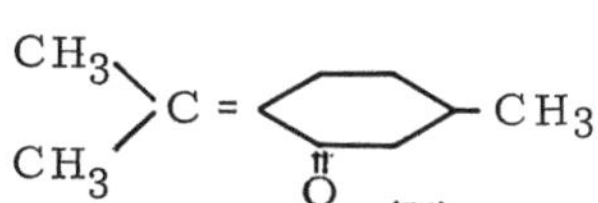
(II)

2) Carbonyl or carboxyl groups are not essential to obtain measurable LD signals. The compounds N-benzylidenemethylamine (III) and N-benzylideneaniline (IV) give large LD signals.

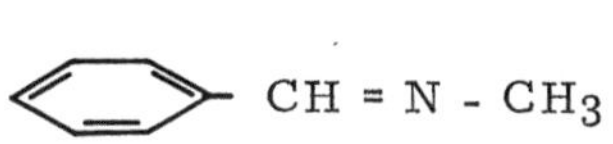
(III)

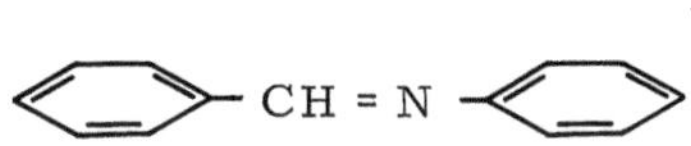
(IV)

3) Compounds that do have cis-trans geometrical isomers do not necessarily give rise to detectable LD signals. Thus, the various retinols, retinyl oximes (at neutral pH) and aliphatic retinyl Schiff bases (at neutral pH) do not give LD signals despite the fact that cis-trans photoisomerization is known to occur in these compounds.

4) Protonation of retinyl oximes and aliphatic retinyl Schiff bases leads to a change in the absorption spectrum and at the same time produces a compound that gives LD signals upon illumination.

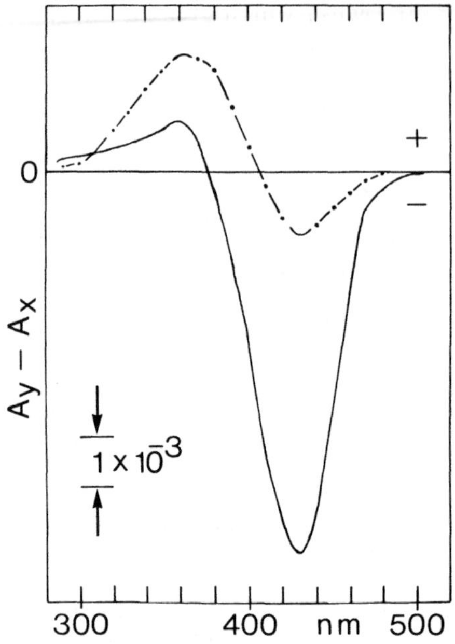

Fig. 7. Linear dichroism spectra of two identical samples of 11-cis retinal (1.8 mM final concentration) in glycerol:ethanol (1:1 v/v) after illumination at 77°K with unpolarized light for 30 min using a high pressure xenon lamp (PEK X-75) and a Corning filter (CS 7-51). The samples were then analyzed for their retinal composition by the method of photoselection and linear dichroism as explained in the text. Solid line: spectrum of sample which was first warmed to 298°K before analyzing the composition. Dashed line: spectrum of sample which was kept at 77°K throughout. All other conditions as in Fig. 4

From this series of photochromic substances that do or do not give rise to LD signals upon illumination at 77°K with linearly polarized light it is evident that cis-trans isomerization is not a prerequisite for the production of detectable signals. It is quite possible that these photochromic changes are due to several mechanisms, only one of which is cis-trans geometrical isomerization around either carbon-carbon or carbon-nitrogen double bonds.

In this connection it is interesting to note that the method of photoselection and linear dichroism indicates that the interpretation of the real chemical events that take place upon illumination at 77°K is not always that easy. It is generally assumed that if, after illumination of retinals at liquid nitrogen temperature, the light is turned off and the solution is warmed to room temperature and then analyzed for its components, the mixture of isomers which is isolated reflects the true composition at 77°K. That this is not so can be easily shown by illuminating two identical solutions of 11-cis retinal at 77°K with intense non-polarized light. One part is then warmed to room temperature in the dark, cooled again to 77°K and then illuminated as usual with linear polarized light to determine its isomer composition, while the other solution is illuminated with linear polarized light without first going through the cycle of warming up and freezing again. As can be seen from Fig.7, the two solutions do not give the same LD signal; in other words, they have a different isomer composition. Thus the warming up in the dark allowed some reactions (s) to take place that did not occur if the solution was kept in the cold. That other thermallyreversible reactions besides cis-trans isomerization take place upon illumination of retinals at 77°K was also suggested by Becker *et al.* (4).

The method of photoselection and linear dichroism can be used to add some new information about the linkage between the chromophore and the protein in rhodopsin. A constant puzzle to workers in the field is the question of whether the aldehyde function of the retinal chromophore is linked to the protein through an aldimine linkage (Schiff base), and if it is, whether or not the nitrogen of the Schiff base is protonated in the native pigment. As reported above, we found that unprotonated Schiff

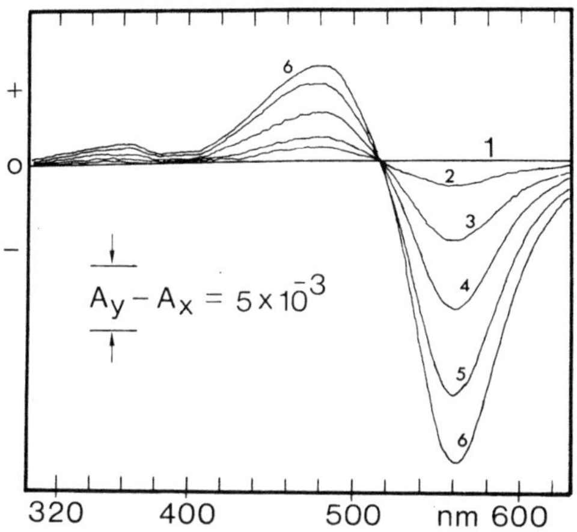

Fig. 8. Linear dichroism spectra of 11-cis visual pigment (final A_{500} = 0.1625) in 0.07 M CTAB, 0.066 M sodium phosphate buffer, pH 7.0, 50% glycerol, light path 1 mm, 77°K. Spectra are instrument tracing, scanning speed 200 nm/min, time constant 300 msec, slit width 0.08 mm. Illumination was at 500 nm with light linearly polarized in x-axis, slit width 0.2 mm. 1) The spectrum before illumination (base line), 2) after 20 sec illumination, 3) after further 1 min illumination, 4) after further 1.5 min, 6) after further 3 min, 6) after further 11 min (total 16 min 50 sec) illumination

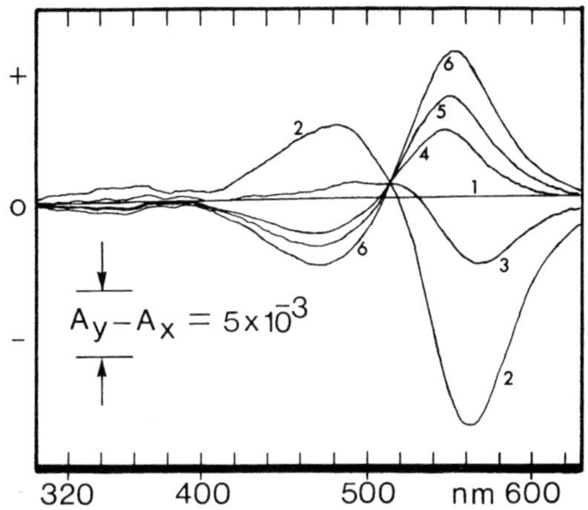

Fig. 9. Linear dichroism spectra of 11-cis visual pigment. Conditions were similar to those described for Figure 8: 1) The spectrum before illumination (baseline), 2) after illuminating for 24 min 50 sec at 500 nm, 3) after illumination at 560 nm for 3 min, 4) after further illumination at 560 nm for 5 min, 5) after further illumination at 560 nm for 5 min, 6) after further illumination at 560 nm for 15 min (total 28 min)

bases do not give detectable LD signals, and only upon protonation of various Schiff bases or retinal can one obtain an LD signal upon illumination at 77°K.

We have observed that illumination of rhodopsin or isorhodopsin at 77°K with linearly polarized light results in the production of large LD signals.

Illumination of rhodopsin solution at 77°K with monochromatic, linearly polarized light produced a certain signal which we call state I (Fig. 8). State I was always the first signal to be obtained by illuminating visual pigments, irrespective of the wavelength of the illuminating light. Varying the wavelength of the illuminating light changed only the efficiency in which state I was produced. With longer illumination at a particular wavelength, say 440 nm, the signal of state I increased up to a certain given magnitude, after which it changed very little with further prolonged illumination. When this preparation was now illuminated with linearly polarized light of wavelength longer than about 560 nm, the LD spectra changed and a new state was produced. We call this state II (Fig. 9). It should be noted that whereas the LD spectra of state I, which was initially produced from the native pigment, had a sharp isosbestic point with the baseline (i. e., the LD "spectra" of the native state which is isotropic, shows a flat baseline), the spectra of state II and I had a sharp isosbestic point at a different place, away from the baseline (Fig. 10). State I and state

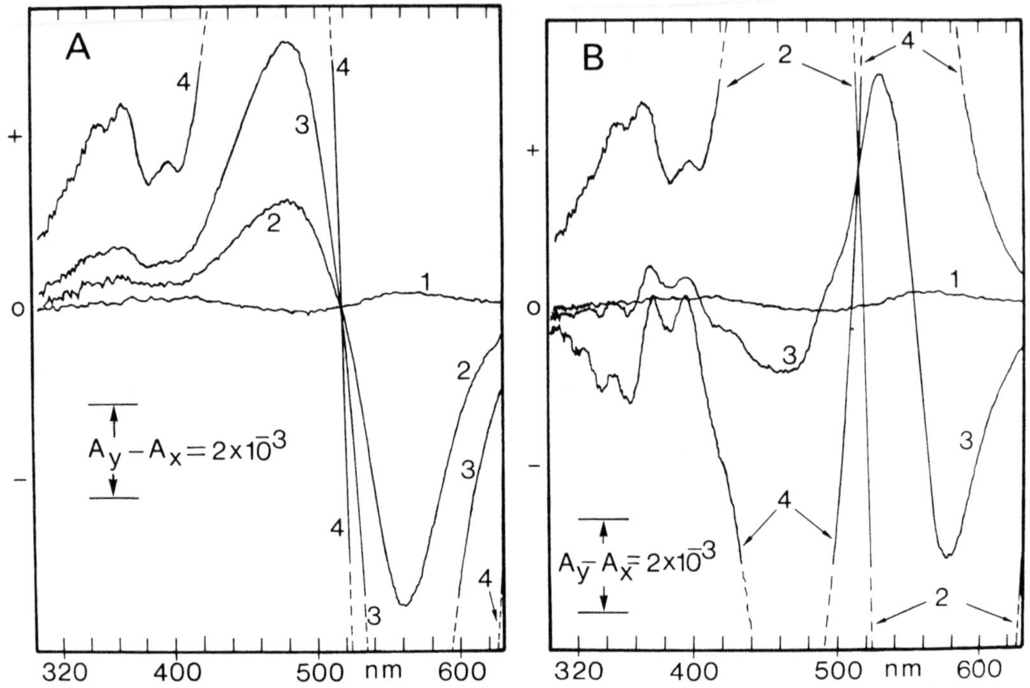

Fig. 10. A. Linear dichroism spectra of 11-cis visual pigment. Conditions were similar to those given in Fig. 8 except for final A_{500} which was 0.45 and time constant of 100 msec: 1) Linear dichroism spectrum of the sample at 77°K before illumination, 2) after 30 sec illumination at 500 nm with a slit width of 0.2 mm, 3) after a further 1 min illumination with a slit width of 0.2 mm, 4) after a further 8.5 min illumination at 500 nm with a slit width of 0.3 mm. B. Linear dichroism of 11-cis visual pigment. Conditions were similar to those described in Fig. 10A: 1) Linear dichroism spectrum of the sample at 77°K before illumination, 2) after illumination at 500 nm with a slit of 0.2 mm for 3.5 min and with a slit of 0.3 mm for 6.5 min, 3) after a further illumination at 560 nm for 7.5 min with a slit of 0.3 mm, 4) after further illumination at 560 for 30 min

---

II were reversibly interconvertible by illumination and this transformation always passed through a sharp isosbestic point.

It is important to note that the LD spectra of state I and II were obtained from frozen ROS fragments without any detergent, and from rhodopsin solutions in digitonin or CTAB. The LD spectra in all these media were identical. Changing the concentration of glycerol did affect the LD spectra by changing the peak position in a minor way.

When the same series of experiments was performed with the 9-cis pigment (isorhodopsin) the same qualitative results were obtained. Native isorhodopsin when illuminated with linearly polarized light of appropriate wavelength gave initially an LD signal that resembled state I of the 11-cis pigment but which was displaced some 10 nm to be blue (Fig. 11). When this state I of isorhodopsin was illuminated with light of wavelength longer than 560 nm, another LD spectrum was produced. State II of the 9-cis pigment was quite different in shape from the corresponding state of

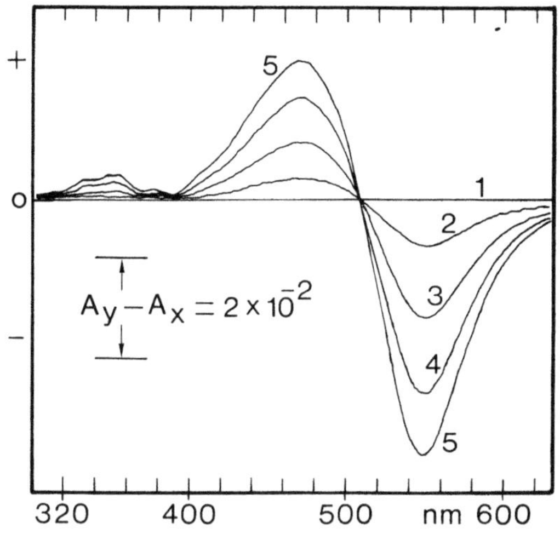

Fig. 11. Linear dichroism spectra of 9-cis visual pigment. Conditions were similar to those given for Fig. 8. Illumination was at 480 nm at a slit width of 0.5 mm: spectra at 77°K 1) before illumination, 2) after 15 sec illumination, 3) after a further 30 sec illumination, 4) after a total of 13 min illumination

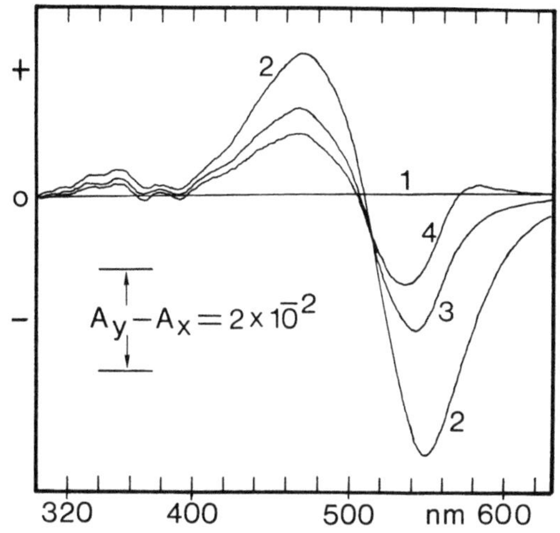

Fig. 12. Linear dichroism spectra of 9-cis visual pigments. Conditions were similar to those described for Fig. 8: Linear dichroism spectra at 77°K 1) of the unilluminated sample, 2) after 13 min illumination at 480 nm, 3) after 1.5 min illumination at 580 nm, 4) after a further 3 min illumination at 580 nm

the 11-cis pigment. Yet it had, similar to the rhodopsin states I and II, a sharp isosbestic point with state I and this isosbestic point of states I and II of the 9-cis pigment was above the baseline (Fig. 12). In other words, similar to the 11-cis pigment, no conditions of illumination were found that could transform either photoproducts of states I or II back into the native pigment.

It is known that in these visual pigment molecules the retinal chromophore is covalently linked to the protein. The fact that both rhodopsin and isorhodopsin give distinct and unique LD spectra upon illumination suggests that the nitrogen of the aldimine bond is protonated.

Further evidence to this point is offered by denaturation of rhodopsin with the detergent sodium dodecyl sulfate (SDS). We have observed that when rhodopsin is denatured with SDS in the dark at pH 6.0, the absorption spectrum shifts instantaneously from 500 to 440 nm. When this solution is frozen and illuminated with linearly polarized light at 440 nm we obtained LD signals that are very similar to the signals observed in protonated aliphatic Schiff bases of retinal. It is important to emphasize that when SDS was added to free retinyl aliphatic Schiff bases there was no change in the absorption spectrum and no LD signals were obtained from these compounds in the presence of SDS at 77°K. We conclude from these experiments that SDS as such does not protonate retinyl Schiff bases at pH 6.0. Thus, the simplest hypothesis to account for the observed shift of the spectra in rhodopsin upon the addition of SDS is to assume that the absorption spectrum at 500 nm is due to at least two factors: 1) A non-covalent interaction between the retinyl chromophore polyene chain and the protein accounting for the shift from 440 nm to 500 nm. This non-covalent interaction is destroyed by denaturation with SDS. 2) The covalent linkage between the aldehyde function and the protein is a protonated Schiff base. The protonation of the SDS-rhodopsin Schiff base is a pre-existing internal protonation (some acidic function of the protein) since neither the medium nor the SDS protonates Schiff bases under our conditions. These observations lend support to various claims in the literature that the protonated Schiff base of retinyl-rhodopsin is not by itself sufficient to account for the shift in absorption spectrum from 380 nm (free retinal) to 500 nm (rhodopsin) and one has to invoke further specific interactions between the polyene chain and the protein (for instance: 5, 6).

Our finding that treatment of rhodopsin (either as sonicated rod outer segment particles or as a complex with digitonin) with SDS leads to immediate and complete loss of the circular dichroism (CD) bands at 500 and 340 nm adds further support for the view that SDS leads to a conformational change in the protein which is expressed as a loss of the specific absorption band at 500 nm. There was also no CD band in the 440 nm absorption peak of SDS-rhodopsin. It is known that the 500 nm CD band is due to induced optical activity in the symmetrical chromophore upon its interaction with the protein (7). In addition, Waggoner and Stryer (8) have found that metarhodopsin I and II, in which the chromophore is already present in the photoisomerized all-trans form, still shows induced CD bands corresponding to the absorption peaks at 480 and 380 nm. Thus, the abolition of the CD band in SDS-rhodopsin, in which the chromophore is still covalently linked to the protein as the 11-cis protonated aldimine (Schiff base) form, indicates that besides the covalent interactions between the chromophore and protein in native rhodopsin, there are other crucially important interactions between the protein and the polyene.

Our experiments concerning the photoselection and linear dichroism of rhodopsin and isorhodopsin indicate that the native pigments cannot be regenerated from the photoproducts by any kind of illumination at $77^{\circ}K$. The very great sensitivity of the LD technique clearly shows that the photoproducts are always different from the native state. Yoshizawa and Wald (1) observed that after illumination of rhodopsin with light of 440 nm at $77^{\circ}K$, about half of the initial material was bleached and half remained native. Yoshizawa and Wald assumed then that this represents also the proportion of the mixture of native and "bleached" rhodopsin which was supposed to be present in illuminated visual pigment at $77^{\circ}K$. We feel that this assumption, namely, that the situation at room temperature after warming represents also the true chemical composition at $77^{\circ}K$, is unwarranted without independent proof. It is obvious that various stable intermediates can be obtained at $77^{\circ}K$ which could be further rearranged or modified upon warming, i.e., a stable intermediate X at $77^{\circ}K$ can decay into 50% 11-cis and 50% all-trans upon warming to $298^{\circ}K$. We feel that the relationship between the chemical nature of the photoproduct at $77^{\circ}K$ to the observed products at $298^{\circ}K$ is really the crux of the whole problem and is a question which remains to be answered by experimental means. Moreover, as shown by our experiments (Fig. 7) the mixtures of illuminated retinals at $77^{\circ}K$ is different from the obtained after warming to room temperature.

It is clear that these results are incompatible with the interpretation given by Yoshizawa and Wald to their observations as being due to a reversible cis-trans photoisomerization of the retinal chromophore. At this time we do not know the nature of the photoproducts responsible for states I and II of the 11-cis and 9-cis visual pigments. Two reports (9, 10) have recently suggested, on the basis of analogy with other carotenoids and theoretical calculations, that the geometrical retinal isomers can exist in several conformationally distinct forms (s-cis and s-trans, or gauche and anti). It was suggested that these conformational forms, which are in rapid equilibrium in solution at room temperature, are stabilized by the interaction of the chromophore with the protein; this gives rise to several distinct species which have different absorption spectra and might thus account for the various intermediates observed at low temperature (9,10). Since both these reports were purely theoretical, we do not know whether our results confirm or deny them. Besides conformational changes of the chromophore suggested above, one should probably take into account the possible cis-trans isomerization of the —C=N— linking the chromophore to the protein, possible tautomeric forms resulting from hydrogen abstraction or addition to the aldimine bond or other homolytic or heterolytic cleavage reactions.

Acknowledgement. This investigation was supported by NIH grants Nos. EY0031 and EY00704.

References

1. YOSHIZAWA, T., G. WALD: Pre-lumirhodopsin and the bleaching of visual pigments. Nature (Lond.) 197, 1279-1286 (1963).
2. YOSHIZAWA, T., G. WALD: Transformations of squid rhodopsin at low temperatures. Nature (Lond.) 201, 340-345 (1964).
3. YOSHIZAWA, T., G. WALD: Photochemistry of iodopsin. Nature (Lond.) 214, 566-571 (1967).
4. BECKER, R.S., K. INUZUKA, J. KING, D.E. BALKE: Comprehensive investigation of the spectroscopy and photochemistry of retinals. II. Theoretical and experimental consideration of emission and photochemistry. J.Am.Chem.Soc. 93, 43-50 (1971).
5. KROPF, A., R. HUBBARD: The mechanism of bleaching rhodopsin. Ann.NY Acad.Sci. 74, 266-280 (1958).
6. IRVING, C.S., G.W. BYERS, P.A. LEERMAKERS: Spectroscopic model for the visual pigments. Influence of microenvironmental polarizability. Biochemistry 9, 858-864 (1970).
7. CRESCITELLI, F., W.F.H.M. MOMMAERTS, T.I. SHAW: Circular dichroism of visual pigments in the visible and ultraviolet spectral regions. Proc.Nat. Acad.Sci.USA 56, 1729-1734 (1966).
8. WAGGONER, A.S., L. STRYER: Induced optical activity of the metarhodopsins. Biochemistry 10, 3250-3254 (1971).
9. HONIG, B., M. KARPLUS: Implications of torsional potential of retinal isomers for visual excitation. Nature (Lond.) 229, 558-560 (1971).
10. SUNDARALINGAM, M., C. BEDDELL: Structures of the visual chromophores and related pigments: A conformational basis of visual excitation. Proc.Nat. Acad.Sci.USA 69, 1569-1573 (1972).

Discussion

G.O. Schenck: Protonation of the Schiff base system, for example by hydrogen bonding with the phenolic or enolic hydroxyl group could be followed (after excitation of the chromophore) by some hydrogen or less probable electron transfer to the Schiff base. Ascorbic acid or tocopherol are example of natural enols and phenols. This question may have only relevance to the photochemistry of rhodopsin.

G. Wald: The method described by Dr. Heller of selective photodichroism was a favorite procedure among classic photochemists half a century ago, notably Fritz Weigert, who with Nakashima also used it with gelatine films of frog rhodopsin. The method had been used earlier with the photographic film. It demands only three conditions: 1) a photochemical change; 2) a directional factor in the reacting molecules - i.e., an axis of orientation; and 3) that the molecules remain fixed in position. The AgBr micelles in the photographic plate fulfill these conditions.

Studies on Intermediates of Visual Pigments by Absorption Spectra at Liquid Helium Temperature and Circular Dichroism at Low Temperatures

Tôru Yoshizawa and Shinri Horiuchi
Department of Biophysics, Faculty of Science, University of Kyoto, Kyoto/Japan
Department of Biology, Osaka Medical College, Takatsuki, Osaka/Japan

One of the fundamental problems in the mechanism of the primary process of visual photoreception is to identify the intermediates in the process of bleaching of visual pigment and to analyze their conformations and configurations. In this paper, some of our investigations concerning this subject are reported.

A. Photoproduct of Cattle Rhodopsin at Liquid Helium Temperature

On absorption of light visual pigment decomposes over a series of intermediates to the final photoproduct, all-trans retinal and opsin. Since the intermediates are too unstable to measure the spectrum at room temperature, low temperature spectrophotometry has been used as a powerful technique to investigate them. More than ten years ago, one of us with his associates found a deep-red intermediate (1), named pre-lumirhodopsin (2), which can be produced by irradiation of cattle rhodopsin at liquid nitrogen temperature (-196 $^{\circ}$C, 77 $^{\circ}$K). The change of rhodopsin to pre-lumirhodopsin has been regarded as the photo-isomerization of the chromophore from 11-cis retinal to the all-trans form.

Afterwards, the comparable pre-lumi intermediates have been found with chicken (3), frog (4), and squid (5) rhodopsins, chicken iodopsin (6) and carp porphyropsin (7). Recently, Cone (8) has succeeded in observing the formation of pre-lumirhodopsin in frog retinas at room temperature by means of irradiation with laser flashes of about 5 nsec duration. It is sure, therefore, that the pre-lumi intermediate is a significant intermediate in the process of visual photoreception.

Now, it is of interest to determine whether or not an earlier intermediate than prelumi intermediate exists. In order to examine this subject, an attempt was made to measure the spectrum at liquid helium temperature (-269 $^{\circ}$C, 4 $^{\circ}$K).

Method. Cattle rhodopsin was extracted with 2% digitonin (M/15 phosphate buffer, pH 6.4) from isolated rod outer segments which had been treated with 4% alum solution and petroleum ether. The extract obtained was mixed with twice the volume of glycerol. For measurement of the absorption spectrum of rhodopsin or its products at liquid helium temperature, two different kinds of cryostats (9) - a metal-cryostat and a double-vacuum glass cryostat - were designed for attachment to a Hitachi recording spectrophotometer (EPS-3T). A xenon lamp (Ushio, DSB 100, 1 KW) or a high pressure mercury lamp (Ushio, HMB-500/B, 500 W) was used as the light source for irradiating the preparation.

Results. The rhodopsin preparation was cooled to near 77 $^{\circ}$K. When only a few cracks were formed in the preparation, the rise of maximal absorbance was about 1.14 on the average. Further cooling such a preparation from 77 $^{\circ}$K to 4 $^{\circ}$K causes only a slight rise (1.03 times) of maximal absorbance. λ_{max} moves from about 498 nm at room temperature to 505 nm at 77 $^{\circ}$K and about 506 nm at 4 $^{\circ}$K.

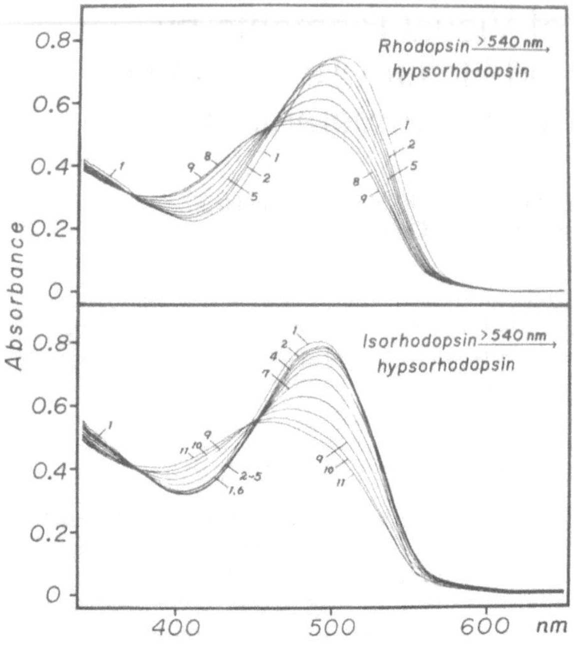

Fig. 1. Course of photoconversion of rhodopsin (above) or isorhodopsin (below) to hypsorhodopsin at 4°K. Above: Curve 1, cattle rhodopsin in a glycerol-water mixture (2:1) at 4°K. Curves 2-9, products of irradiation at wavelengths longer than 540 nm for successive periods of 0.5, 0.5, 1, 2, 4, 8, 16,, and 32 min. The final spectrum (curve 9) represents a mixture of rhodopsin, hypsorhodopsin, and isorhodopsin. Below: For preparing isorhodopsin, the final preparation (curve 9 above) was warmed to 77°K and then irradiated for 512 min. with orange light at wavelengths longer than 580 nm. This preparation is a mixture composed of about 83 per cent isorhodopsin and 17 per cent rhodopsin. After warming to 288°K, the preparation was cooled to 4°K (curve 1). Curve 2-11, products of irradiation at wavelengths longer than 540 nm for successive periods of 0.25, 0.25, 0.5, 1, 2, 4, 8, 16, 32, and 64 min. The final spectrum (curve 11) represents a mixture of rhodopsin, hypsorhodopsin, and isorhodopsin

---

Now the rhodopsin was irradiated with yellow light at 4°K (Fig. 1, above). As the irradiation goes on, the spectrum shifts to shorter wavelenghts, showing a hypsochromic shift, the maximal absorbance simultaneously falling considerably. In the case of isorhodopsin, almost similar results were obtained (Fig. 1, below), except for the short period of the irradiation, in which the direction of the shift of spectrum is different from the case of rhodopsin. This fact indicates that a mixture of rhodopsin and isorhodopsin is formed by the irradiation for the short period. Anyway, the last spectra in both figures show a mixture composing rhodopsin, isorhodopsin and a new intermediate, which we call hypsorhodopsin (curve 3 in Fig. 3).

On the other hand, if one irradiates with blue light instead of yellow light at about 4°K, the spectrum shifts to longer wavelengths. This bathochromic shift corresponds to the change from rhodopsin to a photosteady state mixture (curve 2 to curve 3 in Fig. 2), composed of rhodopsin, isorhodopsin and an intermediate. We call this intermediate bathorhodopsin (curve 4 in Fig. 3).

In order to examine whether or not bathorhodopsin is pre-lumirhodopsin, the mixture was warmed to 77°K and then re-cooled to 7°K for re-measuring the spectrum. Curve 4, thus obtained, perfectly coincides with curve 3 (Fig. 2). This fact indicates that bathorhodopsin does not change on warming to 77°K. Another sample containing bathorhodopsin, was warmed to room temperature. The bathorhodopsin decomposes to the final photoproduct, all-trans retinal and opsin, via lumi- and metarhodopsins. From these experimental results, it is clear that bathorhodopsin is the same photoproduct as pre-lumirhodopsin. Since pre-lumirhodopsin has all-trans

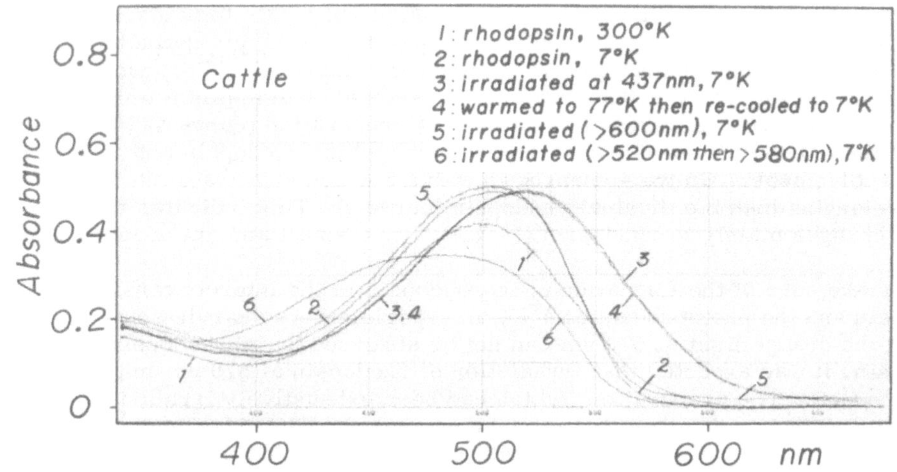

Fig. 2. Photochemical interconversions among rhodopsin, bathorhodopsin, and hypsorhodopsin at 7°K. Curve 1, cattle rhodopsin in a glycerol-water mixture (2:1) at 300°K. After cooling to 7°K (curve 2), the rhodopsin was irradiated at 437 nm for 64 min. The rhodopsin changed to a photosteady state mixture of rhodopsin, bathorhodopsin, and isorhodopsin (curve 3). The mixture was warmed to 77°K, and then re-cooled to 7°K (curve 4). Curve 4 coincides with curve 3, indicating that bathorhodopsin is pre-lumirhodopsin. On irradiation at wavelenghts longer than 600 nm for 32 sec, the spectrum returned to curve 5, which consists mainly of rhodopsin with a trace of isorhodopsin. The preparation was irradiated at wavelengths longer than 520 nm for 64 min., resulting in a mixture of rhodopsin, isorhodopsin, and hypsorhodopsin with a trace of bathorhodopsin. For removing the bathorhosopsin the preparation was irradiated with orange light at wavelengths longer than 580 nm for 10 min at 7°K (curve 6)

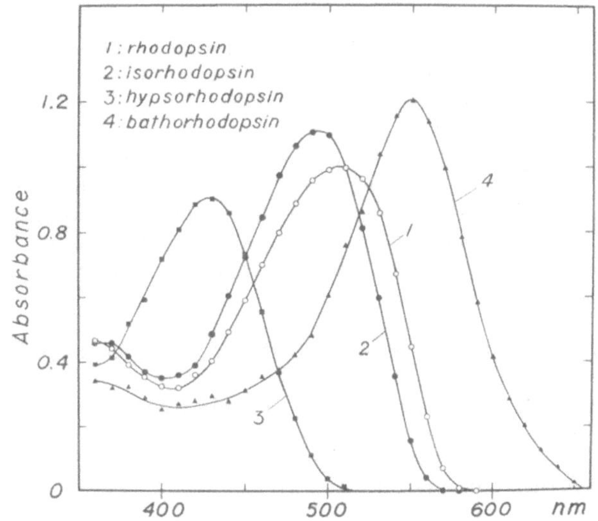

Fig. 3. Absorption spectra of cattle rhodopsin (curve 1), isorhodopsin (curve 2), hypsorhodopsin (curve 3), and bathorhodopsin (curve 4) at 4°K. The maximum absorbance of rhodopsin (λ_{max}: 506 nm) is plotted arbitrarily at 1.0. Isorhodopsin (λ_{max}: 494 nm), hypsorhodopsin (λ_{max}: ca. 430 nm), and bathorhodopsin (λ_{max}: 548 nm) of equivalent concentration possess maximum absorbance about 1.12, 0.91, and 1.22 times, respectively, as high as that of rhodopsin (9)

retinal as the chromophore, the conversion of rhodopsin to bathorhodopsin is due to the photo-isomerization of the chromophore from 11-cis retinal to all-trans. In order to examine whether the all-trans chromophore of bathorhodopsin can be re-isomerized to 11-cis by absorption of a photon at liquid helium temperature or isomerized to other cis forms like 9-cis, the preparation (curve 4) was irradiated with red light at wavelengths longer than 600 nm, which rhodopsin and isorhodopsin scarcely absorb. Curve 4 changed to curve 5 which shows a slight shift to shorter wavelengths than the original rhodopsin (curve 2). This indicates that bathorhodopsin changes mainly to rhodopsin (11-cis) with a small amount of isorhodopsin (9-cis).

To make sure of the formation of isorhodopsin in the interconversion among rhodopsin and the photoproduct at 4°K, an experiment was carried out. For the reason that the orange light at 579 nm can not be absorbed by isorhodopsin and hypsorhodopsin, it was expected that irradiation of rhodopsin at 579 nm may trap these photoproducts. The preparation, which had been exhaustively irradiated under this condition, was converted to a mixture composed of 36% rhodopsin, 58% isorhodopsin, and 6% hypsorhodopsin, according to the evaluation by a procedure similar in principle to that described by Yoshizawa and Wald (5). Judging from the yield of hypsorhodopsin, rhodopsin and bathorhodopsin are converted to isorhodopsin easier than to hypsorhodopsin. It may be inferred, therefore, that the formation of isorhodopsin from rhodopsin passes through bathorhodopsin rather than hypsorhodopsin.

Next experiment to investigate is the photoconversion of hypsorhodopsin. A mixture of rhodopsin, isorhodopsin, and hypsorhodopsin, which had been formed by irradiation with yellow light near 4°K, was irradiated with blue light at 406 nm for successive periods at the same temperature. Since rhodopsin and isorhodopsin have their minimal absorbance near 406 nm, the irradiation for short periods causes the photoconversion of the hypsorhodopsin rather than the rhodopsin and the isorhodopsin. The spectral change by irradiation for short periods forms an isosbestic point at 476 nm, indicating the conversion of hypsorhodopsin to bathorhodopsin.

The thermal conversion of hypsorhodopsin to bathorhodopsin was also examined. In the process of warming the photosteady state mixture containing hypsorhodopsin in the dark, the spectra were repeatedly measured at intervals of about 10°. The spectrum did not change up to 23°K. Above this temperature, the spectrum gradually shifted to longer wavelengths and formed an isosbestic point near 470 nm, close to the isosbestic point mentioned above. This indicates not only that hypsorhodopsin is converted to bathorhodopsin above 23°K in the dark, but also that there is no intermediate between them.

Discussion. All the reactions mentioned above are summarized in Fig. 4. It should be noted that there is no evidence for photochemical conversions of bathorhodopsin into hypsorhodopsin and of hypsorhodopsin into rhodopsin or isorhodopsin. Anyway, this scheme shows that rhodopsin, isorhodopsin, hypsorhodopsin, and bathorhodopsin can be converted into one another, either directly or indirectly, by suitable irradiation.

Now a question arises whether the first photoproduct in the photolysis of rhodopsin is hypsorhodopsin or bathorhodopsin or both. The thermal conversion of hypsorhodopsin into bathorhodopsin strongly suggests that hypsorhodopsin is the first photoproduct. However, considering that the process of the formation of hypsorhodopsin from rhodopsin or isorhodopsin is not simple (Fig. 1), it seems premature to draw this conclusion.

Fig. 4. Bleaching of cattle rhodopsin. Photoreactions are symbolized by wavy lines, thermal reactions by straight lines. λ max for rhodopsin, hypsorhodopsin, isorhodopsin, and bathorhodopsin at -269°C (4°K); lumirhodopsin and metarhodopsin I at -65°C; other pigments near 0°C

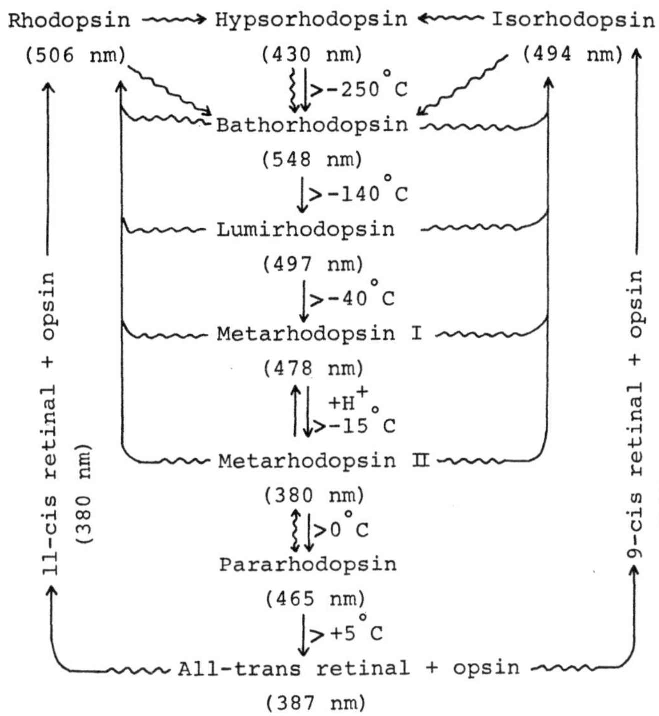

The next problem is in what part of its structure hypsorhodopsin is different from bathorhodopsin. Since warming these photoproducts to room temperature releases all-trans retinal from the opsins, the chromophores of both photoproducts must be trans at the 11-12 double bond. Therefore the difference between both photoproducts may be due to the rotation of single bonds of the retinal chromophore and/or the amino acid residue of opsin combining with or surrounding the chromophore. Other possible interpretations may be related to the next problem.

This problem is how to account for the spectral shift of rhodopsin to hypsorhodopsin. One may consider at least these possible mechanisms: 1) The twist in some single bonds of the side chain retinal; 2) the polarity of the environment surrounding the chromophore; 3) the deprotonation of the nitrogen atom on the Schiff base linkage. We cannot decide yet which mechanism is most suitable in interpretation of the hypsochromic shift of rhodopsin to hypsorhodopsin.

B. Photic Reaction of Chicken Visual Pigments at Liquid Helium Temperature

In order to clarify whether or not the formation of the hypso-intermediate is a phenomenon common to all visual pigments, chicken visual pigments have been examined in cooperation with Mr. Y. Tsukamoto at Osaka University. Chicken iodopsin is one of the most different visual pigments from cattle rhodopsin in many respects.

Iodopsin in 2% digitonin solution, prepared by the procedure described in a previous paper (6), was mixed with an equal amount of glycerol. The absorption spectrum of the iodopsin was intensified by application of the Keilin-Hartree effect — formation of micro-crystals in the medium at low temperature. The low temperature spectrophotometry was carried out by use of the cryostats mentioned above.

On cooling the iodopsin from room temperature to $77^{o}K$, λ_{max} moves from 562 nm to 575 nm. Further cooling to $4^{o}K$ has no effect upon the shift of λ_{max}. On irradiating iodopsin with green light at 540 nm, it changes to a photosteady state mixture composed of iodopsin and bathoiodopsin (the same photoproduct as pre-lumiiodopsin) with a trace of isoiodopsin. Irradiation of the mixture with deep red light at wavelengths longer than 640 nm converts the bathoiodopsin to iodopsin. If one irradiates iodopsin with red light at wavelengths longer than 600 nm, which is absorbed by iodopsin, bathoiodopsin, and isoiodopsin, one can expect the formation of the hypso-intermediate. The irradiation with this light at $4^{o}K$ converted iodopsin to isoiodopsin with an isosbestic point at 562 nm and finally to a steady state mixture of iodopsin and isoiodopsin. This result is quite similar to that obtained at $77^{o}K$. When the steady state mixture was irradiated with orange light at wavelenghts longer than 580 nm, the iodopsin in the steady state mixture slightly increased. Thus we could not confirm the production of the hypso-intermediate by the irradiation of iodopsin. When one begins with isoiodopsin, a similar result is obtained. Thus, the photochemical reaction of iodopsin at $4^{o}K$ is similar to that at $77^{o}K$ and we could not confirm the formation of a hypso-intermediate by irradiation at $4^{o}K$.

Chicken rhodopsin (λ_{max}: 500 nm) hereupon was examined by the same procedure as cattle rhodopsin. The λ_{max} of chicken rhodopsin is at 508 nm at $4^{o}K$. Isorhodopsin was produced by low temperature irradiation. The λ_{max} of the isorhodopsin (488 nm) is at 496 nm at $4^{o}K$. Bathorhodopsin and hypsorhodopsin possess λ_{max} at 550 nm and 430 nm respectively at $4^{o}K$. The spectral changes of interconversion among rhodopsin, isorhodopsin, bathorhodopsin, and hypsorhodopsin are similar to the cattle rhodopsin system. From these experimental results, it seems that hypsorhodopsin is a photoproduct common to rhodopsins.

C. Circular Dichroism of Metarhodopsins

The existence of intermediates in the process of bleaching of visual pigment has been confirmed by low temperature spectrophotometry in the visible range. However, we have as yet little information about the conformation and configuration of the intermediates. The low temperature spectrophotometry of circular dichroism in combination with absorbance gives more information about interactions between the chromophore and opsin in the intermediates. These measurements were performed with a specially designed absorption vessel (9, Fig. 2) attached to an automatic recording spectropolarimeter (JASCO, model ORD/UV-5).

The circular dichroic spectrum of cattle rhodopsin-digitonin solution (in glycerol: water = 2:1) shows optically active bands of the prosthetic group (retinal) at about 498 nm and about 350 nm corresponding to the α- and β-bands of the absorption spectrum. These bands are considered to be induced by the asymmetric structure of the opsin surrounding the prosthetic group (10), but the bleached product of rhodopsin does not show the optically active band of the prosthetic group. On cooling the rhodopsin to about $-70^{o}C$ or $-20^{o}C$, the circular dichroic spectrum rises slightly at the peaks, probably owing to the contraction of the medium. On irradiating rhodopsin with orange light at $-20^{o}C$, both absorption and circular dichroic spectra shift from 500 nm to 482 nm, owing to the conversion of rhodopsin to a steady state

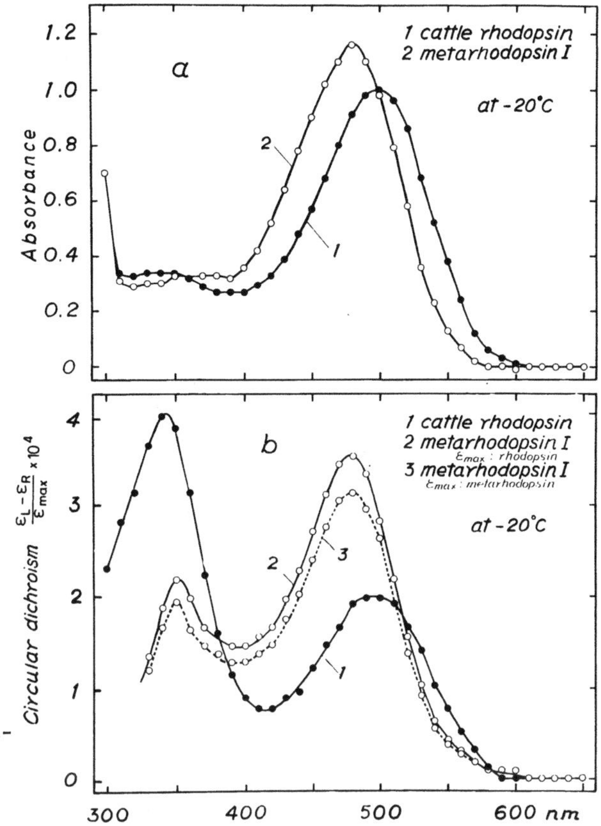

Fig. 5. Absorption spectra (above) and circular dichroic spectra (below) of rhodopsin and metarhodopsin I (MI). The maximum absorbance of rhodopsin (ε_{max}) is plotted arbitrarily at 1.0 and other curves, except for curve 3 in the figure below, are relative to it. Curve 3 was calculated by plotting the maximum absorbance (ε_{max}) of MI at 1.0. Cattle rhodopsin in a glycerol-water mixture (2:1), containing 0.1 M hydroxylamine at pH 6.5 (curve 1), was irradiated with orange light at wavelengths longer than 560 nm for 10 min at -20°C, in order to convert the rhodopsin into a photosteady state mixture composed of rhodopsin, isorhodopsin, and MI. After measuring the absorption spectra and circular dichroic spectra, the preparation was warmed to room temperature to decompose MI to all-trans retinal oxime and opsin. After cooling to -20°C, the spectra were re-measured. The difference spectra before and after the warming represent that of MI. After re-warming to room temperature, the preparation was completely bleached with orange light at wavelengths longer than 520 nm for 10 min and then the spectra were measured at -20°C. The addition of the spectrum of the completely bleached sample to the difference spectrum of MI of equivalent concentration yields the absorption spectrum or circular dichroic spectrum of MI (curve 3)

---

mixture composed of rhodopsin, isorhodopsin, and metarhodospin I. The absorption spectrum of metarhodopsin I, which has been calculated from a set of spectra (11), possesses a prominent band at 478 nm and a small band at 350 nm, corresponding to the α - and β-bands of rhodopsin, respectively (Fig. 5). The circular dichroic spectrum of metarhodopsin I, calculated by the same procedure as the absorption spectrum, also shows two prominent bands, corresponding to the band of the absorption spectrum. Compared with rhodopsin, the circular dichroic spectrum of metarhodopsin I shows about a two fold increase at the α-band and is cut to one half at the β-band.

On warming metarhodopsin I to about -15°C, or irradiating rhodopsin at 0°C, metarhodopsin II (λ_{max}: 380 nm) is produced (12). In Fig. 6 (upper), curve 2 represents the spectrum of metarhodopsin II with a small admixture of pararhodopsin (λ_{max}: 465 nm). The main band of metarhodopsin II (λ_{max}: 380 nm) is optically active (Fig. 6 (lower), curve 2). Warming metarhodopsin II to above 5°C, converts it to pararhodopsin. The formation of pararhodopsin can be confirmed by the dif-

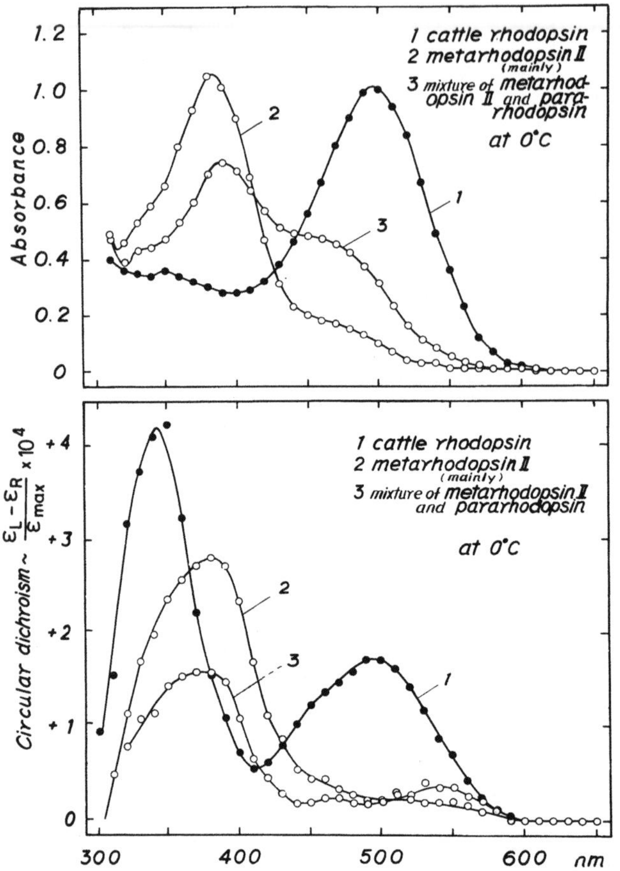

Fig. 6. Absorption spectra (above) and circular dichroic spectra (below) of rhodopsin, metarhodopsin II, and pararhodopsin. The maximum absorbance of rhodopsin (λ_{max}) is plotted arbitrarily at 1.0, and the other curves are relative to it. Cattle rhodopsin in a glycerol-water mixture (2:1) at pH 6.4 (curve 1) was irradiated with orange light at wavelengths longer than 560 nm for 10 min at 0°C, resulting in a mixture of rhodopsin, isorhodopsin, and metarhodopsin II. After measuring the spectrum at 0°C (result a, not shown in the figure), the mixture was warmed to 5°C in the dark. A part of the metarhodopsin II was converted to pararhodopsin. After measuring the spectra at 0°C (result b), the preparation was warmed to room temperature. Then 0.1 M neutral hydroxylamine was added, in order to decompose the intermediates contained in the preparation to the final product, all-trans retinal oxime and opsin. Then the spectra were measured at 0°C (result c). After complete bleaching of residual rhodopsin and isorhodopsin by irradiation at room temperature, the spectra were measured at 0°C (result d). Results (c) and (d) were corrected for dilution by addition of the hydroxylamine. Result (a) or (b) minus corrected result (c) gives the difference spectrum of metarhodopsin II or pararhodopsin with some amount of metarhodopsin II. Addition of the corrected result (d) to the difference spectra of equivalent concentration yields the spectrum of metarhodopsin II (curve 2) or pararhodopsin with some amount of metarhodopsin (curve 3)

---

ference in absorbance at 465 nm between curve 2 and 3 in upper part of Fig. 6. However, in Fig. 6 (lower) (circular dichroic spectra), warming causes little change at 465 nm and reduces the intensity of the band at 380 nm to half. This indicates that pararhodopsin is not optically active. Qualitatively similar results, measured at 3°C, were reported by Waggoner and Lubert (13). From these results, it is inferred that the prosthetic group of metarhodopsin may possess a twisted conformation to some extent.

D. Circular Dichroism of Bathorhodopsin

Recently, Horwitz and Heller (14) reported the results of the circular dichroism of rhodopsin at 77°K. We also have carried out combined measurements of circular

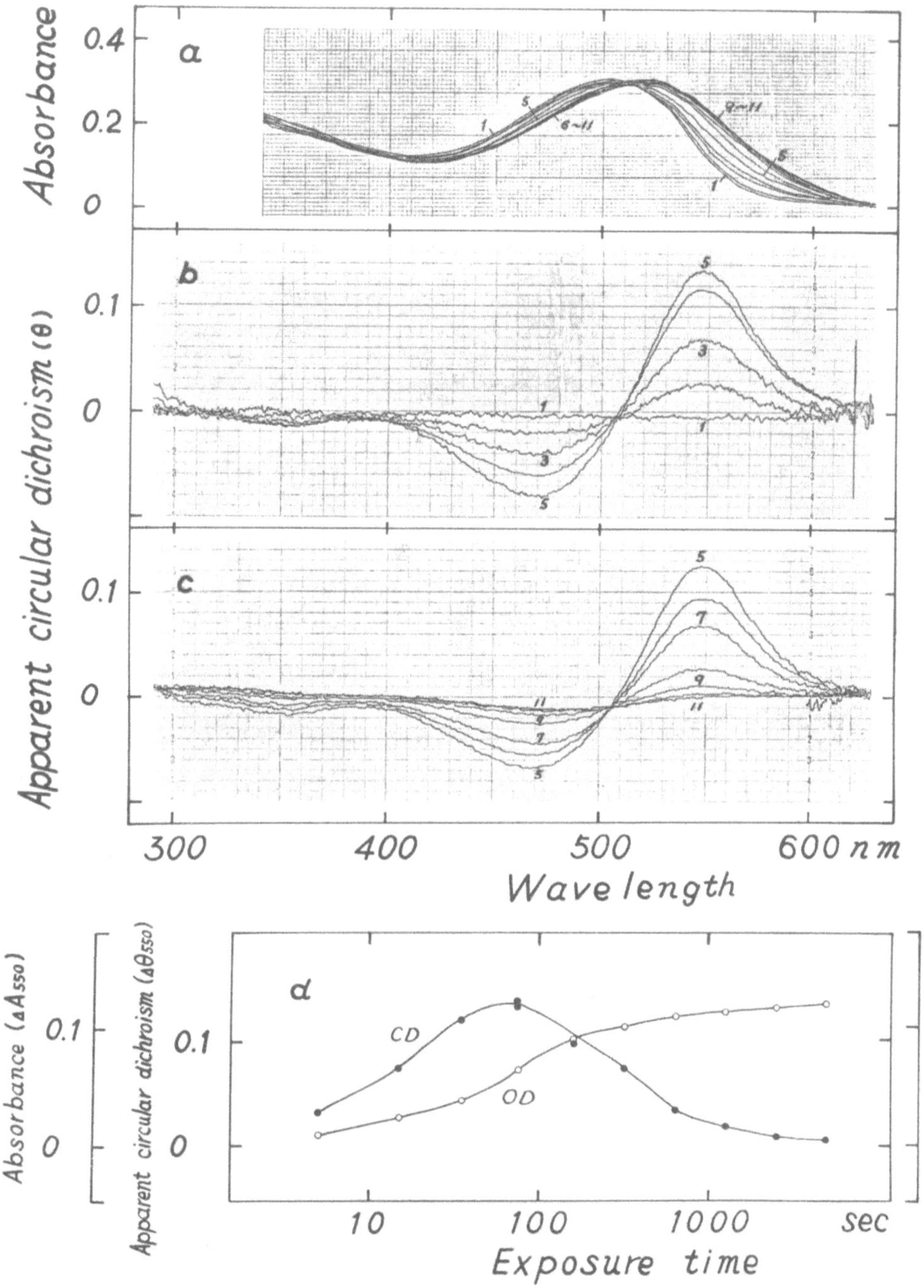

Fig. 7. Spectral changes of absorbance (a) and apparent circular dichroism (b and c) in the conversion of rhodopsin to bathorhodopsin with polarized light. Curve 1: Rhodopsin in glycerol-water mixture (1:2) at 77°K. Curve 2-11: Products of irradiation with polarized light at 437 nm for a total of 5, 15, 35, 75, 155, 315, 635, 1275, 2555, 5115, and 10235 sec (42 min 40 sec). In Fig. d, increments of absorbance (o) and circular dichroism (•) at 550 nm were plotted against exposure time

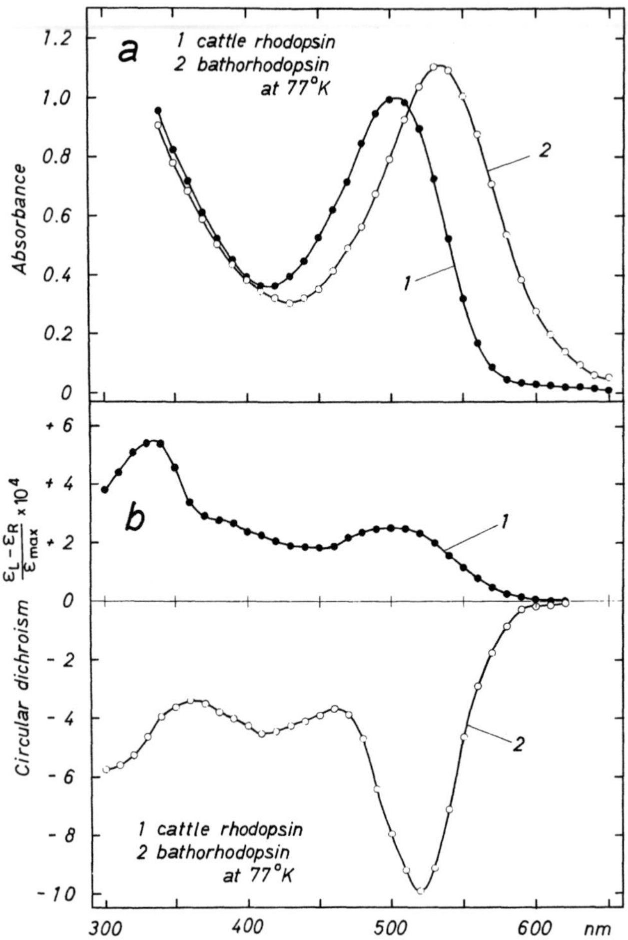

Fig. 8. Absorption spectra (a) and circular dichroic spectra (b) of equivalent concentrations of rhodopsin (curve 1) and bathorhodopsin (curve 2). The maximum absorbance of rhodopsin (ε_{max}) is plotted arbitrarily at 1.0. Absorption and circular dichroic spectra were calculated by a similar procedure to that described by Yoshizawa and Wald (2)

dichroic and absorption spectra at 77°K. We report here the circular dichroic spectra of rhodopsin and bathorhodopsin and some results of rhodopsin irradiated with polarized light. Measurements of circular dichroism and absorbance were made with the automatic recording spectropolarimeter (JASCO, model J-20) and recording spectrophotometer (Hitachi, EPS-3T), respectively. A quartz Dewar with a pair of parallel windows was used as the absorption vessel. Cattle rhodopsin extracted with 2% digitonin solution was mixed with twice the volume of glycerol. In order to get clear samples without any cracks at 77°K, the preparation (66% glycerol) must be cooled rapidly. One of the most noteworthy facts in measurements of circular dichroism at 77°K is that circular dichroic spectra of such a photosensitive pigment as rhodopsin are subject to the influence of the measuring light. With repetition of the measurement of circular dichroism by the usual procedures (scale: 0.002/cm, scan speed: 2 cm/min), signal of apparent circular dichroism continues to increase tremendously, reaches to a peak, and then gradually begins to decrease. This phenomenon was observed by irradiating rhodopsin with monochromatic light from the spectropolarimeter.

If one irradiates rhodopsin with linearly polarized light instead of the light from the spectropolarimeter, more distinct results can be obtained as shown in Fig. 7. The course of formation of bathorhodopsin by irradiation of rhodopsin (curve 1) with linearly polarized light at 437 nm is shown in Fig. 7a and the increment of absorbance at 550 nm (o) in Fig. 7d. The change of apparent circular dichroism (Fig. 7b and c; • in part d) in the same preparation exhibits not only a quite different course from that of the formation of bathorhodopsin (Fig. 7a; o in part d), but also an extraordinarily high intensity at 550 nm (about 70 times that of rhodopsin at 500 nm at room temperature). Similar changes were observed when rhodopsin was irradiated with monochromatic light from the spectropolarimeter, but only a little change with natural light. Even though the photosteady state mixture, which was formed by exhaustive irradiation with natural blue light at 437 nm, was irradiated with circularly or linearly polarized light at the same wavelength, the change of circular dichroism was scarcely observed.

These experimental results may be interpreted as follows: On irradiating rhodopsin for short periods with linearly polarized light at $77^{\circ}K$, only the molecules which have some electric vector oriented in parallel with the plane of polarized light may undergo photochemical reaction, resulting in the formation of bathorhodopsin. Since the molecule may not move in such a rigid solvent as glycerol-water at $77^{\circ}K$, all the bathorhodopsin molecules resulting have electric dipole inevitably oriented in the same direction. Consequently, the preparation develops into an oriented structure. According to a report by Disch and Sverdlik (15), the apparent circular dichroic spectrum of an oriented system can differ considerably from its true spectrum, because of the effect of linear dichroism and birefringence. It is clear that the apparent circular dichroic spectrum which appears through the irradiation with linearly polarized light is caused by the oriented structure. It may be inferred that the similar circular dichroic spectrum induced by light from the spectropolarimeter may be attributed to the same origin as by irradiation with linearly polarized light, since the light from the spectropolarimeter consists of polarized light with various degrees of ellipticity.

We tried to measure the circular dichroic spectrum of bathorhodopsin. Rhodopsin was exhaustively irradiated with natural blue light at 437 nm at $77^{\circ}K$, resulting in a photosteady state mixture composed of bathorhodopsin and rhodopsin with a small admixture of isorhodopsin. If one subtracts the spectra of the rhodopsin and the isorhodopsin contained in the steady state mixture from that of the mixture, one can get the absorption and circular dichroic spectra of bathorhodopsin. In the calculation, we neglected the formation of isorhodopsin because of lack of its exact circular dichroic spectrum at $77^{\circ}K$. In addition, it is rather difficult to get the exact circular dichroic spectrum of rhodopsin at $77^{\circ}K$, because a little change of rhodopsin by the measuring light produces a large signal of apparent circular dichroism, as shown in Fig. 7. For these reasons, the spectra shown in Fig. 8 give only qualitative information. Bathorhodopsin exhibits a negative circular dichroic signal in the main band. One may suppose that the change from a positive signal to a negative one indicates a reversal in direction of strain of the retinal chromophore, owing to the photoisomerization from the 11-cis configuration to the all-trans.

Acknowledgement. We express sincere thanks to Prof. G. Wald for reading the manuscript, to Dr. F. Tokunaga for useful discussion, and to Miss S. Furuyama for preparation of the manuscript.

References

1. YOSHIZAWA, T., Y. KITÔ: Chemistry of the rhodopsin cycle. Nature (Lond.) 182, 1604-1605 (1958).
2. YOSHIZAWA, T., G. WALD: Pre-lumirhodopsin and the bleaching of visual pigments. Nature (Lond.) 197, 1279-1286 (1963).
3. TSUKAMOTO, Y., S. HORIUCHI, T. YOSHIZAWA: Photoconversion of chicken visual pigments at liquid helium temperature. Dobutsugakuzasshi 80, 406 (1971).
4. TOKUNAGA, F., T. YOSHIZAWA: unpublished observation.
5. YOSHIZAWA, T., G. WALD: Transformations of squid rhodopsin at low temperatures. Nature (Lond.) 201, 340-345 (1964).
6. YOSHIZAWA, T., G. WALD: Photochemistry of iodopsin. Nature (Lond.) 214, 566-571 (1967).
7. YOSHIZAWA, T., S. HORIUCHI: Intermediates in the photolytic process of porphyropsin. Exp.Eye Res. 8, 243-244 (1969).
8. CONE, R.A.: Rotational diffusion of rhodopsin in the visual receptor membrane. Nature New Biology 236, 39-43 (1972).
9. YOSHIZAWA, T.: The behaviour of visual pigments at low temperatures. In: Handbook of Sensory Physiology, Vol. VII/1, ed. H.J.A. DARTNALL, pp. 146-179. Springer, Berlin-Heidelberg-New York (1972).
10. SHAW, T.: The circular dichroism and optical rotatory dispersion of visual pigments. In: Handbook of Sensory Physiology, Vol. VII/1, ed. H.J.A. DARTNALL, pp. 180-199. Springer, Berlin-Heidelberg-New York (1972).
11. KROPF, A., R. HUBBARD: The mechanism of bleaching rhodopsin. Ann.NY Acad.Sci. 74, 266-280 (1958).
12. MATTHEWS, R., R. HUBBARD, P.K. BROWN, G. WALD: Tautomeric forms of metarhodopsin. J.Gen.Physiol. 47, 215-240 (1963).
13. WAGGONER, A., L. STRYER: Induced optical activity of the metarhodopsins. Biochem. 10, 3250-3254 (1971).
14. HORWITZ, J., J. HELLER: Circular dichroism of native and illuminated bovine visual pigment$_{500}$ at 77°K in the 620- to 320-nm region. Biochem. 10, 1402-1409 (1971).
15. DISCH, R., D. SVERDLIK: Apparent circular dichroism of oriented systems. Anal.Chem. 41, 82-86 (1969).

Discussion

G. Wald: Is it possible that isorhodopsin is the precursor of hypsorhodopsin? This would explain why 540 nm, which isorhodopsin absorbs, forms much hypsorhodopsin, whereas 579 nm, which isorhodopsin hardly absorbs, forms very little hypsorhodopsin.

T. Yoshizawa: We carried out an experiment which compared rhodopsin with isorhodopsin in the rate of formation of hypsorhodopsin by irradiation with light at wavelengths longer than 520 nm. It was shown that the formation of hypsorhodopsin from rhodopsin is faster than that from isorhodopsin, even in due consideration of the difference in absorption spectra between rhodopsin and isorhodopsin and of the wavelength of the actinic light. Therefore, this fact does not support the view that hypsorhodopsin can be directly formed only from isorhodopsin.

J. Heller: In your figures you showed the circular dichroism of rhodopsin at 77°K. Dr. Horwitz and myself found that you can determine the circular dichroism only of native, unilluminated rhodopsin. Illumination of rhodopsin which is embedded in rigid media at 77°K always leads to photoselection and consequently linear dichroism (LD). The LD signal is recorded by the circular dichroism spectrophotometer as a CD signal. We believe that the CD spectra of illuminated rhodopsin at 77°K are actually LD spectra. The only sure way of distinguishing between CD and LD at 77°K is by rotating the sample in the spectrophotometer. CD signals which are due to molecular properties will not be affected, while LD signals change their sign upon rotation of the sample. We have previously published a paper on the CD of illuminated rhodopsin at 77°K (14) where the signals are actually due to linear dichroism.

T. Yoshizawa: In connection with an answer to your questions, I would like to offer our opinion to your paper (14). I certainly know that you described that CD spectrum of rhodopsin (in CTAB) has disappeared by cooling to 77°K and furthermore that the extraordinarily large signal was brought by irradiation of rhodopsin at 77°K irrespective of polarized light, light from the spectropolarimeter and natural light. Our experiment shows that rhodopsin-digitonin glycerol mixture exhibits circular dichroism at 77°K . The conflict between your CD spectrum and ours may be explained by three possible reasons as follows: 1) Difference in detergent for extraction of rhodopsin (digitonin and CTAB). 2) Presence of cracks in preparation at 77°K. If the cracks are formed in the direction of parallel to light path, CD signal will reduce in the height. If the cracks are not parallel, measuring light will not only be dispersed by them, but also causes a change in ellipticity to a certain extent. On the other hand, our preparation has no cracks at 77°K, because the preparation was rapidly cooled to 77°K by means of dropping it into a pre-cooled absorption vessel. 3) Disappearance of CD signal of rhodopsin by superposition of the signal due to ordered system and/or CD signal of bathorhodopsin formed by measuring light (Fig. 7). Concerning the measurement of the preparation of low absorbance, one cannot neglect this possibility (our preparation: OD = 1.2). The second problem is irradiation of rhodopsin at 77°K. In the case of measurement of rhodopsin preparation irradiated at 77°K with light containing linear polarized light, by which the preparation can be transformed from random structure to ordered one, it is sure - as you stated - that the signal containing linear dichroism (apparent circular dichroism, your LD signal) is recorded by the circular dichroism spectrophotometer as CD signal. Our experiment showed in Fig. 7 has been carried out in order to confirm this. Thus, we realized that, even though one irradiates preparation at 77°K with linear polarized light, exhaustive irradiation brings the disappearance of the apparent CD signal, indicating the transformation of the ordered structure (which had been formed by the irradiation for short period, curve 1-5) into the random structure (photosteady state mixture) containing rhodopsin, isorhodopsin, and bathorhodopsin. If one wants to measure the circular dichroism of bathorhodopsin, one must use the preparation containing bathorhodopsin without any trace of ordered structure. In order to get such a preparation, we think that there are at least two ways. One is to use the photosteady state mixture produced by the exhaustive irradiation, and the other is to rotate the preparation continuously during the irradiation at 77°K under considering both the velocity of rotation and the intensity of light. By using the former method, we have carried out the experiment on bathorhodopsin (Fig. 8). As a matter of fact, the measurement of CD spectrum of bathorhodopsin involves some errors similar to that of rhodopsin, because the irradiation with polarized light at different wavelengths again transforms from random structure into an ordered structure.

Interconversion of Metarhodopsins

T. P. Williams, B. N. Baker, and D. J. Eder
Institute of Molecular Biophysics, Florida State University, Tallahassee, FL 32306/USA

When the highly colored visual pigments of the eye absorb light, they communicate this fact to the membrane of the receptor cell in which they reside. This process of converting the energy of absorbed light into the initial response of the membrane is called transduction, and, at the present time, the study of transduction is a popular one in many laboratories, including ours. We have chosen to study the pigments, on one hand, and the receptor response, on the other, hoping that we can ascertain what it is that links them in a cause-effect relationship. The receptor response under investigation is the one which is elicited from the rod-dominated eye of the albino rat. We picked this rod response because the visual pigment of rods, rhodopsin, can readily be obtained and studied *in vitro*.

If one is interested in those pigment reactions which might be directly involved in transduction, he obviously must choose to study those reactions which occur fast enough to precede, or be synchronous with, the onset of the membrane response. The metarhodopsins fulfill this requirement, and we have been studying them for several years.

Matthews *et al.* (1) originally disclosed some of the interesting properties of the metarhodopsins. Not the least interesting of these was the rather large, positive change in entropy which occured when metarhodopsin I (MI, λ_{max} 478 nm) was converted, in the dark, to metarhodopsin II (MII, λ_{max} 380 nm). Along with this interesting entropy change, they showed that a proton was *added* to the molecule as MI became MII. If retinal, the aldehyde of vitamin A is linked to opsin, the visual protein, by a Schiff base, why should MII absorb at shorter wavelengths than MI? Blue-shifts in the spectra of Schiff bases occur upon *deprotonation*. Another item emerged from the Matthews paper: An extrapolation of their data to (mammalian) physiological temperature showed that the $t_{\frac{1}{2}}$ for production of MII was *ca.* 1 msec - fast enough to be associated with transduction. Experiments in our lab have long ago confirmed this rate (2). Having done so, we then began to chemically dissect the MI-MII interconversion.

Many, if not all, workers in this field agree that *opsin* is the pigment moiety which is directly linked with transduction. The action of light on retinal probably just provides the trigger for the protein reaction. Yet, when we study spectral changes which occur, we are focussing on retinal and its immediate environs. Changes which go in other parts of opsin, remote from retinal, may not register as spectral changes at all. How can *these* changes be measured?

A method we have employed extensively involves photoreversal of bleaching. This is a process whereby light regenerates rhodopsin from certain of the intermediates involved in bleaching. The action of light on the original pigment is the conversion of 11-cis retinal to all-trans. The shape of all-trans signals to opsin that light has been absorbed and there ensues a series of complex "dark" reactions. If, early in the lifetimes of some of these all-trans intermediates, another photon is absorbed, the all-trans retinal may become 11-cis or 9-cis. The former yields a rhodopsin which is indistinguishable from the starting material (3,4,5). The latter results in isorhodopsin. How much of these two pigments, relative to each other, is produced upon photoreversal? And what is the mechanism whereby they are produced? We have studied these problems and what has emerged provides a picture of the opsin reactions which take place.

Baumann and Ernst (6) have shown that photoreversal of the very early intermediates (probably no later than the pre-lumirhodopsin stage) yields only isorhodopsin. We have found that photoreversal of later intermediates gives quite a different result. For example, photoreversal of MI produces a 60/40 mixture of rhodopsin/isorhodopsin. Photoreversal of MII, on the other hand, results in a 93/7 mixture of rhodopsin/isorhodopsin. These results are summarized in Table I. The "moles" of the various substances are normalized by setting the starting concentration of rhodopsin at 100.

Table I. Photoreversal from MI and MII: Percent isorhodopsin (ISO) produced at 37°C

Photoreversal from	Total Moles of Pigment Regenerated	Moles R	Moles ISO	% ISO
MI alone	12.7	7.6	5.1	40
MI + MII	26.6	20.5	6.1	23
MII alone	13.9	12.9	1.0	7

These results taken along with those of Baumann and Ernst suggest that there is a continuous change in the conformation of opsin, starting with one which selects the 9-cis isomer exclusively and ending (in MII) with one which accepts only 11-cis.

The mechanisms whereby these photoregenerated pigments are produced provides additional insight into the opsin reactions. For example, when MI is irradiated with a 2 msec photoreversing flash, it produces the 60/40 mixture of rhodopsin and isorhodopsin as fast as the light is absorbed (7). The experiments were done as follows: Digitonin solutions of cattle rhodopsin were irradiated with orange light to produce MI and MII. A blue flash, selective for MI, was given to the mixture and the optical density changes were followed on the sub-millisecond time scale. Optical density build-up, corresponding to stable pigment accumulation, followed the integrated quantal input and was complete when the flash was over, i.e. in 2 msec The results at 5°C are shown in Fig. 1.

Temperature changes in the range 5°C to 18°C had no effect. Thus, the conformation of opsin in the "MI state" is capable of accepting 9-cis or 11-cis and, when it does, the spectrum of the rhodopsin/isorhodopsin mixture appears immediately. We were able to resolve time to 50 microseconds, and therefore the term "immediately" has this qualification attendant to its meaning. Despite this qualification, the situation here is very different from that encountered in the reversal of MII.

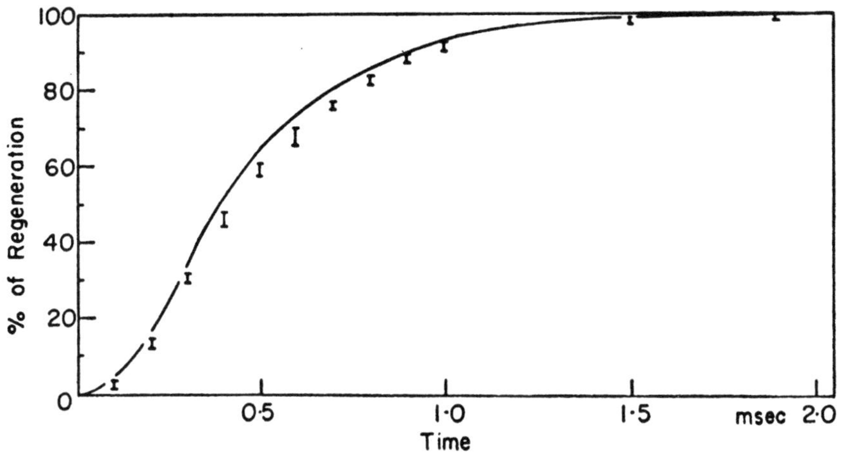

Fig. 1. Rate of production of rhodopsin and isorhodopsin from MI

In these experiments (8), MI and MII were produced as before and this time a UV flash, selective for MII was given. Two kinds of spectral changes occured: A fast one, producing P470 (pararhodopsin?), and a slower one, producing rhodopsin. These spectral changes are shown in Fig. 2.

The ratio of P470 to rhodopsin was always 3.5/1.0, and the P470 appeared as fast as the light was absorbed. Not so for rhodopsin: It appeared during the course of a dark reaction. Rhodopsin was produced from a material whose partial difference spectrum is shown in Fig. 3, and whose λ_{max} is that of the MII from which it has just been produced.

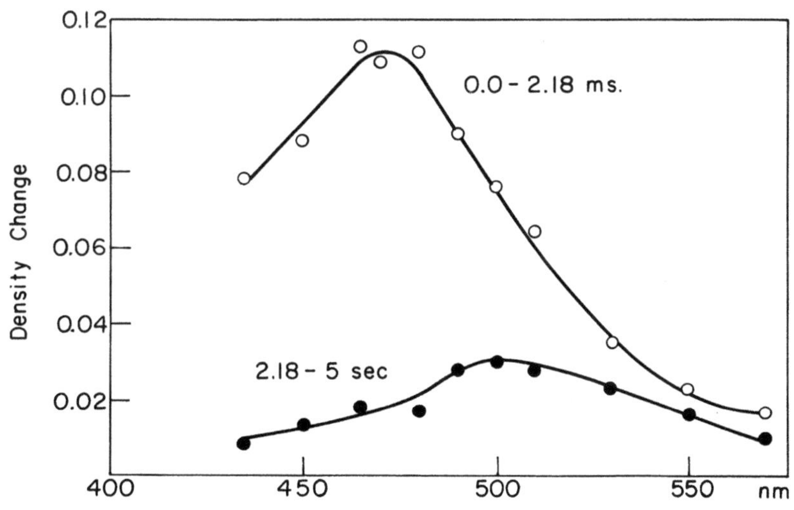

Fig. 2. Spectral changes which occur when MII is irradiated with 2 msec flash at 5°C and pH 6.5

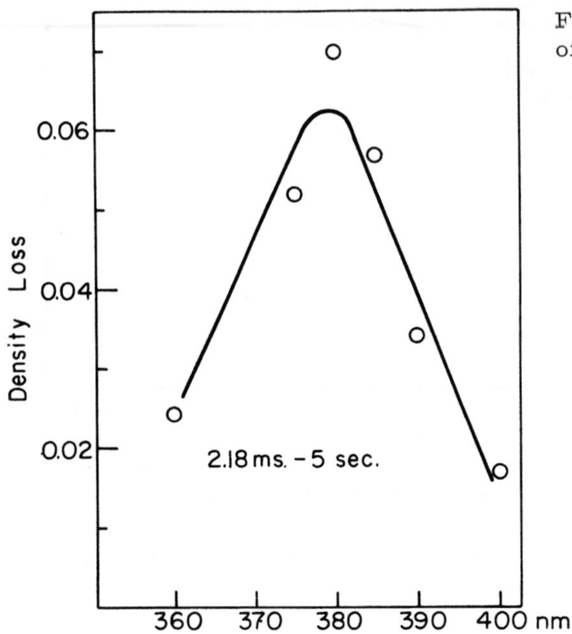

Fig. 3. Partial difference spectrum of the precursor of rhodopsin

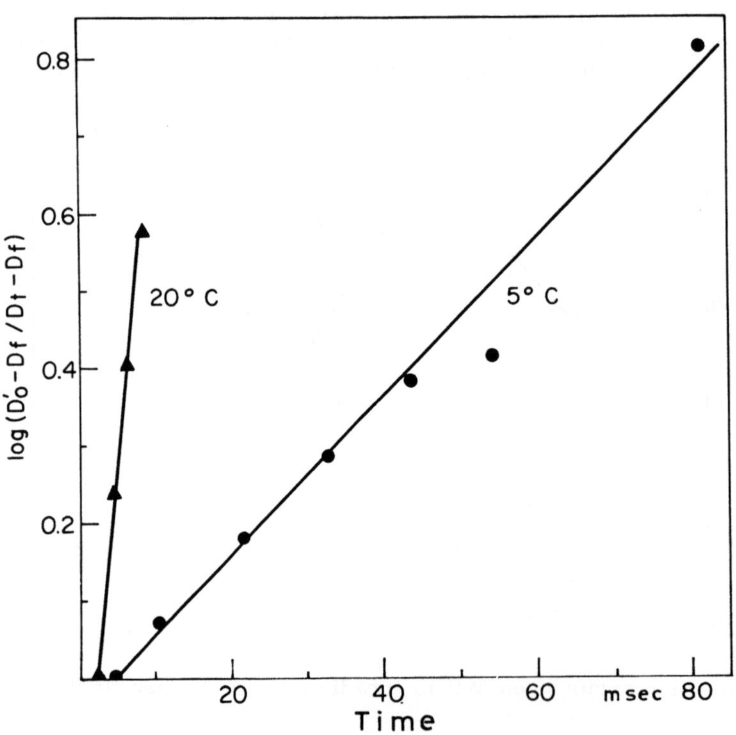

Fig. 4. Integrated first-order rate plot of production of rhodopsin from MII

In other words, the flash isomerizes MII to a form which must contain the 11-cis isomer because, in the dark, it produces rhodopsin. The rate at which this compound (MII-prime) decays to rhodopsin is shown to be both first-order and temperature dependent in Fig. 4. The activation energy for this production of rhodopsin is 18 kcal/mole and this implies that an appreciable internal barrier must be traversed in the process (as opposed to a simple protonation, for example).

The photoreversal mechanisms of MI and MII then are quite different: When MI absorbs light, the opsin immediately accepts 9- or 11-cis and the photoregenerated pigments appear as fast as the light is absorbed. When MII absorbs light, the formation of 11-cis occurs immediately and, within a few milliseconds this induces the opsin to attain the rhodopsin configuration.

In summary, we have found that, by studying photoreversal of the metarhodopsins, we can gain insights into the opsin reactions. Such insights may well provide a means of understanding transduction which years of investigation of spectral changes have failed to do.

References

1. MATTHEWS, R.G., R. HUBBARD, P.K. BROWN, G.WALD: Tautomeric forms of metarhodopsin. J.Gen.Physiol. 38, 269-315 (1963).
2. WILLIAMS, T.P., S.J. BREIL: Kinetic measurements on rhodopsin solutions during intense flashes. Vision Res. 8, 777-786 (1968).
3. WILLIAMS, T.P.: Induced asymmetry in the prosthetic group of rhodopsin. Vision Res. 6, 293-300 (1966).
4. BAKER, B.N., T.P. WILLIAMS: Thermal decomposition of rhodopsin, photoregenerated rhodopsin and P470. Vision Res. 8, 1467-1469 (1968).
5. SHICHI, H.: Spectrum and purity of bovine rhodopsin. Biochemistry 9, 1973-1977 (1970).
6. BAUMANN, CH., W. ERNST: Formation of isorhodopsin in isolated frog retinae by intense nanosecond flashes. J.Physiol. (Lond.) 210, 156-157P (1970).
7. BAKER, B.N., T.P. WILLIAMS: Photolysis of metarhodopsin I: Rate and extent of conversion to rhodopsin. Vision Res. 11, 449-458 (1971).
8. WILLIAMS, T.P.: Photolysis of metarhodopsin II. Rate of production of P470 and rhodopsin. Vision Res. 8, 1457-1466 (1968).
9. ARDEN, G.B., H. IKEDA, I.M. SIEGEL: Effects of light adaptation on the early receptor potential. Vision Res. 6, 357-371 (1966).

Discussion

G.O. Schenck: May I raise a "dirty" question on a "dirty" solution of rhodopsin which is described as especially suitable. In scientific language "dirty" should be clearly defined (meaning independent of context) or much better generally avoided. It is not clear whether the dirt came from the janitor (etc.) or whether "more dirty" is associated with "more natural"? May I suggest "less purified" for want of a better definition.

T.P. Williams: I shall be happy to agree with that.

G.P. Arden: Photoreversal from MII or MIII gives apparently equal isorhodopsin and normal rhodopsin; this was in rat, with plenty of pararhodpsin, and with control by early receptor potential (9).

T.P. Williams: We cannot photoreverse MIII. However, with really high intensities, it is possible that some of our MIII was photoconverted to MII (1). If so, and if this MII absorbed additional light, you could get the same ratio of rhodopsin to isorhodopsin as if you were irradiating MII. It would thus appear that MIII was reversible.

Temperature Dependence of Slow Thermal Reactions during the Bleaching of Rhodopsin in the Frog Retina

Ch. Baumann and R. Reinheimer
W. G. Kerckhoff-Institut, Max-Planck-Gesellschaft, Bad Nauheim/W. Germany

The photolysis of rhodopsin is followed by a sequence of thermal reactions, and various intermediates appear before stable breakdown products are formed. In the retina, the early intermediates, i.e., prelumi-, lumi- and metarhodopsin I (MI) are very short lived. Metarhodopsin II (MII) and its successors, on the other hand, decay slowly. The kinetics of these slow reactions have recently been studied in the frog retina (1) where two parallel pathways for the decay of MII were demonstrated. Rate constants at 21 °C were determined for these two reactions, for the decay of metarhodopsin III (MIII) and for the reduction of all-trans-retinal.

The present paper deals with the effect of temperature on the decay rates. This type of study provides further data to test the model used to describe earlier results (1). It also yields information about changes of the molecular structure of rhodopsin during the process of bleaching. Such changes have been extensively studied in rhodopsin solutions but the results of different groups are not in good agreement (for a review see 2). In addition, it may be questioned whether the results from solutions are always relevant to the situation in intact photoreceptors.

Method

Isolated retinas of frogs (Rana esculenta) were incubated in a transparent perfusion chamber and investigated spectrophotometrically. Absorbances and absorbance changes on bleaching were measured in the visible and the near ultraviolet region. Bleaching was carried out by means of flashes from a xenon filled discharge tube (90% light dissipation within 1.5 msec). The flash light was filtered so that only wavelengths between 490 and 620 nm were transmitted. It was calculated that approximately 10 photons were absorbed per molecule of rhodopsin per flash (cf. 3). In spite of substantial photoreversal effects, one flash bleached at least $\frac{2}{3}$ of the pigment. Further details about the method have been published elsewhere (1).

The temperature of the retina was varied by cooling or heating the perfusate which was a buffered frog Ringer. The preparations were perfused at constant rates (5 to 7 ml/min). Before entering the perfusion chamber, the perfusate passed a heat exchanger where it reached the desired temperature. Temperatures were measured inside the perfusion chamber in the immediate vicinity of the preparation by means of a thermocouple. During the course of one experiment, the actual temperature did not deviate by more than 0.2 °C from the set temperature (i.e. 10, 15, 20, 25, or 30 °C).

Theory

The data of this study will be analysed in terms of a scheme consisting of four reactions:

1) Conversion of metarhodopsin II (A) into metarhodopsin III (B);
2) Hydrolysis of metarhodopsin II (A) into all-trans-retinal (C) and opsin;
3) Decay of metarhodopsin III (B) to all-trans-retinal (C) and opsin;
4) Reduction of all-trans-retinal (C) into all-trans-retinol (D).
Schematically, this may be written as

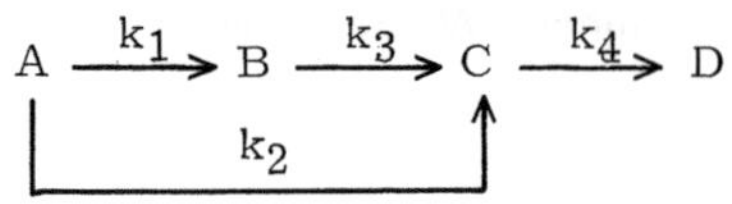

where k_1, k_2, and k_3 are simple rate constants. k_4 is the rate constant for the enzymatic formation of D. Opsin is not shown because it does not absorb appreciably at any of the wavelengths used to monitor the concentrations of A, B, C, or D.

Reactions prior to the decay of metarhodopsin II do not interfere with the present measurements because they are too fast. At room temperature, the MI to MII conversion reaches virtual completion within a few milliseconds (cf. 4). MII, on the other hand, decays comparatively slowly, and the fraction of it that decays during the flash period of 1.5 msec is negligibly small. Therefore, immediately after the flash, all the bleached pigment is present in the form of MII. As the molar absorbance coefficients of rhodopsin and MII are known to be equal (4), the temporary increase of absorbance at 380 nm (near the λ_{max} of MII) will be equal to the permanent absorbance loss at 500 nm due to the bleaching of rhodopsin. This was observed in all the experiments and is important to the formulation of the initial conditions (see below).

The conversion of MII into MIII and the decay of MIII are both probably pure conformational changes and therefore first-order processes. If water is present in excess, the hydrolysis of MII will also obey first-order kinetics. The enzyme-controlled reduction of retinal into retinol was analysed in two independent ways, viz. in terms of a zero order and in terms of a first-order reaction. Only the latter was found to be compatible with the experimental findings. All the reactions were therefore assumed to be first-order.

The reaction scheme is represented by the following set of simultaneous differential equations

$$\frac{dA}{dt} = -(k_1+k_2)A; \quad \frac{dB}{dt} = k_1A-k_3B; \quad \frac{dC}{dt} = k_2A+k_3B-k_4C; \quad \frac{dD}{dt} = k_4C. \qquad 1$$

The initial conditions are $A = A_o$; $B = C = D = 0$ if $t = 0$, where t is the time in seconds and A_o the concentration of A at the commencement. It is convenient to set A_o equal to unity. In practice this means that all the data have to be normalized with respect to the MII concentration at the beginning (see also 1). The solution of eqn. 1 becomes then:

$$A = e^{-(k_1+k_2)t} \qquad 2$$

$$B = \frac{k_1}{k_3-(k_1+k_2)}\left(e^{-(k_1+k_2)t}-e^{-k_3t}\right) \qquad 3$$

$$C = \frac{k_1k_3+k_2(k_3-k_1-k_2)}{(k_3-k_1-k_2)(k_4-k_1-k_2)}\,e^{-(k_1+k_2)t} + \frac{k_1k_3}{(k_1+k_2-k_3)(k_4-k_3)}\,e^{-k_3t} + \frac{k_1k_3+k_2(k_3-k_4)}{(k_1+k_2-k_4)(k_3-k_4)}\,e^{-k_4t} \qquad 4$$

$$D = 1-(A+B+C) \qquad 5$$

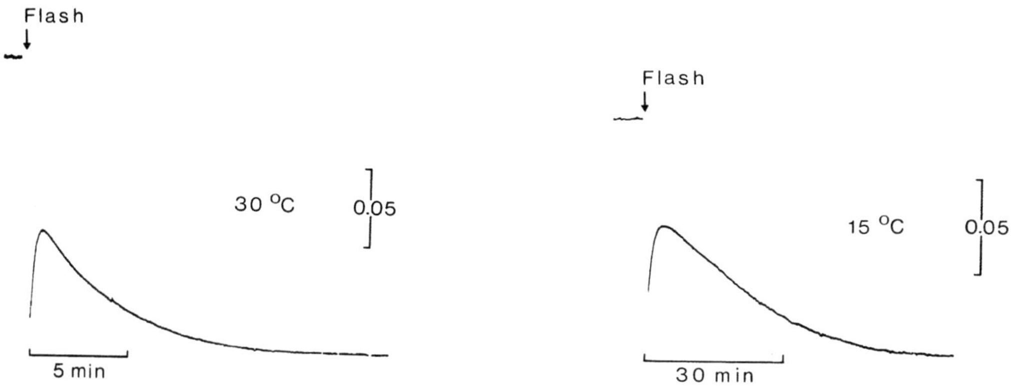

Fig. 1. Time course of absorbance changes in isolated frog retinas after exposure to a strong bleaching flash. Wavelength: 480 nm

---

These equations describe the relative concentrations of the four reactants as functions of t, the time after the flash.

Results

Range of temperatures studied. At temperatures above 30°C, frog rhodopsin begins to decompose spontaneously, i.e., it is bleached without having absorbed any light (5). This temperature was therefore not exceeded in the present study. Limitations of technique imposed a lower boundary on the range. At 10°C, several of the rate constants become smaller than $10^{-3}sec^{-1}$, and their accurate evaluation at this temperature requires extremely stable measuring conditions (e.g. constancy of the light source, stability of the preparation, constant temperature of the perfusate

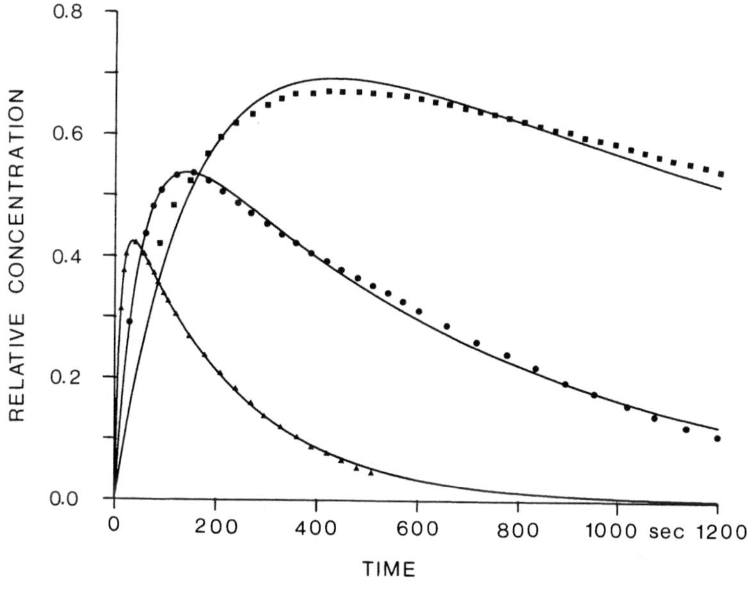

Fig. 2. Formation and decay of MIII at 30°C (triangles), at 20°C (circles) and at 10°C (squares). The curves were calculated according to equation (3) with the respective rate constants of Table I

Table I. Temperature dependence of k_1, k_2, and k_3

Temperature ^{o}C	k_1 sec^{-1}	k_2 sec^{-1}	k_3 sec^{-1}
10	5.55×10^{-3}	9.71×10^{-4}	4.71×10^{-4}
15	8.47×10^{-3}	3.30×10^{-3}	8.25×10^{-4}
20	1.31×10^{-2}	6.69×10^{-3}	1.45×10^{-3}
25	2.88×10^{-2}	2.39×10^{-2}	2.47×10^{-3}
30	4.35×10^{-2}	4.37×10^{-2}	4.50×10^{-3}

etc.). As the data obtained at $15^{o}C$ and $10^{o}C$ did not suggest that the reactions at lower temperatures were qualitatively very different from those at higher ones, no efforts were made to work at temperatures below $10^{o}C$.

Evaluation of k_1, k_2, and k_3. According to eqn. 3, the relative concentration of M III (B) depends on three rate constants viz. k_1, k_2, k_3. If a signal can be recorded that reflects the presence of B over a sufficiently long time, then the constants can be calculated. Fig. 1 shows absorbance changes at 480 nm following a strong bleaching flash. These absorbance changes are exclusively due to the presence of M III because none of the other reactants (A, C, and D) absorbs measurable quantities of light at this particular wavelength. Recordings like those of Fig. 1 thus provide the possibility of computing k_1, k_2, and k_3. The data have to be normalized with respect to the concentration of A at the beginning. In addition, corrections have to be made for differences between the absorbance coefficients of A and B and for the fact that the absorbance change recorded at 480 nm is slightly lower than at 465 nm where the absorbance of M III is maximum (cf. 1). Data corrected in this way are shown in Fig. 2. The triangles refer to the experiment of Fig. 1 ($30^{o}C$); two other sets of data are from experiments carried out at $20^{o}C$ (circles) and $10^{o}C$ (squares), respectively. The curves running through the data points are graphical representations of eqn. 3. The optimum fit was achieved by applying a least squares method, i.e. the three parameters of eqn. 3 were modified until the deviations between the points and the curve (strictly speaking the error sum of squares) became minimum. A BMDX 85 computer program (6) was used for this purpose. Those parameters that gave the best fit are summarized in Table I.

Each row in the table refers to a single experiment which was selected for the computation because it was free of technical imperfections. Recordings from another five experiments (not shown) had to be discarded since they were either spoiled by air bubbles running through the perfusion chamber, or the preparations contained relatively little rhodopsin.

The curves in Fig. 2 fit the data fairly well but there are some deviations. At $30^{o}C$ (triangles), the agreement between the curve and the data points is almost perfect. At $20^{o}C$ (circles), the agreement is still very good, though some small deviations can be seen around 600 and 1200 sec after the flash. These deviations, if expressed as absorbances, amount to approximately 0.002 and are still within experimental error. The situation is somewhat different at $10^{o}C$ where some systematic deviations occur. They are greatest during the first two minutes after the flash. Then, the relative concentration of B is larger than eqn. 3 predicts. This is also illustrated in Fig. 1 (right) where the same effect was observed at $15^{o}C$. At higher tem-

Fig. 3. Arrhenius plots of k_1 (triangles), of k_2 (squares), and of k_3 (circles) within the range of 10^oC ($3.53 \cdot 10^{-3}$ $^oK^{-1}$) to 30^oC ($3.30 \cdot 10^{-3}$ $^oK^{-1}$)

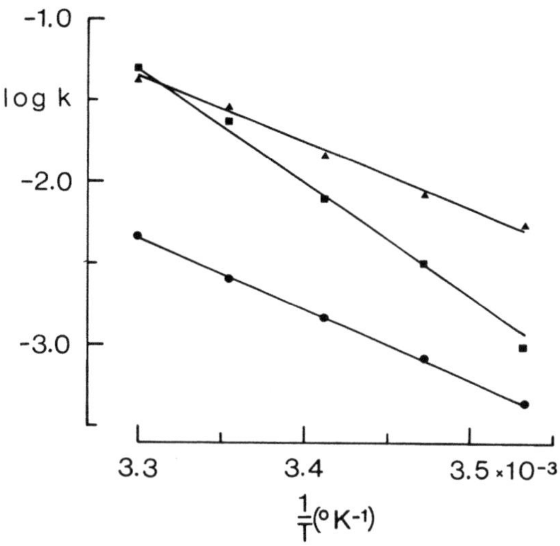

peratures (20^oC and more), the absorbance (λ = 480 nm) recorded immediately after the flash is very low and appears to start at the level that is finally reached when MIII has disappeared (Fig. 1, left; see also Baumann, 1, Fig. 4). At 10 and 15^oC, on the other hand, the absorbance measured shortly after the flash is definitely larger than it is at the end of the experiment, and the first observable change is a fall, not a rise of absorbance. The effect continues for some 10 to 20 seconds; thereafter, the formation of MIII becomes the dominant process and the absorbance changes in accordance with eqn. 3. The agreement is less perfect than it is at the higher temperatures but it is possible to analyse the data on the basis of eqn. 3 provided that data recorded during the first two minutes or so are discarded. This is the reason why the first datum of the 10^oC experiment (Fig. 2) corresponds to 90 sec after the flash and not earlier.

The process that causes the initial fast absorbance decrease at 480 nm cannot be explained on the basis of the present data. Were the conversion of MI (λ_{max}=478 nm) into MII (λ_{max}=380 nm) responsible, then the absorbance at 380 nm should rise in conjunction with the fall at 480 nm. But even at 10^oC, the absorbance at 380 nm falls from the very beginning of the recordings (i.e. ca. 2 sec after the flash).

Influence of temperature on k_1, k_2, and k_3. Table I shows a marked temperature dependence of the rate constants. This is illustrated in Fig. 3, where the logarithms of the constants are plotted against the reciprocal of the absolute temperature. It is obvious that straight lines may be used to fit the data. This means that the Arrhenius law holds within the temperature range studied and the slope of the lines can be expressed as $-E_a/2.303$ R, where E_a is the activation energy and R is the gas constant. Values of E_a are given in Table II. In addition to the empirically determined activation energy E_a, the free energy of activation, $\Delta F^{\ddagger}$, the enthalpy of activation, $\Delta H^{\ddagger}$, and the entropy of activation, $\Delta S^{\ddagger}$, were calculated from the following formulas:

$$\Delta F^{\ddagger} = -RT \ln \frac{k_r h}{kT}; \quad \Delta H^{\ddagger} = E_a - RT; \quad \Delta S^{\ddagger} = \frac{\Delta H^{\ddagger} - \Delta F^{\ddagger}}{T};$$

where k_r is the reaction rate constant, T the temperature (^oK), h Planck's constant, and k the Boltzmann constant.

Table II. Kinetic parameters of the decay of MII and of MIII

Reaction	E_a kcal/mole	$\Delta F^{\ddagger}$ kcal/mole	$\Delta H^{\ddagger}$ kcal/mole	$\Delta S^{\ddagger}$ cal/mole·°K
Conversion of MII into MIII	18.2	19.6	17.6	- 7
Hydrolysis of MII into retinal and opsin	32.8	20.1	32.2	+ 41
Decay of MIII	19.1	20.9	18.5	- 8

$\Delta F^{\ddagger}$, $\Delta H^{\ddagger}$, and $\Delta S^{\ddagger}$ are parameters having meaning in the context of the Absolute Reaction Rate Theory. This theory originally was not meant to be applied to macromolecular processes. However, from various studies with proteins (enzymes in particular) it is known that large enthalpies and entropies of activation are indicative of major conformational changes (7,8,9). If the present data are considered in this way, it is only the hydrolysis of MII that requires a comparatively large conformational change of the molecule. The MII to MIII transition and the decay of MIII, on the other hand, are accompanied by very little conformation changes (see also Discussion).

Evaluation and temperature dependence of k_4. The conversion of retinal into retinol is governed by the constant k_4. The conversion process is mathematically described by eqns. 4 and 5. The former gives the time course of the decay of retinal (C), and the latter describes the formation of retinol (D). As the constants k_1, k_2, and k_3 are already known, k_4 may be considered as the only unknown parameter of eqns. 4 and 5. If the concentration changes of either C or D could be measured without interference due to the presence of the other reactants, k_4 could be evaluated by fitting eqns. 4 and 5 to the data. Unfortunately, the real situation is somewhat more complicated, as wavelengths preferably absorbed by C and D are also absorbed by the other reactants. Thus, isolation of C (or D) on the basis of absorbance measurements is impossible under the conditions of the present experiments. Nevertheless, k_4 can be determined if the proportions are known by which the various reactants contribute to the absorbance at a particular wavelength. At 380 nm, where the absorbance of retinal is maximum, the contributions of A, B, C, and D stand in the ratio 1:0.25:1:0.02 (cf. 1). Therefore

$$E' = A + 0.25\,B + C + 0.02\,D, \qquad 6$$

where E' is proportional to absorbance to 380 nm and A, B, C, and D are given by eqns. 2 to 5. Adjusting the experimental data for use in eqn. 6 just requires a normalization with respect to the concentration of A at the beginning. After this, the absorbance changes recorded at 380 nm provide the basis for the computation of k_4 (computer program as quoted above).

The computational procedure can be simplified by ignoring the concentration of A in eqn. 6. The decay of A is relatively fast and, a few minutes after the flash, its concentration becomes very small indeed (cf. eqn. 2). At 10°C, A approaches a figure < 0.001 after some 1,100 sec. k_4 was therefore computed on the basis of data read between the 1,200th and the 4,800th second after the flash. Fig. 4 illustrates

Fig. 4. E' of eqn. 6 plotted versus time (curve) and compared with absorbance changes at 380 nm (squares). Temperature: 10°C

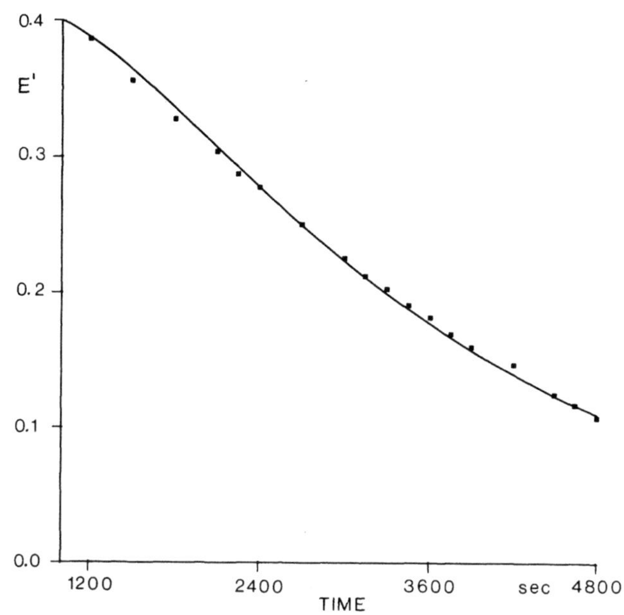

the agreement between eqn. 6 and the data of one experiment (10°C) in which k_4 was found to be equal to $9.08 \cdot 10^{-4} sec^{-1}$. In another experiment carried out at the same temperature, k_4 was $1.004 \cdot 10^{-3} sec^{-1}$. At higher temperatures, the following figures (± standard error of the mean) were obtained: $(2.52 \pm 0.26) \cdot 10^{-3} sec^{-1}$ (21°C; 9 expts.) and $(5.63 \pm 0.17) \cdot 10^{-3} sec^{-1}$ (30°C; 6 expts.).

Fig. 5 shows the effect of temperature upon k_4. It is obvious that, within the temperature range considered, the data obey the Arrhenius law. The activation energy (E_a) is 15.1 kcal/mole. This figure lies within the range normally found in the case of enzyme reactions. However, an unambiguous interpretation of the activation process to which E_a belongs is not possible. Enzyme-controlled reactions are known to proceed in two steps. First, enzyme (E) and substrate (S) form a complex (X). This is a reversible reaction: $E + S \rightleftharpoons X$. The second step is the breakdown of the enzyme-substrate complex resulting in the formation of products ($X \longrightarrow$ products). At low substrate concentrations (when first-order kinetics are encountered) E_a can either be the activation energy for the initial complex formation ($E + S \longrightarrow X$) or it is the activation energy for the second step ($X \longrightarrow$ products) plus the total energy increase for the first step ($E + S \longrightarrow X$) (10). It must be emphasized that this interpretation is valid only if, as in the present case, the Arrhenius law applies directly to the apparent first-order rate constant (k_4). For further details the reader is referred to Laidler's textbook on enzyme kinetics (10; pp. 200-202).

Discussion

The experimental results of this study can be satisfactorily explained in terms of the kinetic model for four slow reactions of rhodopsin bleaching outlined on page 90. However, the possibility must be considered that other reaction schemes may also lead to an adequate quantitative description of the experimental findings. It was of

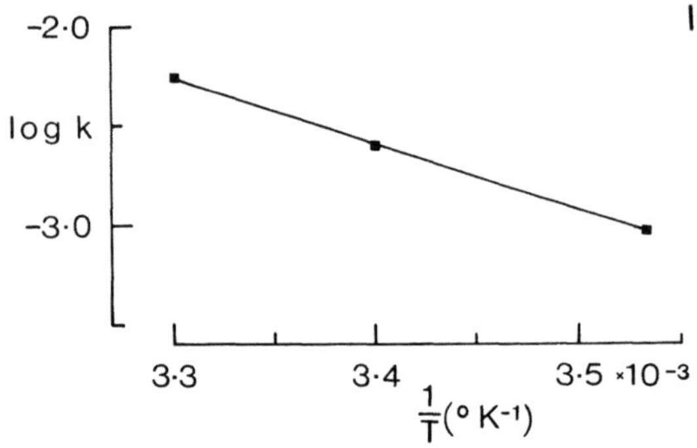

Fig. 5. Arrhenius plot of the apparent rate constant (k_4) for the enzyme-controlled conversion of retinal into retinol

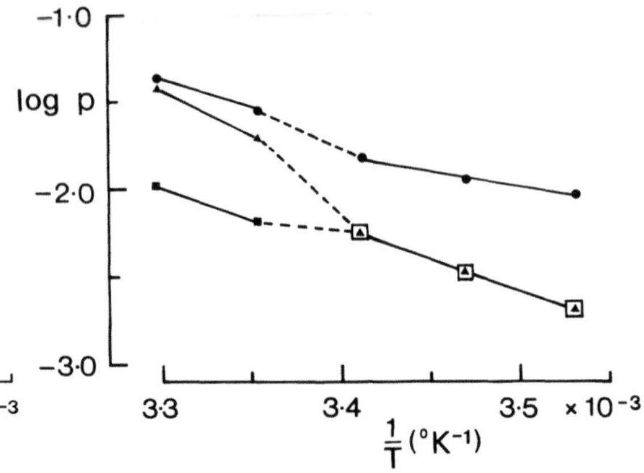

Fig. 6. Arrhenius plot of p_1 (circles), of p'_1 (triangles), and of p_2 (filled and plain squares). The dotted lines connect corresponding sets of data obtained at higher temperatures (left) with those from lower ones (right)

particular interest to use the present data to test one scheme (1) which may be written as

$$B \underset{p'_1}{\overset{p_1}{\rightleftharpoons}} A \xrightarrow{p_2} C \xrightarrow{p_3} D$$

where A is MII, B is MIII, C is retinal, and D is retinol. The p's are first-order rate constants. It can be shown that, for the given initial conditions

$$A = 1;\ B = C = D = 0 \quad \text{at time } t = 0,$$

the change of B with time t is

$$B = \frac{p_1}{s_1 - s_2}\left(e^{s_1 t} - e^{s_2 t}\right) \qquad 7$$

where s_1 and s_2 are the roots of the equation

$$s^2 + s\,(p_1 + p'_1 + p_2) + p'_1 p_2 = 0. \qquad 8$$

The computation of p_1, p'_1, and p_2 is analogous to that of k_1, k_2, and k_3, i.e., it is based on a least squares fit. To give an example: at 30°C eqn. 7 fits the data when $p_1 = 4.35$, $p'_1 = 3.79$, $p_2 = 1.04$ ($\cdot 10^{-2}\ sec^{-1}$). This seems to be plausible, but if the findings obtained at different temperatures are compared with one other, an odd picture emerges. This is illustrated in Fig. 6 where the logarithms of p_1, p'_1, and p_2 are plotted versus the reciprocal of the absolute temperature in the range between 10°C and 30°C. None of the sets of data can be fitted by a single straight line because at the two higher temperatures (30 and 25°C) the slope is obviously different from that at the lower ones (10 to 20°C). In addition, at 10, 15, and 20°C p'_1 and p_2 are exactly equal to each other and would consequently have identical activation energies. Equality of p'_1 and p_2 as a condition of an optimum agreement between the data and eqn. 7 has also been found in previous experiments carried out at 21°C (1).

There are thus several reasons to consider this model as implausible.
1) It shows a non-uniform behaviour when tested at different temperatures which

Table III. Activation entropies of corresponding thermal reactions studied in the retina (this work) or in rhodopsin solutions (cf. column Ref.)

Reaction	$\Delta S^{\ddagger}$/retina cal/mole·°K	$\Delta S^{\ddagger}$/solution cal/mole·°K	References
Conversion of MII into MIII	- 7	- 50	Ostroy et al. (11)
Hydrolysis of MII into retinal and opsin	+ 41	- 7	Matthews et al. (4)
Decay of MIII	- 8	+ 48	Ostroy et al. (11)

makes analysis of the data in terms of activation energies impossible.

2) The precise equality of p_1' and p_2 required for an optimum fit is another odd property of the model, and it makes it all the more improbable as this particular result was obtained at all temperatures up to 21°C.

3) The goodness of fit, i.e., the agreement between the experimental findings and the theoretical curves was significantly worse than it was in the model considered in the "Results" section.

Reasons for the rejection of two other schemes were given in a previous paper (1). The most plausible model for interpreting all results remains one in which MII can enter two different pathways of decay, viz. conversion into MIII and hydrolysis into retinal and opsin. The two reactions clearly differ in their kinetic parameters (Table II). The former has an enthalpy of activation ($\Delta H^{\ddagger}$) of approximately 18 kcal/mole and a small negative entropy of activation ($\Delta S^{\ddagger} = -7$ cal/mole·°K). However, the thermal barrier of the hydrolysis is much larger ($\Delta H^{\ddagger} = 32$ kcal/mole) and, consequently, the activation entropy must also be fairly large ($\Delta S^{\ddagger} = 41$ cal/mole·°K), otherwise the reaction would be too slow to be observable at physiological temperatures.

The fact that $\Delta H^{\ddagger}$ and $\Delta S^{\ddagger}$ both are large has still another consequence: it suggests that a fairly large conformational change of opsin accompanies the hydrolysis. This leads to the conclusion that the only covalent bond between retinal and opsin is probably difficult to break unless it is exposed to the hydrolysing agent (whatever this may be) by an unfolding of the protein. The conversion of MII into MIII and the decay of MIII, on the other hand, have kinetic parameters which indicate that in neither of the two reactions major conformational changes are involved (see Table II).

The activation entropies are different for the two pathways of MII decay which both lead to the formation of retinal and opsin. The difference is difficult to explain if it is maintained that $\Delta S^{\ddagger}$ parallels the behaviour of ΔS, the entropy change, for the latter does not depend on the reaction pathway but only on the initial and final reactants. It may be, however, that in the retina, as in solution (11), MIII is first converted into N-retinylidene-opsin. Then, by a rapid hydrolysis (cf. 12,13) retinal is formed. This means that one of the pathways is not completely described yet and that a comparison of the two pathways in terms of the various $\Delta S^{\ddagger}$ would be premature.

The thermal reactions have been studied earlier in rhodopsin solutions, and the kinetic parameters found there may be compared to those obtained in the retina. This is done in Table III. The comparison is confined to $\Delta S^{\ddagger}$ because the values of $\Delta F^{\ddagger}$ of all the reactions the table refers to are clustered around 21 kcal/mole so that a large positive figure of $\Delta S^{\ddagger}$ necessarily indicates a large $\Delta H^{\ddagger}$ and a negative figure of $\Delta S^{\ddagger}$ indicates a small $H^{\ddagger}$ (smaller than 21 kcal/mole; cf. p. 93).

The table shows clearly that none of the data obtained in the retina is in agreement with its respective counterpart determined in solution. One might ascribe this to species differences because the solutions were all prepared from cattle retinas. On the other hand, data measured in excised rat eyes (14) are in reasonably good agreement with those from the frog retina. This applies to the conversion of MII into MIII for which a $\Delta S^{\ddagger}$ of – 10 cal/mole· °K was found. The corresponding figure of Table III (– 7 cal/mole· °K) is very similar. The comparison looks different, however, when the decay of MIII is considered. In the rat retina, this is an extremely slow reaction which will need over an hour for virtual completion at 37 °C. The kinetic data given by Ebrey (14) are consistent with a slow reaction ($\Delta H^{\ddagger}$ = 32 kcal/mole; $\Delta S^{\ddagger}$ = 30 cal/mole· °K) but they are certainly different from the corresponding frog data. So in this particular case, a species difference is obvious. Nonetheless, Table III suggests that, in certain details, rhodopsin situated in the outer segments reacts differently from rhodopsin in solution.

Acknowledgement. This work was supported by the Deutsche Forschungsgemeinschaft.

References

1. BAUMANN, CH.: Kinetics of slow thermal reactions during the bleaching of rhodopsin in the perfused frog retina. J.Physiol. 222, 643-663 (1972).
2. ABRAHAMSON, E.W., J.R. WIESENFELD: The structure, spectra, and reactivity of visual pigments. In: Handbook of Sensory Physiology, Vol. VII/1, ed. H.J.A. DARTNALL, Springer, Heidelberg (1972).
3. BAUMANN, CH.: Flash photolysis of rhodopsin in the isolated frog retina. Vision Res. 10, 789-798 (1970).
4. MATTHEWS, R.G., R. HUBBARD, P.K. BROWN, G. WALD: Tautomeric forms of metarhodopsin. J.Gen.Physiol. 47, 215-240 (1963).
5. LYTHGOE, R.J., J.P. QUILLIAM: The thermal decomposition of visual purple. J.Physiol. 93, 24-38 (1938).
6. DIXON, W.J.: Biomedical Computer Programs. X-series Supplement. University of California Press, Berkeley (1970).
7. LUMRY, R., H. EYRING: Conformation changes of proteins. J.Phys.Chem. 58, 110-120 (1954).
8. BRAY, H.G., K. WHITE: Kinetics and Thermodynamics in Biochemistry. Churchill, London (1957).
9. JOLY, M: A Physico-Chemical Approach to the Denaturation of Proteins. Academic Press, London (1965).
10. LAIDLER, K.J.: The Chemical Kinetics of Enzyme Action. Oxford University Press, Oxford (1958).
11. OSTROY, S.E., F. ERHARDT, E.W. ABRAHAMSON: Protein configuration changes in the photolysis of rhodopsin. II. The sequence of intermediates in thermal decay of cattle metarhodopsin in vitro. Biochim.Biophys.Acta 112, 265-277 (1966).
12. ABRAHAMSON, E.W., S.E. OSTROY: The photochemical and macromolecular aspects of vision. In: Progress in Biophysics, Vol. 17, ed. J.A.V. BUTLER and H.E. HUXLEY, Pergamon Press, Oxford (1967).

13. BRIDGES, C.D.B.: Studies on the flash photolysis of visual pigments - IV. Dark reactions following the flash-irradiation of frog rhodopsin in suspensions of isolated photoreceptors. Vision Res. 2, 215-232 (1962).
14. EBREY, T.G.: The thermal decay of the intermediates of rhodopsin in situ. Vision Res. 8, 965-982 (1968).

Discussion

G. Wald: Your observation that the reaction MII → retinal + opsin has both a large heat of activation (ca. 33 kcal/mole) and a large entropy of activation (+41 e.u.) probably means that a lot more is happening than simple hydrolysis. When Rowena Matthews in our laboratory looked for the step in rhodopsin bleaching that exposes 2 -SH groups, it turned out to be that hydrolysis. It seems to include therefore a considerable rearrangement ("opening up") of the protein, opsin.

T.G. Ebrey: Matthews et al. (4) found, I believe, an isosbestic point in the conversion of MII → pararhodopsin in a digitonin solution at 3 °C. Could you find conditions that would give such an isosbestic point?

Ch. Baumann: No, we could not because we were not dealing with a two-component system.

S.L. Bonting: Could compound C in your scheme be aldimine-bound rather than free retinaldehyde, in view of our findings (a) that after MII decay little free retinaldehyde appears (and rather slowly), and (b) that retinoldehydrogenase is able to reduce aldimine-linked retinaldehyde to free retinol.

Ch. Baumann: In its natural environment, retinal is probably always bound to other molecules. Our data, however, are compatible with the absorption spectrum of retinal.

Metarhodopsin III

R. A. Weale
Department of Visual Science, Institute of Ophthalmology, London WC1 H9QS/Great Britain

The photic isomerization of rhodopsin is followed by a number of dark reactions (1) some of which are also photosensitive (2). Information is available on their various half-lifes, which, as expected, vary with temperature. However, they also depend on the environment of the parent pigment, in particular whether it is in solution or in situ.

One of the less labile products, called by Lythgoe and Quilliam (3) transient orange, absorbs maximally at $\lambda \approx 470$ nm, and is easily detected even in the intact eye. E.g. cat eyes were studied by fundus reflectometry (4,5) and difference spectra were obtained to measure changes caused by various chromatic radiations. If the sequence was concluded with a strong white bleaching light then the resulting difference spectrum bore no resemblance to the result obtained by prolonged bleaching (Fig. 1). Note the difference in the spectral distribution: under no conditions was it possible to obtain a final ΔD_{max} greater than about 40 per cent of the overall ΔD_{max}.

When the regeneration of rhodopsin was studied in the human eye (6) a similar situation was observed. The retina having been bleached, difference spectra were recorded (7) as a function of various times of dark-adaptation. Their spectral variation and time-course could be described by the hypothesis that rhodopsin regener-

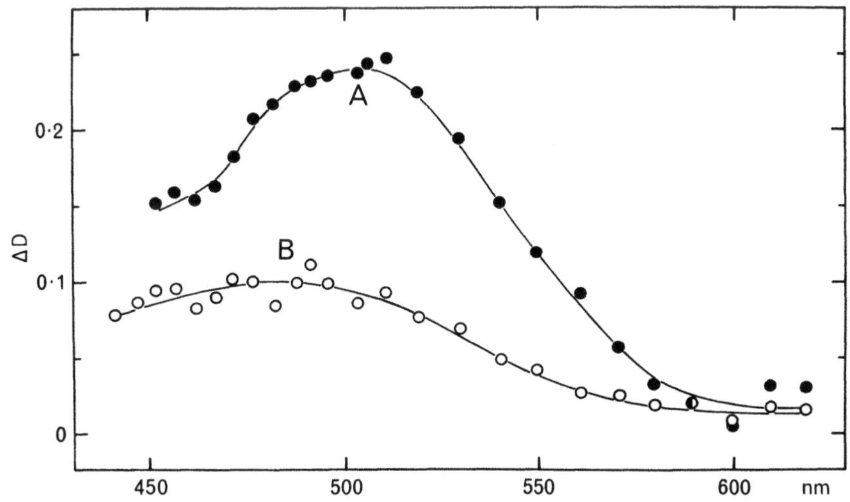

Fig. 1. Difference spectra obtained as a result of bleaching the retina of the cat (N=4). A: sum of repeated fractional bleaches with monochromatic light and a terminal exposure to white light; B: density difference due to the terminal exposure only (after Weale, 5)

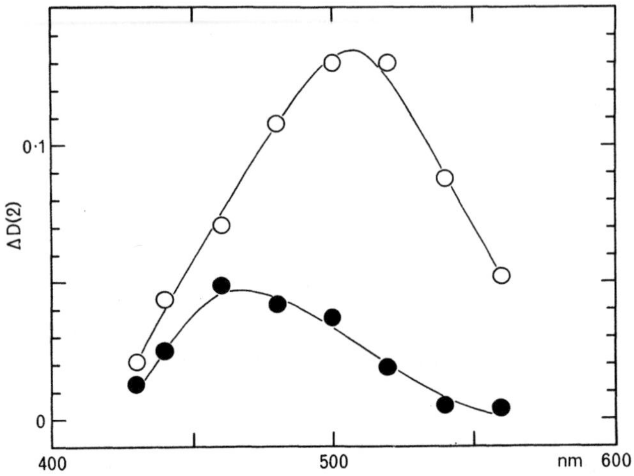

Fig. 2. Regeneration difference spectra (double transit of light) obtained for man. The full circles represent the density difference which metarhodopsin III would achieve if it did not decay: the time-constants of formation and decay are 1 and 2 min respectively. The empty circles represent the density difference due to the regeneration of rhodopsin with a time-constant of 4 min (Ripps and Weale, 7)

ates with a time-constant of 4 min, and that an additional pigment (metarhodopsin III) with $\lambda_{max} \approx 470$nm is formed and decays with time constants of 1 and 2 min respectively. The ratio of the ΔD_{max} values (Fig. 2) is approximately 0.3 but a consideration (8) of the relevant rate-constants yields a value of approximately 0.5. The gist of these observations has now been confirmed with Rushton's apparatus (9).

However, the ratio of the extinction coefficients of the two pigments (1,10) is more nearly equal to unity. A possible explanation for the low ΔD_{max} value of metarhodopsin III is that the dipoles of its molecules are oriented much less favourably for absorbing light than are those of rhodopsin. E.g. if they included with the rod axis an average angle of $\sim 48^{o}$ as compared with the 75° or so for rhodopsin (11) then the above results would be explained. Credence to this view is lent by Denton's observation (12) that the fully bleached dipole includes with the rod axis an angle of fewer than 45°.

It is of obvious interest for our understanding of the behaviour of the rhodopsin molecule following irradiation whether the individual products are oriented in specific ways, whether the change-over is gradual, or whether perhaps other explanations have to be sought.

The hypothesis of the rotation of the dipoles was tested as follows. Ranae temporariae were dark-adapted over-night, and pithed in deep-red light. Their eyes were enucleated, the retinae removed, and single rod dispersions prepared in deep-red light as previously described (13). A Zeiss polarizing microscope was used to measure the transmissivity of the rod with monochromatic light polarized in planes parallel and perpendicular to the rod axis; the transmitted light was incident on a photo-multiplier (14), the resulting response being displayed on a cathode-ray oscillograph and recorded on film.

In these experiments, density values were determined for the dark-adapted rod, the intensity of the measuring light being kept so low that it could not cause any significant bleaching (Fig. 3). Thereupon the rod was exposed for 5 sec to a white light intense enough to bleach all the rhodopsin in 2-3 sec. After this the density was measured repeatedly during an over-all period of 3-4 min, care being taken not to expose the rod to light except during the measurement. Average data ($\sigma_M \sim$ 0.006) obtained with each of the three measuring wavelengths are shown in Fig. 3

Fig. 3. Changes in the optical density of frog rods. Circles: electric vector perpendicular to the axis of the rod; squares: electric vector parallel. The density differences refer to the density measured after the rods had been exposed to an intense 5-sec flash of white light W. Full symbols: dark-adapted state; empty symbols: data obtained after the bleaching exposure. The measuring wavelengths are shown in each section

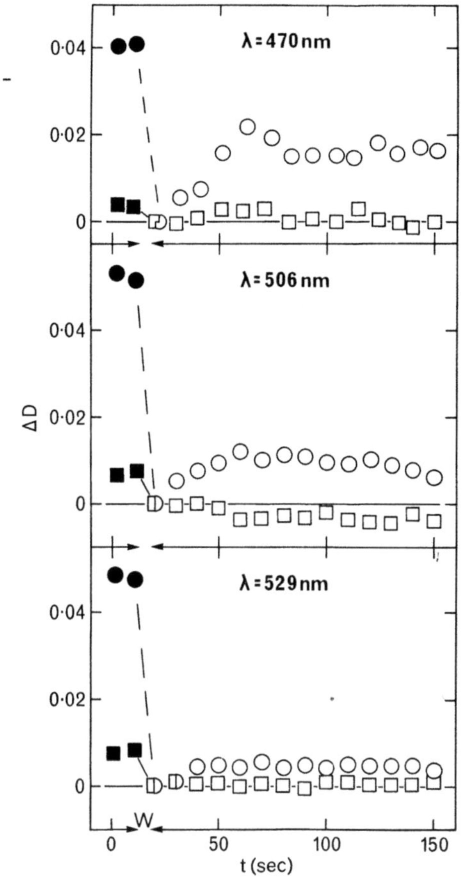

and represent density differences referred to the values obtained immediately after completion of the exposure to the strong light. They lead to the following conclusions:

1) The density rises following exposure of the dark-adapted rod to light with a time-course strictly comparable with that observed by Baumann (15) for metarhodopsin III.

2) For the electric vector perpendicular to the axis of the rod, the ratio of ΔD at 470 nm (which is near its maximum (13)) to that of the parent pigment as shown by the data for λ =506 nm is approximately 0.38.

3) The increase in density after bleaching is confined almost entirely to the plane of polarization perpendicular to the rod axis, and bears much the same relation to the density change parallel to that axis as is true of the corresponding values obtained from the dark-adapted rod.

4) It follows that the spatial distributions of the dipoles of rhodopsin and metarhodopsin III are very similar.

This, then, disproves the above hypothesis, namely that the dipole of metarhodopsin III includes with the rod axis a mean angle considerably different from that believed to be included by this axis and the dipole of the parent rhodopsin molecule.

It would seem, therefore, that, if the extinction coefficients of rhodopsin and metarhodopsin III are nearly equal, but the density differences are very discrepant, one has to postulate the existence of a shunt (8). On this view, about one half or more of the molecules - the fraction will vary inter alia with body temperature - of metarhodopsin II follow a route through metarhodopsin III. The function of this component which has received little attention so far is obscure. Ernst and Kemp (16) have shown that the time-course of its decay correlates strongly with a process of visual adaptation (in the rat): the presence of metarhodopsin III appears to delay dark-adaptation. It may be noted that the spectral position of its absorption band is such as to favour absorption of the light of the sky, and it remains to be seen whether its decay to retinol may not be the rate-limiting factor of one of the basic elements in the regeneration of rhodopsin.

References

1. ABRAHAMSON, E.W., S.E. OSTROY: The photochemical and macromolecular aspects of vision. Progr.Rep. in Biophys. and Mol.Biol. 17, 179-215 (1967).
2. HUBBARD, R., A. KROPF: The action of light on rhodopsin. Proc.Nat.Acad. Sci. (Wash.) 44, 130-139 (1958).
3. LYTHGOE, R.J., J.P. QUILLIAM: The relation of transient orange to visual purple and indicator yellow. J.Physiol. (London) 94, 399-410 (1938).
4. WEALE, R.A.: A photographic method for in vivo studies of visual pigments. Proc.XX Internat.Congress Physiol.Brussels, 951 (1956).
5. WEALE, R.A.: Observations on photochemical reactions in living eyes. Brit. J.Ophthal. 41, 461-474 (1957).
6. WEALE, R.A.: On an early stage of rhodopsin regeneration in man. Vision Res. 7, 819-827 (1967).
7. RIPPS, H., R.A. WEALE: Rhodopsin regeneration in man. Nature (Lond.) 222, 775 (1969).
8. BAUMANN, C.: Kinetics of slow thermal reactions during the bleaching of rhodopsin in the perfused frog retina. J.Physiol. (Lond.) 222, 643-663 (1972).
9. ALPERN, M.: Rhodopsin kinetics in the human eye. J.Physiol. (Lond.) 217, 447-471 (1971).
10. MATTHEWS, R.G., R. HUBBARD, P.K. BROWN, G. WALD: Tautomeric forms of rhodopsin. J.Gen. Physiol. 47, 215-240 (1963).
11. LIEBMAN, P.: In situ microspectrophotometric studies on the pigments of single retinal rods. Biophys.J. 2, 161-178 (1962).
12. DENTON, E.J.: The contribution of the orientated photo-sensitive and other molecules to the absorption of whole retina. Proc.Roy. Soc.London,Ser.B. 150, 78-94 (1959).
13. WEALE, R.A.: On the linear dichroism of frog rods. Vision Res. 11, 1373-1385 (1971).
14. WEALE, R.A.: in preparation.
15. BAUMANN, C.: Sehpurpurbleichung und Stäbchenfunktion in der isolierten Froschnetzhaut. Pflügers Arch.ges.Physiol. 298, 44-60 (1967).
16. ERNST, W.J.K., C.M. KEMP: The growth of the P III photoresponse of isolated rat retina and its dependence on the disappearance of the metarhodopsins. In: "The Visual System, Neurophysiology, Biophysics and their Clinical Applications". Ed. G.B. Arden. In: Adv.exp.Med.and Biol., Vol. 24. Plenum Press, New York (1972).

The Effect of Urea upon the Photosensitivity of Bovine Rhodopsin

Aubrey Knowles
MRC Vision Unit, University of Sussex, Falmer, Brighton, BN1 9QY, Sussex/Great Britain

Urea denatures proteins by a weakening of the hydrogen bonding that leads to a loss of secondary structure. Several workers have commented upon the effect of high concentrations of urea upon the thermal stability of visual pigments, for example, Hubbard (1) finds bovine rhodopsin to be "stable" in 8M urea while Girsch and Rabinovitch (2) report a half-life of 11 min for the thermal bleaching of bovine rhodopsin in 6M urea at 25°C. Ostroy, Erhardt and Abrahamson (3) found that urea increases the rate of decay of the bleaching intermediates on the irradiation of bovine rhodopsin at 3°C. Apparently the denaturation of the protein has some effect upon the binding of the chromophoric group and it is of interest to see whether this weakening would affect the quantum yield of photobleaching of rhodopsin.

Bovine rod outer segments (ROS) were separated by sucrose flotation and the rhodopsin extracted with 2% aqueous digitonin. In the experiments described in this paper the extract was buffered with sodium borate and the temperature maintained at 25 ± 0.5°C. The technique developed by Dartnall (4,5) was used to measure photosensitivities, that is, the product of the quantum yield of photobleaching (γ) and the extinction coefficient of the visual pigment (ε), and the results are given in Table I.

The absolute values of the photosensitivities are based on the value for bovine rhodopsin in 20 mM hydroxylamine being the same as the mean value found by Dartnall (5) for rhodopsins (A_1 - based pigments) from several species. This was checked by measuring the intensity of the source with a calibrated thermopile. The quantum yields are calculated from the value of $4.06 \cdot 10^4$ M^{-1} cm^{-1} for the extinction coefficient of bovine rhodopsin at 499 nm (6).

The increase of photosensitivity on the addition of hydroxylamine to the pigment is similar to that found with other species (5) and is apparently due to an increase of quantum yield. The nearly identical increase on the addition of urea also indicates an increase of quantum yield since there is no evidence to suggest that the extinction coefficient has increased. Measurement of the first-order thermal decay of bovine rhodopsin in 4M urea at pH 8.4 and 25°C gave a lifetime of $8.5 \cdot 10^4$s., which is

Table I. The photosensitivity of bovine rhodopsin in aqueous digitonin at pH 8.4 and 25°C

Additive	Photosensitivity $\varepsilon \cdot \gamma$ cm^{-1} $einstein^{-1}$	Quantum yield γ mole $einstein^{-1}$
None	$2.15 \cdot 10^4$	0.53
5.4 M urea	$2.68 \cdot 10^4$	0.66
20 mM hydroxylamine	$2.74 \cdot 10^4$	0.67

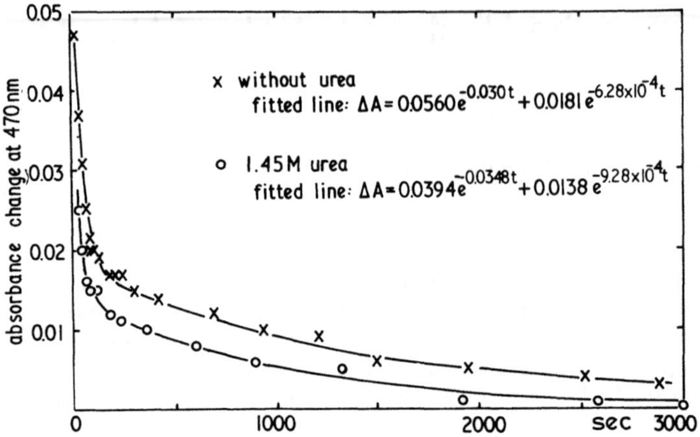

Fig. 1. Decay of transient 470 nm absorbance following flash irradiation of bovine rhodopsin at pH 8.4 and 25°C, alone and with 1.45 urea

considerably longer than the value of Girsch and Rabinovitch (2), and so the decay is sufficiently slow that it will not appreciably affect the photosensitivity determination.

The simultaneous addition of hydroxylamine and urea caused no further increase in quantum yield, and since both give the same limiting value, it seems possible that both act by the same mechanism. It has been thought that the action of hydroxylamine is to prevent regeneration of the pigment molecule by reaction with the chromophoric group as soon as it comes into a reactive condition. Ball et al. (7) showed that retinal does not react with urea, and I have confirmed that this reaction does not occur in the digitonin micelle, and so there is no reaction of the chromophoric group with urea analogous to that with hydroxylamine.

The method of calculation of the photosensitivity compensates for the absorption of stable photoproducts of the bleaching reaction but not slowly-decaying intermediates in the sequence of dark reactions leading to the stable products:

$$\text{Rhodopsin} \xrightarrow{h\nu} X \rightarrow Y \rightarrow Z \xrightarrow[\text{slow}]{} \text{Products}$$

If the rate of reaction of Z is slow compared with the rate of the absorption of light, the dark reaction will become the rate-limiting step in the reaction sequence and determine the quantum yield. Recent experiments by Dartnall (8) show that if the irradiation of rhodopsin without hydroxylamine is interrupted by dark periods, the dark reactions can "catch up" and the apparent quantum yield rises to the hydroxylamine value. This suggests that the action of the hydroxylamine is to increase the rate of the rate-limiting dark reaction. Such an increase in the rate of decay of intermediates of frog rhodopsin in the receptor on the addition of hydroxylamine was noted by Bridges (9).

On the other hand, Johnson (10) has argued that hydroxylamine cannot increase the rate of decay of intermediates preceding M II.

In the photosensitivity determinations, the rhodopsin at the start of the experiment is absorbing about $1.25 \cdot 10^{-11}$ einstein sec^{-1}, and assuming a quantum yield of 0.67, this corresponds to a rate for the primary photoreaction of $0.80 \cdot 10^{-11}$ mole sec^{-1}. The concentration of the pigment is about $1.5 \cdot 10^{-5}$ M and the cell volume $0.39 \cdot 10^{-3}$ and so this reaction rate corresponds to a first-order reaction of rate constant of $1.4 \cdot 10^{-3}$ sec^{-1}. Thus a dark reaction of rate constant less than this value, that is,

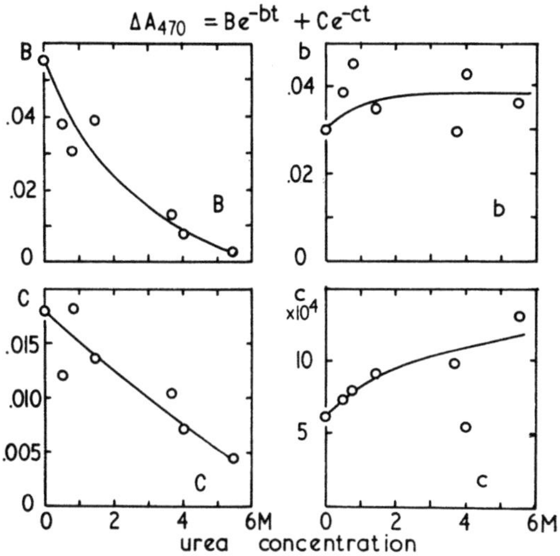

Fig. 2. The effect of urea upon the parameters for the decay of the transient 470 nm absorption following flash irradiation of bovine rhodopsin at pH 8.4 and 25 °C

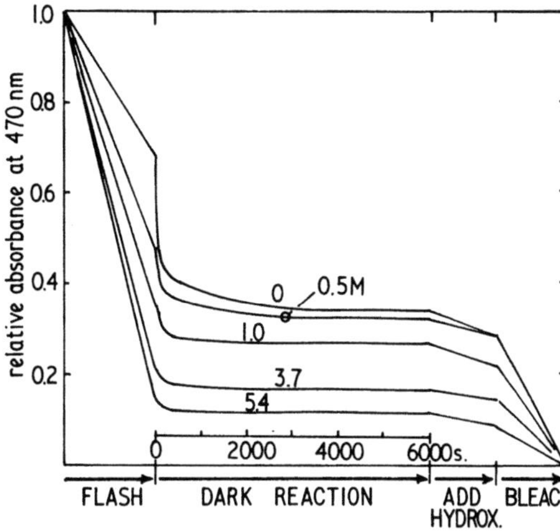

Fig. 3. The effect of urea upon the changes in the 470 nm absorbance of bovine rhodopsin at pH 8.4 and 25 °C caused by flash irradiation

an intermediate of half-life greater than 500 sec will reduce the apparent quantum yield. Since these photosensitivity values are based on observations at 500 nm, anomalous results will only be obtained if the slow intermediate has an extinction coefficient at 500 nm comparable with that of the original pigment.

A search for such an intermediate has been made using the simple flash-photolysis technique of Bridges (11). The rhodopsin extract was exposed to a 400 J flash of about 4 msec half-life from a photographic flash unit. Changes in absorbance from about 20 sec after the flash were followed by means of a manual spectrophotometer. The flash was passed through a Wratten No. 9 filter which transmits less than 1% below 470 nm. The absorption spectrum measured immediately after the flash has a maximum at about 480 nm (12) that is primarily due to pararhodopsin (PR), (alternatively named Transient Orange or M_{465} (13, 3, 14). This decays over a period of about one hour to give the alkaline form of N-retinylidene opsin (λ_{max} 365 nm). I have followed the decay of this intermediate at wavelengths from 440 to 500 nm, and at all wavelengths the process seems to be biphasic and apparently composed of two first-order reactions differing in rate by a factor of at least 50. Fig. 1 shows the decay at 470 nm together with a fitted line of the form:

$$A = Be^{-bt} + Ce^{-ct}$$

representing two simultaneous first-order processes.

This could mean that there are two species with similar absorption spectra decaying at the same time, or that a single species has a complex decay process involving the formation of an equilibrium, but no positive conclusion about this has been reached so far. For the purpose of this discussion, it will be assumed that only one species is present. The slower rate constant (c) has the value $6.28 \cdot 10^{-4}$ sec^{-1} and could thus be the rate determining factor. The decay at 500 nm is essentially simi-

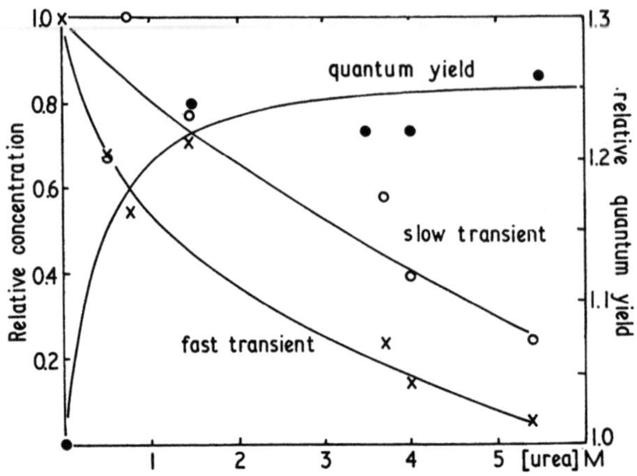

Fig. 4. Relative transient yields and quantum yields of photobleaching for the photolysis of bovine rhodopsin at pH 8.4 and 25°C

lar, but the overall change is not so great as at 470 nm and so the observations at the latter wavelength will be used in this discussion.

The Effect of Urea upon the Transient

When urea is added, the two pre-expotential factors (B and C) fall and the decay rates increase slightly. Thus less of the transient is generated and it reacts more rapidly. The parameters for a range of urea concentrations are shown in Fig. 2.

The increase in the rates (b and c) is probably not significant and so the decline in the initial yields (B and C) is the more important feature. This suggests that urea acts upon some intermediate preceding PR.

The decay of PR is placed in the overall picture of the bleaching of the rhodopsin in Fig. 3 which shows all the steps in the experiment.

The flash is of sufficient intensity that all of the pigment molecules absorb at least one photon. A certain fraction undergoes photoregeneration and so the absorbance measured at the start of the dark reaction is due to a mixture of regenerated pigment and PR. At the end of 6000 sec, when effectively all of the transient was thought to have decayed, hydroxylamine was added so that the solution contained only regenerated pigment and retinal oxime. The extract was then exposed to a continuous orange light to bleach completely the remaining pigment. These curves show that the decay of PR is equivalent to a high proportion of the total change at 470 nm when no urea is present, and the relative size of this change is diminished by the urea. A further effect of the urea is to reduce the regenerated pigment seen after the addition of hydroxylamine. This falls from 29% to 9% at the highest urea concentration. If urea prevents photoregeneration during the lifetime of the flash, it must degrade an intermediate of lifetime less than 4 msec. Thus the apparent photosensitivity is enhanced by the rapid decomposition of an early intermediate which lowers the yield of production of PR, rather than by any major increase in the decay of PR. This is in accordance with the low-temperature experiments of Ostroy et al. (3), although these workers did not comment on the yield of PR. The photosensitivity reaches its limiting value at 1.5 M urea (Fig. 4) which roughly corresponds to the rapid drop in the yield of the fast component (B).

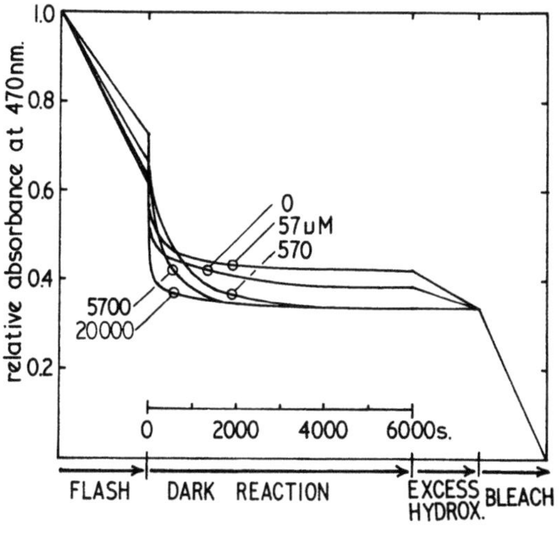

Fig. 5. The effect of hydroxylamine upon the changes in the 470 nm absorbance of bovine rhodopsin at pH 8.4 and 25°C caused by flash irradiation

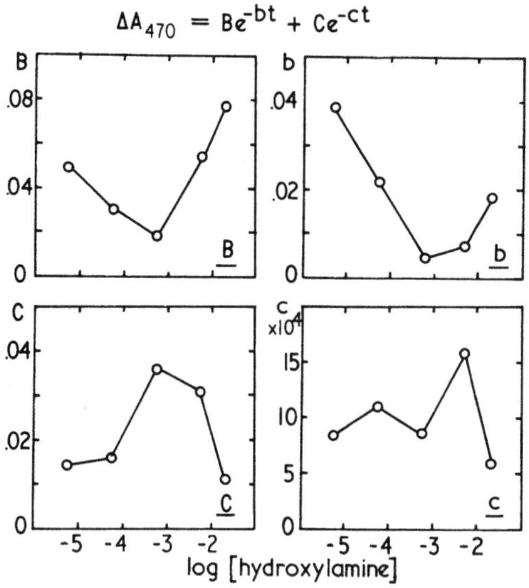

Fig.6. The effect of hydroxylamine upon the parameters for the decay of the transient 470 nm absorption following flash irradiation of bovine rhodopsin at pH 8.4 and 25° C

The Effect of Hydroxylamine upon the Transient

The effect of hydroxylamine, though superficially similar, has some major differences. Fig. 5 shows the sequence of changes at 470 nm. The dark reaction is again biphasic but the shape of the curve changes as the concentration is increased.

Further hydroxylamine was added at the end of the dark period to bring the concentration up to 20 mM and this caused a further change even when 57 μM hydroxylamine was already present, although this is an eight-fold excess over the pigment concentration. All of the solutions showed the same degree of photoregeneration which is in accordance with Johnson's finding (10) that hydroxylamine does not affect the rhodopsin $\rightleftharpoons$ MI equilibrium formed during the course of the flash. The decay parameters show some dramatic changes with hydroxylamine concentration (Fig. 6), the yield (B) and rate constant (b) of the fast component passing through a minimum at intermediate concentrations, while those of the slow component (C and c) pass through a maximum.

These effects apparently cancel so that the quantum yield with 570 μM hydroxylamine - an 80-fold excess - is the same as that with no hydroxylamine. However, with 20 mM hydroxylamine the rate constants (b and c) are less than those at very low hydroxylamine concentration, while the yields (B and C) are quite similar. An interesting feature is the vast range of concentration over which these changes occur. Hydroxylamine is a small molecule that should readily enter the digitonin micelle, though its polar nature might inhibit this. A reaction only involving oxime formation with the chromophoric group would, therefore, only require a stoichometric ratio of 1:1. The fact that no change in photosensitivity is seen at 80:1, and changes are still seen at 2600:1 suggests that the action of the hydroxylamine is upon the protein part of the molecule, though with bovine rhodopsin this does not lead to bleaching in the dark, or prevent photoregeneration. It should be noted that other visual pigments such as the green-rod pigment of the frog, are completely degraded by hydroxylamine in the dark (15).

Conclusions

Urea increases the photosensitivity, or better, increases the overall rate of photobleaching of bovine rhodopsin by degrading the earlier intermediates in the bleaching sequence and so by-passing the longer-lived species such as PR. The decay of PR is not greatly accelerated.

The effect of hydroxylamine is more complex, but appears to involve a denaturation of the protein rather than a simple reaction of the chromophoric group. These experiments were conducted on the assumption that the primary effect of the two reagents is upon the photo-excited molecule since the thermal bleaching of the intact pigment molecules is slight during the period of the irradiation. To minimise the thermal reaction, the reagent was added immediately before the irradiation. However, another dimension is added to the problem if either hydroxylamine or urea is added some time before the start of the irradiation, for then the photochemistry is quite different. Thus a denaturation takes place in the dark that does not diminish or shift the 500 nm absorption maximum, but modifies the binding of the chromophoric group so that the molecule "falls apart" more readily when excited by light. While the rate of the dark reactions may be modified by denaturation of the protein, it should be borne in mind that the overall quantum yield must be limited by the initial act of isomerization of the chromophoric group. If the denaturation is pursued to some point where the chromophoric group behaves more like free 11-cis retinal, one might expect the quantum yield to again fall towards the value of 0.2 found by Kropf and Hubbard (16) for the photoisomerization of 11-cis retinal.

References

1. HUBBARD, R.: Absorption spectrum of rhodopsin: 500 nm absorption band. Nature (Lond.) 221, 432-435 (1969).
2. GIRSCH, S.J., B. RABINOVITCH: The bleaching of rhodopsin in the dark. Biochem.Biophys.Res.Comm. 44, 550-556 (1971).
3. OSTROY, S.E., F. ERHARDT, E.W. ABRAHAMSON: Protein configuration changes in the photolysis of rhodopsin. II. The sequence of intermediates in the thermal decay of cattle metarhodopsin in vitro. Biochem.Biophys.Acta 112, 265-277 (1966).
4. DARTNALL, H.J.A.: The spectral variation of the relative photosensitivities of some visual pigments. In: "Visual problems of colour", National Physical Laboratory Symposium No. 8, pp.121-148, H.M.S.O., London (1958).
5. DARTNALL, H.J.A.: The photosensitivities of visual pigments in the presence of hydroxylamine. Vision Res. 8, 339-358 (1968).
6. WALD, G., P.K. BROWN: The molar extinction of rhodopsin. J.Gen.Physiol. 37, 189-200 (1953).
7. BALL, S., F.D. COLLINS, P.D. DALVI, R.A. MORTON: Studies on vitamin A. II. Reactions of retinene with amino compounds. Biochem.J. 45, 304-307 (1949).

8. DARTNALL, H.J.A.: Photosensitivity. In: Handbook of Sensory Physiology, Vol. VII 1, pp. 122-145, Springer-Verlag, Berlin (1972).
9. BRIDGES, C.D.B.: Studies on the flash-photolysis of visual pigments - IV. Dark reactions following the flash-irradiation of frog rhodopsin in suspensions of isolated photoreceptors. Vision Res. 2, 215-232 (1962).
10. JOHNSON, R.H.: Absence of effect of hydroxylamine upon production rates of some rhodopsin photo-intermediates. Vision Res. 10, 897-900 (1970).
11. BRIDGES, C.D.B.: Studies on the flash-photolysis of visual pigments. I: Pigments present in frog rhodopsin solutions after flash-irradiation. Biochem. J. 79, 128-134 (1961).
12. BRIDGES, C.D.B.: Studies on the flash-photolysis of visual pigments. III. Interpretation of the slow thermal reactions following flash-irradiation of frog rhodopsin solutions. Vision Res. 2, 201-214 (1962).
13. MATTHEWS, R.G., R. HUBBARD, P.K. BROWN, G. WALD: Tautomeric forms of metarhodopsin. J.Gen.Physiol. 47, 215-240 (1963).
14. WALD, G.: The molecular basis of visual excitation. Nature (Lond.) 219, 800-807 (1968).
15. DARTNALL, H.J.A.: The visual pigment of the green rods. Vision Res. 7, 1-16 (1967).
16. KROPF, A., R. HUBBARD: The photoisomerization of retinal. Photochem. Photobiol. 12, 249-260 (1970).

Discussion

W.T. Mason: I would like to comment on the disparity between chemical treatment of rhodopsin in the disc membranes *in situ* and similar treatment of rhodopsin in detergent micellar preparations. We have examined vertebrate and invertebrate rhodopsin solubilized in detergent micellar suspensions (Mason, W.T., R.S. Fager, E.W. Abrahamson, submitted for publication) by electron microscopy. We were interested to find that these rhodopsin micelles exhibit an extremely high electron-density. This fact may reflect on the retardation and extended time course of chemical treatments of micellated visual pigment relative to the time course of similar treatments of visual pigment, in the intact disc membrane. The high electron density of the detergent micelle surrounding rhodopsin may well inhibit the approach of an attacking chemical such as urea, hydroxylamine, etc., to the rhodopsin, whereas in the intact disc membrane the rhodopsin molecule may be easily accessible and therefore a shorter time course of treatment observed.

A. Knowles: I am very interested to hear this. My experience is that hydroxylamine enters the digitonin micelle rather slowly, and the same may be true for urea. I had assumed that this effect was due to the difficulty experienced by these polar molecules in penetrating the hydrophobic interior of the micelle.

III. Regeneration of the Pigments

Interrelations of Visual Pigments and "Vitamins A" in Fish and Amphibia

C. D. B. Bridges
Department of Ophthalmology, New York University Medical Center, New York, NY 10016/USA

In frogs and fish, although perhaps not in mammals (1,2), the pigment epithelium plays a decisive role in visual pigment regeneration (see Baumann, 3, for review). This is illustrated in Fig. 1 for Rana clamitans, a frog where the retina may be separated easily and cleanly from the rest of the pigmented layers of the eye. As the figure shows, the isolated retina at 20° regenerates only about 11% of the pigment bleached (98%) by a $\frac{1}{2}$ hr. exposure to yellow light of $\lambda > 460$ nm (cf. Wald and Hubbard, 4; Ewald and Kühne, 5). On the other hand, after a similar bleach the opened eye cup regenerates 76-95% (compare ref. 6), while a retina stripped away from its pigment epithelium then replaced and bleached regenerates as much as 55-88% (cf. Kühne, 7). The amount regenerated appears somewhat lower when the retina is separated, bleached and then replaced, but it is still considerably more than in the isolated retina.

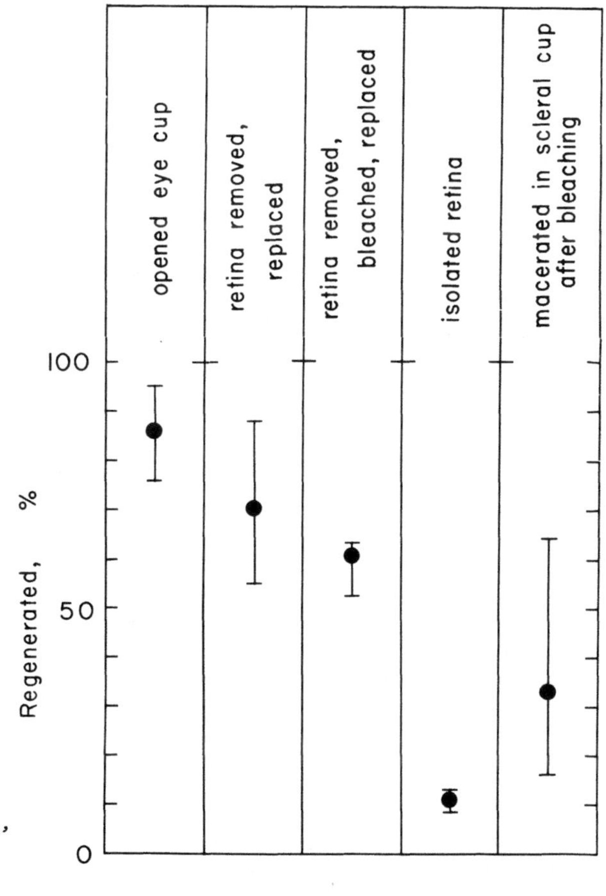

Fig. 1. Regeneration in the eyes of Rana clamitans. Regeneration is expressed as a percentage of the amount bleached by a $\frac{1}{2}$ hr exposure to yellow light of $\lambda > 460$ nm. The contralateral eye was used as an unbleached control in all experiments. Dissection and incubation was in Ringer-phosphate, pH 7.8 (Sickel, 26); the temperature was 20°C

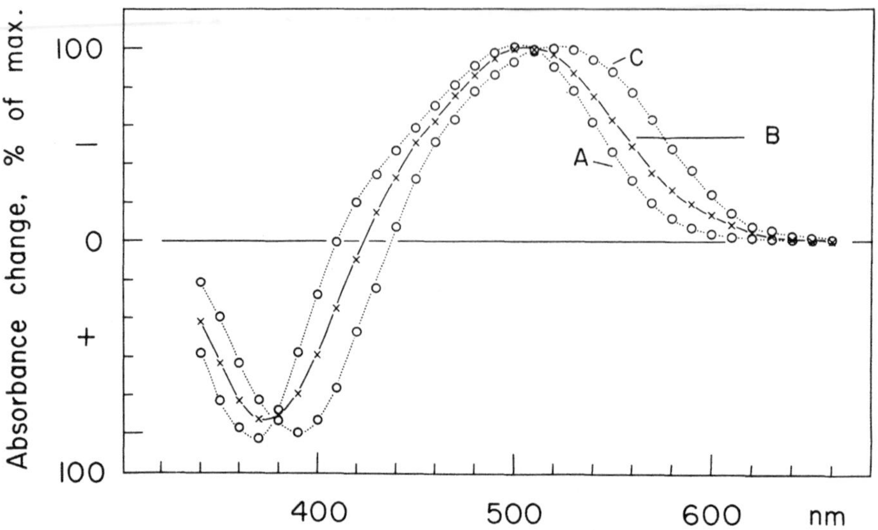

Fig. 2. Regeneration of rhodopsin in a goldfish retina laid in contact with frog (Rana pipiens) pigment epithelium. Difference spectra are measured in the presence of 10 mM hydroxylamine. All conditions as for Fig. 1.
A. Pigment regenerated in a R. pipiens retina on a R.pipiens pigment epithelium.
B. Pigment regenerated in a goldfish retina on frog pigment epithelium.
C. Pigment regenerated in a goldfish retina on goldfish pigment epithelium

Under some circumstances the pigment epithelium determines the composition of pigment regenerated. As shown in Fig. 2, a goldfish (Carassius auratus) retina separated from its own eyecup then laid into that of a frog and bleached regenerates mainly rhodopsin (71%), although it had originally contained only porphyropsin. A similar observation has also been made by Reuter et al. (8), where the rhodopsin-containing ventral retina of the bullfrog was laid in contact with that part of the pigment epithelium normally apposed to the dorsal porphyropsin-containing retina. On bleaching mainly porphyropsin was regenerated. These experiments show that there is a transfer of visual pigment prosthetic group (retinol in the case of rhodopsin or 3-dehydroretinol in the case of porphyropsin) between pigment epithelium and retina, as established in the rat by Dowling (9; see also Wald, 10). Experiments where the retina has been detached and replaced as in Fig. 1 show that specialized intercellular connexions are not necessarily implicated in regeneration. Since retinol and its 3-dehydro derivative are not water-soluble and yet must pass over the aqueous barrier separating the pigment epithelium from the retina, there may be a retinol-binding protein carrier similar to that responsible for retinol transport in the blood (11, 12).

Fig. 1 also shows that while the system survives removal and replacement of the retina fairly well, more extensive damage produced by crudely macerating the illuminated tissues in the scleral cup with a glass rod can reduce the amount regenerated to between 16 and 64%, thus showing that the pigment epithelium cells (perhaps their membranes) must remain relatively intact for regeneration to occur at a level higher than that of the isolated retina.

It is therefore pertinent to ask whether the pigment epithelium is primarily responsible for the pronounced shifts of visual pigment composition that are frequently observed in fishes and frog tadpoles. Such instances are exemplified by the light-

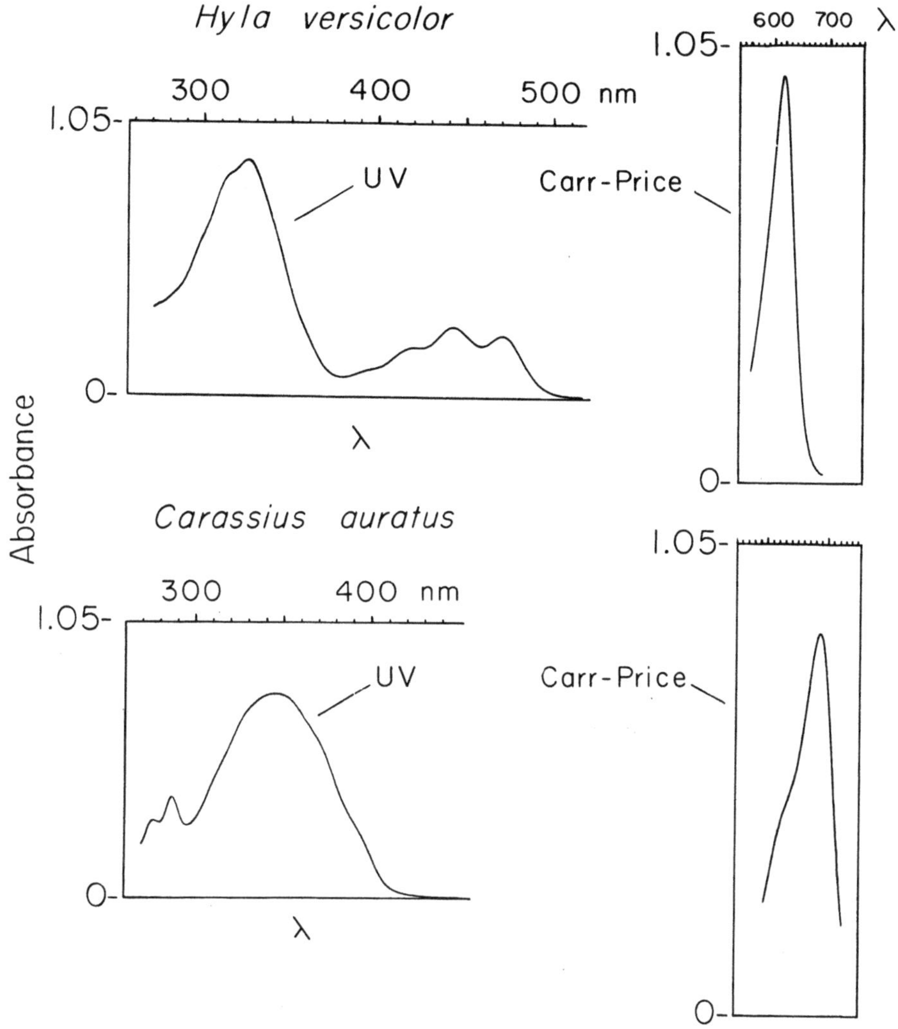

Fig. 3. Retinols extracted from the dark-adapted eyes of tree frogs (Hyla versicolor) and goldfish (Carassius auratus). The dark-adapted eyes (after removal of cornea, iris, lens and, in the case of the goldfish, the sclera) were freeze-dried, then extracted with light petroleum b.p. 37-51°C. The oils were saponified in 0.33 M ethanolic KOH for $1\frac{1}{2}$ hr at 55°C, then extracted with light petroleum, washed and dried. The UV spectra are recorded in light petroleum. The Carr-Price reaction is carried out in the usual way by taking in an optical cell a 0.2 ml aliquot of the sample dissolved in dry chloroform, adding two drops of acetic anhydride then squirting in 2 ml of chloroform saturated with antimony trichloride (sometimes the quantities were reduced by using semi-micro cells). Spectra are measured on a Cary 14 spectrophotometer. Since all operations were in a dark-room, the visual pigment remained unaffected by the freeze-drying and petroleum extraction. They could therefore be analysed by soaking the dry tissue in McIlvaines's pH 4.6 buffer, centrifuging, washing in distilled water and extracting with 2% digitonin

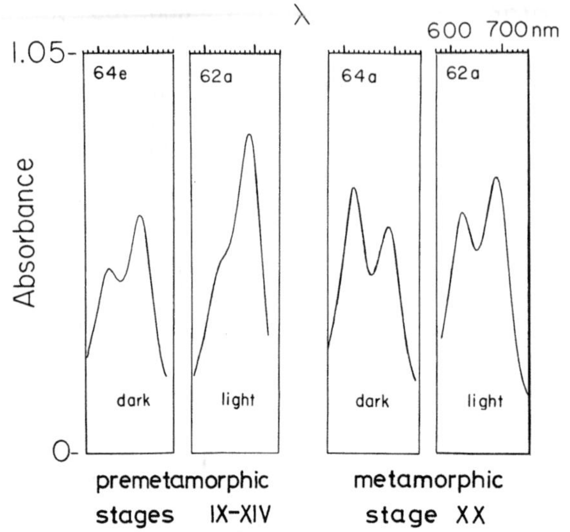

Fig. 4. Carr-Price spectra of retinols stored in the eyes of R. clamitans tadpoles. All groups were kept in incubators at 24°C: some were illuminated from above with white light (77 ft. c. at the water surface), others were in complete darkness for about 3 weeks. Metamorphosis was rapidly induced in 6-8 days by adding thyroxine to the water (36 μ g/ litre). Methods were as outlined in Fig. 3

stimulated conversion of porphyropsin to rhodopsin in the cyprinid fish Scardinius (13,14) and the reverse in ranid tadpoles (15,16), or the induction by thyroid hormone of porphyropsin synthesis at the expense of rhodopsin in kokanee salmon and the cyprinid Richardsonia (17,18). The action of thyroid hormone is reversed in tadpoles, where it causes a switch from porphyropsin to rhodopsin (concurrently with metamorphosis: Wilt, 19). The locus of action of these stimuli is only poorly understood, although some evidence suggests that light and thyroid hormone act directly on the ocular tissues and not via some central mechanism such as the pituitary (20,14).

The primary source of prosthetic groups for visual pigment synthesis is undoubtedly the pigment epithelium. In tadpoles and adults of R. clamitans and R. catesbeiana and in adults of R. pipiens, R. palustris, R. sylvatica and Hyla crucifer the quantities of retinols stored in the pigment epithelium range from 0.9 to 2.2 molar equivalents of retinal visual pigment. This ratio varies with different animals - while the albino rat appears to be exceptional in having no retinol stored in its ocular tissues (9), in the present study the goldfish Carassius auratus was found to possess no less than 41 molar equivalents of retinal visual pigment in the form of 3-dehydroretinol (ester) in the pigment epithelium.

The composition of the retinol supplies in the pigment epithelium is clearly related to the composition of the visual pigments found in the retina (see Bridges, 21, for review). Thus adult Hyla versicolor have only rhodopsin in the retina and only retinol in the pigmented ocular tissues, whereas goldfish have only 3-dehydroretinol stored in the eye and have only porphyropsin in the retina. Examples of analyses of retinols from the eyes of these species are illustrated in Fig. 3. The goldfish apparently has no rhodopsin in its retina at any time, but there are many freshwater fishes that switch from rhodopsin to porphyropsin and back again under a variety of circumstances (21). One such fish is the cyprinid Scardinius, where Bridges and Yoshikami (22) showed that the pigment epithelium retinols of specimens kept in the dark were dominated by 3-dehydroretinol, while those of specimens kept in the light had mainly retinol. The visual pigment of the light fish is retinol-based (rhodopsin) while that of the dark fish has 3-dehydroretinol as prosthetic group (porphyropsin).

Fig. 5. Carr-Price spectra of retinols stored in bullfrog eyes at various stages of development. Here, the eyes have been sectioned into dorsal and ventral segments before processing as described in Fig. 3. The adults were used immediately upon arrival (early December). Metamorphosis was natural (July-August samples) and took place at 25°C in light and dark incubators

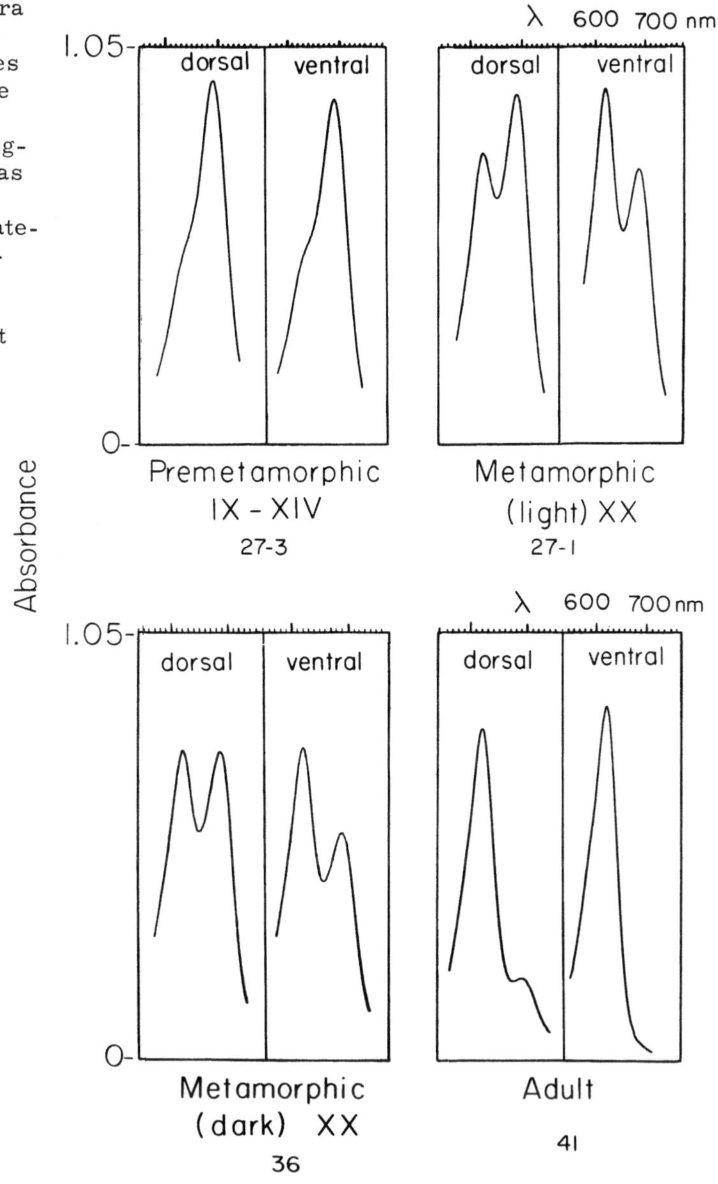

As in fish, light also alters the rhodopsin-porphyrhopsin balance in the eyes of tadpoles - high intensities favour porphyropsin and darkness favours rhodopsin (15); the effect is probably due to light absorbed in the rods (16). Fig. 4 shows observations with R. clamitans tadpoles similar to those described above for Scardinius. Tadpoles kept in the light have nearly pure 3-dehydroretinol, as shown by the prominent Carr-Price peak at 693 nm, whereas those in darkness exhibit the peak at 620 nm characteristic of the presence of retinol. This peak becomes more prominent at metamorphosis (forelimb emergence) in tadpoles kept in the light, and overshadows the 3-dehydroretinol peak at 693 nm in tadpoles brought to metamorphosis in the dark. Visual pigment analyses from the corresponding retinas show that, for example, the premetamorphic light tadpoles had nearly pure porphyropsin, while the dark ones had only 41%.

A similar situation is found in tadpoles of Rana catesbeiana, as Fig. 5 illustrates. The adult frog examined has 32% porphyropsin in its dorsal retina and 24% 3-dehydroretinol in the corresponding pigmented layers. Neither porphyropsin nor 3-dehydroretinol is found in the ventral ocular tissues. This confirms the findings on this species by Reuter et al. (8). Premetamorphic tadpoles kept in the light have nearly pure porphyropsin and nearly pure 3-dehydroretinol stocks, irrespective of retinal region. As illustrated by the Carr-Price blues of Fig. 5, the dorso-ventral difference becomes noticeable at the time of forelimb emergence (although it may be observable in R. catesbeiana tadpoles kept in the dark - these results will be dealt with in detail in a future publication).

On the basis of the regeneration experiments and analyses of stored retinols discussed above it is likely that the retina synthesizes visual pigment according to the availability of the requisite prosthetic group in the pigment epithelium, much of which is often in the necessary 11-cis configuration (23). The composition of these stores may be manipulated in several ways. The cells of the pigment epithelium could preferentially sequester one or other of the retinols from the circulation, or there might be an enzyme system promoting conversion of retinol to 3-dehydroretinol operating wholly within the eye (i.e. a "retinol 3,4 dehydrogenase"). Some evidence supports the latter possibility (24,25). Recently, Bridges and Delisle (to be published) have investigated several populations of smelt (Osmerus eperlanus mordax) isolated in Canadian lakes since the retreat of the ice and the shoaling of the Champlain Sea at the end of the Wisconsin glaciation some 7,000-17,000 years ago. Surprisingly, some have lost the ability to manufacture porphyropsin, their retinas containing only rhodopsin. On the other hand, other groups of the identical species of smelt from similar neighbouring lakes and the St. Lawrence River have nearly pure porphyropsin. The livers of all these fish contain abundant supplies of the 3-dehydroretinol porphyropsin precursor, so it appears that the defect is not associated with any dietary deficiency. Either the ocular tissues are unable to aquire and utilise this liver 3-dehydroretinol, or they cannot synthesize it de novo, perhaps due to a genetic loss of the critical dehydrogenase.

Acknowledgement. This investigation was supported by PHS Research Grant No 5 RO 1 EY 00461 from the National Eye Institute.

References

1. AMER, S., M. AKHTAR: Studies on a missing reaction in the visual cycle. Nature New Biol. 237, 266-267 (1972).
2. FUTTERMAN, S.:, M.H. ROLLINS: A catalytic role of dihydroriboflavin in the geometrical isomerization of all-trans retinal. This volume, pp. 123-130.
3. BAUMANN, CH.: The regeneration of renewal of visual pigment in vertebrates. In: Handbook of Sensory Physiology, vol. VII/1, ed. H.J.A. DARTNALL, pp. 395-416. Springer, Berlin (1972).
4. WALD, G., R. HUBBARD: The synthesis of rhodopsin from vitamin A_1. Proc. Nat. Acad. Sci. (Wash.) 36, 92-102 (1950).
5. EWALD, A., W. KÜHNE: Untersuchungen über den Sehpurpur. Untersuch. Physiol. Inst. Heidelberg 1, 248-290 (1878).
6. REUTER, T.: The synthesis of photosensitive pigments in the rods of the frog's retina. Vision Res. 6, 15-38 (1966).

7. KÜHNE, W.: Zur Photochemie der Netzhaut. Untersuch. Physiol. Inst. Heidelberg 1, 1-14 (1878).
8. REUTER, T., R.H. WHITE, G. WALD: Rhodopsin and porphyropsin fields in the adult bullfrog retina. J. Gen. Physiol. 58, 351-371 (1971).
9. DOWLING, J.E.: Chemistry of visual adaptation in the rat. Nature (Lond.) 188, 114-118 (1960).
10. WALD, G.: Carotenoids and the visual cycle. J. Gen. Physiol. 19, 351-371 (1971).
11. KANAI, M., A. RAZ, D.S. GOODMAN: Retinol-binding protein: the transport protein for vitamin A in human plasma. J. Clin. Invest. 47, 2025-2044 (1968).
12. PETERSEN, P.A.: Characteristics of a vitamin A-transporting protein complex occurring in human serum. J. Biol. Chem. 246, 34-43 (1971).
13. DARTNALL, H.J.A., M.R. LANDER, F.W. MUNZ: Periodic changes in the visual pigment of a fish. In: Progress in Photobiology, ed. B. CHRISTENSEN and B. BUCHMANN, pp. 203-213. Elsevier, Amsterdam (1961).
14. BRIDGES, C.D.B., S. YOSHIKAMI: The rhodopsin-porphyropsin system in freshwater fishes. 1. Effects of age and photic environment. Vision Res. 10, 1315-1332 (1970).
15. BRIDGES, C.D.B.: Reversible visual pigment changes in tadpoles exposed to light and darkness. Nature (Lond.) 227, 956-957 (1970).
16. BRIDGES, C.D.B.: Action of light on the visual pigments of amphibian larvae. Abstracts 6th Intern. Congr. Photobiol., Bochum, Germany, No. 203 (1972).
17. BEATTY, D.D.: Visual pigment changes in juvenile kokanee salmon in response to thyroid hormones. Vision Res. 9, 855-864 (1969)
18. ALLEN, D.M.: Photic control of the proportions of two visual pigments in a fish. Vision Res. 11, 1077-1112 (1971).
19. WILT, F.H.: The differentiation of visual pigments in metamorphosing larvae of Rana catesbeiana. Develop. Biol. 1, 199-233 (1959).
20. WILT, F.H.: The organ specific action of thyroxin in visual pigment differentiation. J. Embryol. Exp. Morph. 7, 556-563 (1959).
21. BRIDGES, C.D.B.: The rhodopsin-porphyropsin visual system. In: Handbook of Sensory Physiology, vol. VII/1, ed. H.J.A. DARTNALL, pp. 417-480. Springer, Berlin (1972).
22. BRIDGES, C.D.B., S. YOSHIKAMI: The rhodopsin-porphyropsin system in freshwater fishes. 2. Turnover and interconversion of visual pigment prosthetic groups in light and darkness - role of the pigment epithelium. Vision Res. 10, 1333-1345 (1970).
23. HUBBARD, R., A.D. COLMAN: Vitamin-A content of the frog eye during light and dark adaptation. Science (Wash.) 130, 977-978 (1959).
24. NAITO, K., F.H. WILT: The conversion of vitamin A_1 to retinene$_2$ in a freshwater fish. J. Biol. Chem. 237, 3060-3064 (1962).
25. OHTSU, K., K. NAITO, F.H. WILT: Metabolic basis of visual pigment conversion in metamorphosing Rana catesbeiana. Develop. Biol. 10, 216-232 (1964).
26. SICKEL, W.: Respiratory and electrical responses to light stimulation in the retina of the frog. Science (Wash.) 148, 648-651 (1965).

[illegible] (1961).
[illegible]

A Catalytic Role of Dihydroriboflavin in the Geometrical Isomerization of all-trans Retinal

Sidney Futterman and Martha H. Rollins
Department of Ophthalmology, University of Washington, School of Medicine, Seattle, WA 98195/USA

The mechanism of visual pigment regeneration during dark adaptation has remained obscure despite sporadic study in a number of laboratories (1-7). It has, however, been clearly established (8) that the apoprotein of rhodopsin reacts spontaneously with 11-cis retinal to regenerate the visual pigment. If 11-cis retinal arises through the direct geometrical isomerization of all-trans retinal, it would then seem reasonable to look for the occurence of such a reaction in photoreceptor outer segments where the Schiff base, metarhodopsin II, and possibly free all-trans retinal occur transiently as a result of visual pigment bleaching. Our recent studies (9,10) indicate some of the metabolic requirements for rapid complete regeneration of photosensitive pigment in preparations of bleached bovine photoreceptors incubated with all-trans retinal.

Methods

Outer segments of photoreceptor cells were prepared in a cold room at 3°C under red photographic darkroom light by homogenizing 80 dark-adapted fresh bovine retinas in 40 ml of 0.154 KCl, diluting with additional KCl solution to approximately 240 ml, centrifuging at 600 xg for 5 minutes to remove nuclei and at 1600 xg for 10 minutes to sediment a crude photoreceptor cell outer segment fraction.

Zinc treated outer segments (9) were prepared from a suspension of outer segments in 6 ml of KCl solution by dilution to approximately 40 ml using 10% $ZnSO_4$ and maintaining the mixture in the dark for 5 minutes at 3°C. The zinc treated protein was recovered by centrifugation at 12.000 xg for 2 minutes and washed twice with 0.154 M KCl, once with isotonic phosphate buffer of pH 7.4 and once again in 0.154 M KCl by dispersing in approximately 40 ml solution and centrifuging as before. The washed sediment was suspended in 6 ml of either 0.154 M KCl or isotonic phosphate buffer of pH 7.4.

The rhodopsin present in 0.3-ml portions of outer segment or zinc treated outer segment suspensions was bleached by exposure to light from three 100-watt incandescent bulbs at a distance of 15 inches for 10 minutes at room temperature. After chilling at 0°C for 2 minutes, supplements were added under red light and reaction mixtures (1.0 ml) were incubated anaerobically in the dark with shaking at 37°C to effect regeneration of rhodopsin. For addition to reaction mixtures, all-trans retinal (5 mg) was dissolved in 0.5 ml of ethanol containing 5% Triton X-100 and diluted with water to 5 ml.

Rhodopsin was extracted from reaction mixtures by the addition of 2 ml of 2% Triton X-100 followed by centrifugation at approximately 30,000 xg for 10 minutes. The rhodopsin content of the supernatant fluid was calculated from differences between the absorbancy at 500 nm before and after bleaching as described above for 5 minutes. The percentage of bleached rhodopsin that was generated during subsequent

Table I. Regeneration of visual pigment in preparations of bleached photoreceptor outer segments and zinc treated photoreceptor cell outer segments

Reaction mixtures containing either bleached photoreceptor cell outer segments or bleached zinc treated photoreceptor cell outer segments, 352 nmoles of all-trans retinal, 0.7 ml of isotonic phosphate buffer of pH 7.4, and 10 µmoles of either EDTA or DTT as indicated, were incubated in air or under nitrogen in the dark for 2 hours and then analyzed for visual pigment as described in the text.

Additions	% Regeneration			
	Outer segments		Zinc treated outer segments	
	Aerobic	Anaerobic	Aerobic	Anaerobic
None	15	101	20	25
EDTA	17	93	19	36
DTT	61	92	33	85
EDTA + DTT	67	95	70	92

incubation in the dark was calculated as follows:

$$\% \text{ regeneration} = \frac{100\,(\text{"bleached and incubated" } \Delta OD_{500\ nm} - \text{"bleached" } \Delta OD_{500\ nm})}{\text{"unbleached and incubated" } \Delta OD_{500\ nm} - \text{"bleached" } \Delta OD_{500\ nm}}$$

Results

When photoreceptor outer segments containing approximately 75 nmoles of rhodopsin were bleached and incubated for 2 hours in the dark at 37°C with 352 nmoles of all-trans retinal, little regeneration occurred under aerobic conditions (Table I). Under anearobic conditions no additional supplements were required and the regeneration of pigment was complete in 2 hours. Both dithiothreitol (DTT) and ethylenediaminetetracetic acid (EDTA) were required for effective regeneration of visual pigment under aerobic conditions. After treatment of outer segments with zinc sulfate for the purpose of inactivating any enzymes and removing any cofactors which might participate in the process of pigment regeneration, the need for supplements remained virtually unchanged except for a pronounced requirement for DTT under anaerobic conditions.

A comparison of the capacity of a variety of reducing agents to support the regeneration of visual pigment in zinc treated preparations of outer segments (Table II) indicated that dithiols were more effective than monothiols. The possible participation of the naturally occuring dithiol, thioctic acid, in the physiological process of visual pigment regeneration was virtually eliminated by observations that 3 mM

Table II Reducing agents capable of stimulating regeneration of visual pigment in preparations of bleached zinc treated photoreceptor cell outer segments

Reaction mixtures containing zinc treated photoreceptor cell outer segments, 352 nmoles of all-trans retinal, 0.7 ml of isotonic phosphate buffer of pH 7.4, 3 μmoles of EDTA, and 10 μmoles of reducing agent as indicated, were incubated for 2 hours and then analyzed for visual pigment.

Additions	% Regeneration	
	Anaerobic[a]	Aerobic
None	28	22
2-Mercaptoethanol	36	25
Glutathionine	54	21
Cysteine	63	69
DTT	85	74
2,3-Dimercaptopropanol	92	81
DL-6,8-Thioctic acid	87	79
AET[b]	92	53
Ascorbate	77	46

[a]No marked stimulation of regeneration of rhodopsin was observed using α-tocopherol, thiourea, or butylated hydroxyanisole.

[b]2-Aminoethylisothiouronium bromide hydrobromide.

arsenite or 1 mM iodoacetate failed to inhibit the regeneration of bleached outer segment preparations incubated for 2 hours under nitrogen without supplements. However, arsenite addition did completely inhibit the regeneration of visual pigment supported by the addition of DTT to zinc treated outer segments. These findings suggested that added thiols might replace or bypass an endogenous reducing agent that participated in pigment regeneration. Although α-tocopherol is present in photoreceptor outer segments (11), the experiments (Table II) also eliminated the possibility that this reducing agent might be a direct participant in the process of visual pigment regeneration.

Since it was apparent that a complete mechanism for isomerization of retinal was present in crude preparations of outer segments, an attempt was made to establish the metabolic requirements for regeneration of visual pigment by shortening the incubation period to 25 minutes. It was assumed that under these conditions, endogenous supplements would be present in concentrations insufficient to effect maximal reaction rates and one could therefore demonstrate the activity of added supplements (Table III). The basic supplements required were NAD, a flavin, magnesium ion and an oxidizable substrate. With retinol as substrate, NAD and NADP were

Table III. Supplements capable of promoting regeneration of bleached visual pigment

Reaction mixtures containing bleached photoreceptor outer segments, 352 nmoles of all-trans retinal, 0.7 ml of isotonic phosphate buffer of pH 7.4, either 3 μmoles of L-lactate, 10 μmoles of succinate or 3.5 μmoles of retinol and 1 μmole of NAD, 0.1 μmole of riboflavin, or 5 μmoles of $MgCl_2$ as indicated were incubated for 25 minutes and analyzed for visual pigment.

Omission	% Regeneration			
	No substrate	With retinol	With lactate	With succinate
NAD	19	70	60	89
None	50	79	80	85
Riboflavin	16	16	18	25
None	57	71	77	83
$MgCl_2$	31	68	72	85
None	48	69	70	81

equally effective. Riboflavin could be replaced by FMN or FAD. An unambiguous cation requirement could not be observed when an oxidizable substrate was added, apparently because efficient non-enzymatic reduction of flavin occured when high concentrations of reduced pyridine nucleotides were present. These preparations of photoreceptor outer segments contained elipsoid mitochondria (12) and it seemed likely that succinate was reducing the mitochondrial electron transport flavoprotein which in turn reduced the added riboflavin. In more recent experiments in which mitochondrial contamination has been eliminated by the sucrose flotation technique, succinate was no longer capable of stimulating visual pigment regeneration, and dithiothreitol was needed even under anaerobic conditions, presumably to reduce an endogenous flavin.

The rate of visual pigment regeneration was profoundly influenced by the addition of supplements (Fig. 1). Under anaerobic conditions with no additions other than excess all-trans retinal, a lag of about 30 minutes occured before visual pigment regeneration began and approximately 2 hours were required to complete regeneration. The addition of riboflavin shortened the lag considerably and greatly increased the rate of regeneration. The addition of dihydroriboflavin abolished the lag period entirely. When a suitable preformed chromophore was supplied, as for example, 9-cis retinal, visual pigment regeneration was essentially complete in approximately 5 minutes. The rate of visual pigment regeneration, therefore, seemed to be limited by the rate at which the underlying metabolism necessary to produce a reduced flavin catalyst could be accomplished.

The regeneration of rhodopsin was maximal in the vicinity of pH 7.0 to pH 7.5 when bleached preparations were incubated with all-trans retinal (Fig. 2). No similar

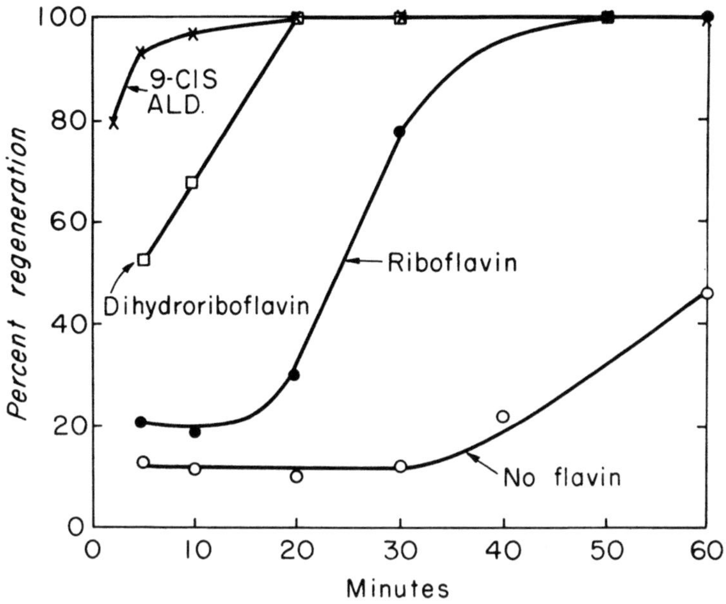

Fig. 1. Rate of visual pigment regeneration in bleached photoreceptor cell outer segments incubated with all-trans retinal in the presence and absence of flavin supplements and with 9-cis retinal. Reaction mixtures, containing bleached outer segments, 0.7 ml of isotonic phosphate buffer of pH 7.4, 352 nmoles of either all-trans retinal (□, ○, ●) or 9-cis retinal (x) and where indicated, 0.1 μmole of riboflavin or dihydroriboflavin, were incubated and analyzed as described in the text

pH optimum was observed when outer segment preparations were incubated with all-trans retinal in the presence of dihydroriboflavin or with 9-cis retinal. It was apparent, therefore, that a suitable chromophore could combine efficiently with the visual pigment apoprotein and that all-trans retinal could isomerize effectively in the presence of a suitable catalyst over an extended range of pH. These results indicated that the pH optimum observed for regeneration of bleached outer segments incubated with all-trans retinal under anaerobic conditions and without other supplements probably reflected the pH requirements for the metabolic reactions responsable for the reduction of an endogenous flavin catalyst.

A large excess of all-trans retinal was required to obtain effective visual pigment regeneration when no other supplements were present (Fig. 3). In contrast, when the dihydroriboflavin catalyst was present, all-trans retinal was used with approximately the same efficiency for visual pigment regeneration as a suitable geometrical isomer, 9-cis retinal.

Discussion

The overall process of visual pigment regeneration has now been resolved into three steps. These can be viewed as supportive metabolism that reduced endogeneous flavin catalyst, a catalytic geometrical isomerization of all-trans retinal, and the recombination of a chromophore molecule with visual pigment apoprotein. As depicted below, retinol itself, which can serve as the reducing agent, is oxi-

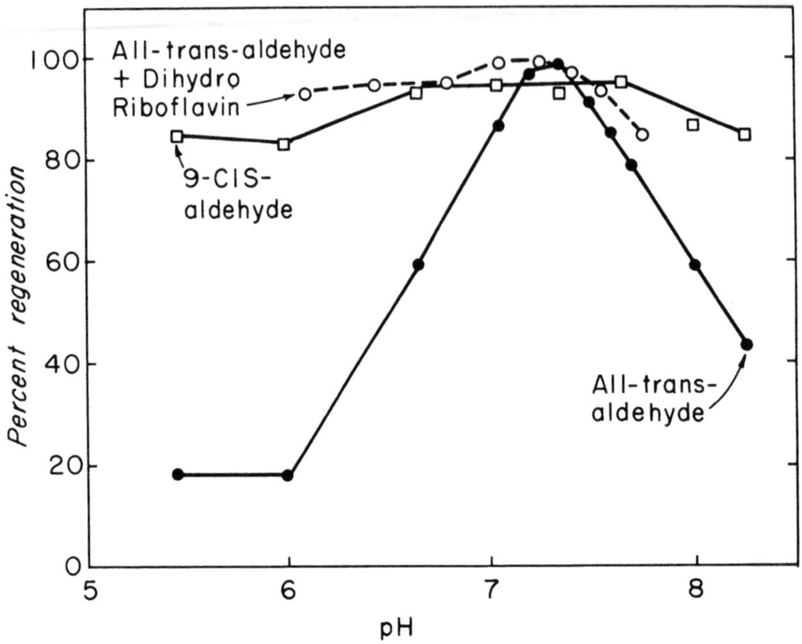

Fig. 2. Effect of pH on regeneration of visual pigment in bleached photorecptor cell outer segments incubated with all-trans retinal in the presence and absence of dihydroriboflavin and with 9-cis retinal. Reaction mixtures, containing bleached outer segments, 0.3 ml of 0.154 MKCl, 0.5 ml of isotonic phosphate buffer and, as indicated, 352 nmoles of all-trans retinal or 9-cis retinal and 0.1 µmole of dihydroriboflavin were incubated for 2 hours and analyzed for visual pigment. The pH of each reaction mixture was determined after spectral measurements were made

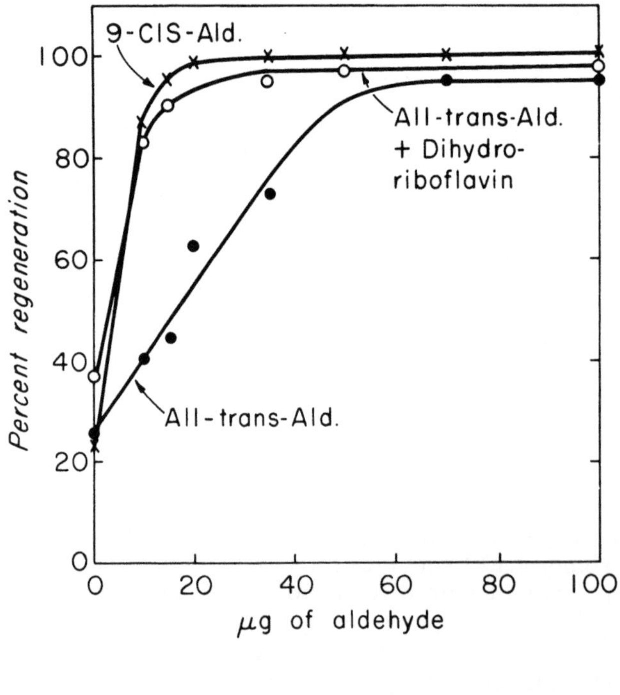

Fig. 3. Efficiency of utilization of retinal in regeneration of visual pigment in bleached photoreceptor cell outer segments. Reaction mixtures, containing bleached outer segments, 0.7 ml of phosphate buffer of pH 7.4, all-trans or 9-cis retinal and, where indicated, 0.1 µmole of dihydroriboflavin, were incubated for 2 hours

dized by the retinal reductase of the photoreceptor organelle (9) and generates reduced pyridine nucleotides. A magnesium ion dependent enzymatic transfer or non-enzymatic transfer of electrons from reduced pyridine nucleotides to flavin produces a reduced flavin derivative which serves as an isomerization catalyst. The catalyst effectively isomerizes all-trans retinal to a geometrical isomer capable of serving as a visual pigment chromophore. In the final step the chromophore spontaneously and rapidly combines with apoprotein to regenerate visual pigment.

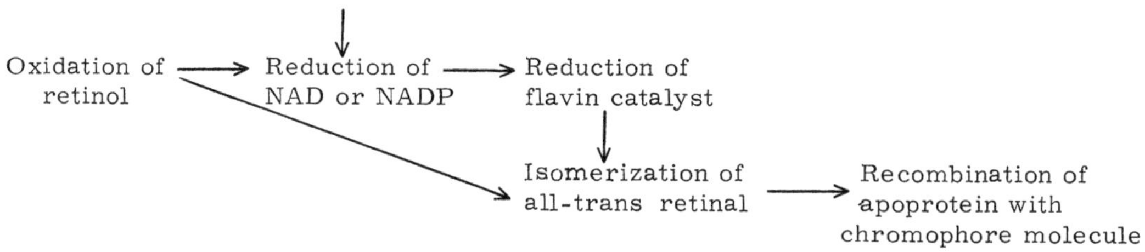

The utilization of all-trans retinal for complete reconversion of apoprotein to photosensitive pigment by bleached preparations of bovine photoreceptor cell outer segments has thus been demonstrated and the involvement of reduced flavin as a geometrical isomerization-catalyst has been implicated in the mechanism of regeneration.

Acknowledgement. This investigation was supported by the PHS Research Grant No. EY 00343 from the National Eye Institute.

References

1. HUBBARD, R., G. WALD: The mechanism of rhodopsin synthesis. Proc. Nat. Acad. Sci. 37, 69-79 (1951).
2. BLISS, A.F.: Properties of the pigment layer factor in the regeneration of rhodopsin. J. Biol. Chem. 193, 525-531 (1951).
3. COLLINS, F.D., J.N. GREEN, R.A. MORTON: Studies in rhodopsin 7. Regeneration of rhodopsin by comminuted ox retina. Biochem. 56, 493-498 (1954).
4. HUBBARD, R.: Retinene isomerase. J. Gen. Physiol. 39, 935-962 (1956).
5. OSTAPENKO, A., A.L. BERMAN, R.N. ETINGOV: Determination of rhodopsin and its resynthesis in retinal homogenate. Biokhimiya 34, 1028-1033 (1969).
6. PLANTE, E.O., B. RABINOVITCH: Enzymes in the regeneration of rhodopsin. Biochem. Biophys. Res. Commun. 46, 725-730 (1972).
7. AMER, S., M. AKHTAR: Studies on a missing reaction in the visual cycle. Nature (Lond.) 237, 266-267 (1972).
8. BROWN, P.K., G. WALD: The neo-b isomer of vitamin A and retinene. J. Biol. Chem. 222, 865-877 (1956).
9. FUTTERMAN, S., M.H. ROLLINS, E. VACANO: Mechanism of visual pigment regeneration in the visual cycle. Federation Proc. 29, 471 (1970).
10. FUTTERMAN, S., M.H. ROLLINS: Role of dihydroflavin in regeneration of rhodopsin in vitro. Federation Proc. 30, 1197 (1971).
11. DILLEY, R.A., D.G. MC CONNELL: Alpha-tocopherol in the retinal outer segment of bovine eyes. J. Membrane Biol. 2, 317-323 (1970).
12. FUTTERMAN, S., A. HENDRICKSON, P.E. BISHOP, M.H. ROLLINS, E. VACANO: Metabolism of glucose and reduction of retinaldehyde in retinal photoreceptors. J. Neurochem. 17, 149-156 (1970).

Discussion

W. de Grip: As to the pH-influence on the regeneration velocity of opsin with 11-cis retinaldehyde, J.P. Rotmans in our laboratory, also found a rather broad curve between pH 5 and 7.5, in agreement with your results with 9-cis retinaldehyde and contradictory to the results of Dr. Akhtar.

Active Site and Enzymological Studies on the Components of the Visual Cycle

M. Akhtar, S. Amer, and M. D. Hirtenstein

Department of Physiology and Biochemistry, University of Southampton, Southampton SO9 5NH/ Great Britain

Developments in the biochemistry of vision over the last quarter century may be summarized by the equations 1-3 of Scheme 1. The first reaction in the Scheme involves the combination in a dark reaction of the visual protein opsin with 11-cis retinal to form the photolabile complex rhodopsin, λ_{max} 450-560 nm (the absorption maxima are species dependent).

On absorption of light, rhodopsin participates through a process whose mechanism is not yet understood in the transmission of impulse responsible for the visual sensation. The knowledge of the biochemical changes accompanying the absorption of light by rhodopsin is at an advanced stage of understanding. The changes culminate in the formation of all-trans retinal and opsin (equation 2, Scheme I) through a number of intermediates. For the completion of the cycle one needs a molecular process which may regenerate 11-cis retinal from all-trans-retinal (equation 3, Scheme I) (reviews: 1,2). We describe recent progress made in our laboratory on the biochemistry of the reactions 1 and 3, of the cycle in Scheme I.

Scheme I

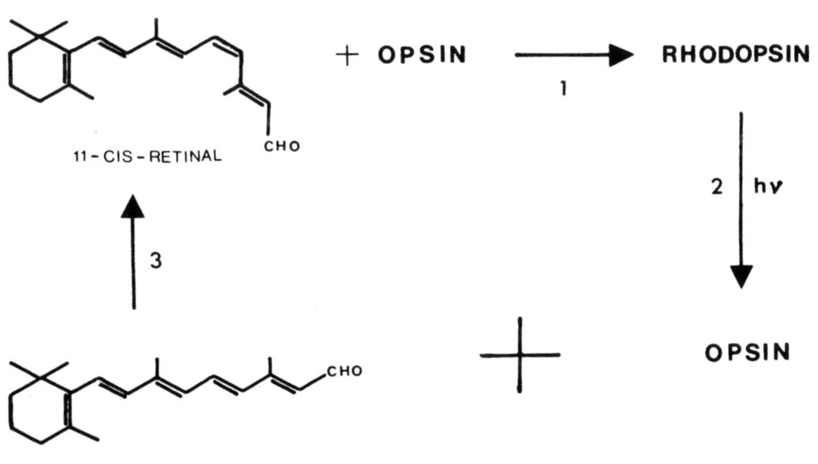

The Active Site of Rhodopsin

The presence of Schiff base linkage in rhodopsin was originally proposed by Pitt et al. (3) and convincing evidence in support of this hypothesis was provided by our work (4) and also that of the Harvard group (5). It was shown that although rhodopsin itself is resistant to attack by sodium borohydride, one of the derivates, metarhodopsin II, readily reacts with sodium borohydride forming a reduced derivative of rhodopsin (4,5).

Table I. Radiochemical data from the degradation, according to Scheme II, of reduced rhodopsin derivatives

	Percentage of total radioactivity present at each stage from	
	^3H Dihydro-metarhodopsin II	TCA* denatured and reduced ^3H rhodopsin
ε-N-retinyl-lysine	68.3	68.3
ε-N-perhydroretinyl-lysine	68.4	72.1
α, ε, -bis-DNP, ε-PHRL**	81.5	85.3

* = trichloroacetic acid; ** = α, ε, bis-dinitrophenyl-ε-N-perhydroretinyl-lysine.

The new derivative was formulated as dihydrometarhodopsin II. Further developments led to the synthesis of labelled rhodopsin which contained tritiated retinyl moiety at the active site. Using this labelled material it was established (6,7) that at the metarhodopsin II stage, the retinyl moiety is linked to an ε-amino group of lysine. Similar results were reported by Bownds (8) and subsequently by other workers.

The direct extrapolation of this observation suggested that an ε-amino group of lysine is also involved in the chromophore binding in native rhodopsin (6,7,8). This simple deduction was however questioned by observations from several laboratories showing that the denaturation of rhodopsin followed by treatment with sodium borohydride resulted in the isolation of N-retinylphosphatidylethanolamine. This suggested involvement of a lipid amino group at the active site of native rhodopsin. Latter systematic study however revealed that the lipid linked retinyl moiety was isolated only from 'crude preparations' in which rhodopsin was still attached to membrane fractions. Samples of highly purified rhodopsin gave only protein bound retinyl moiety (9,10).

Thus the chemical events leading to the isolation of a lipid bound retinyl moiety are irrelevant to the nature of chromophore binding site (9-12). However the conditions and circumstances under which the lipid derivative is formed have led to the suggestion (9) that the retinylidene chromophore in rhodopsin and metarhodopsin I is present in a lipophilic environment and in metarhodopsin II in a hydrophilic environment (for a detailed discussion see ref. 9). Some predictions of the hypothesis are supported by work on model systems (13,14,15).

Identification of the Chromophore Binding Sites

The availability in our laboratory of preparations of highly purified rhodopsin prepared from opsin and labelled 11-cis retinal has permitted quantitative work to be undertaken on the active site using two reduced derivatives of rhodopsin. Labelled rhodopsin was prepared and purified as described previously (9). Samples of the material were either reduced with sodium borohydride in the presence of a denaturant to give "reduced denatured rhodopsin" or exposed to light in the presence of sodium borohydride to give dihydrometarhodopsin II. In each case more than 90% of the original radioactivity was found to be protein bound. Both the derivatives were subjected to the reaction sequence established previously (6,7) and outlined in Scheme II.

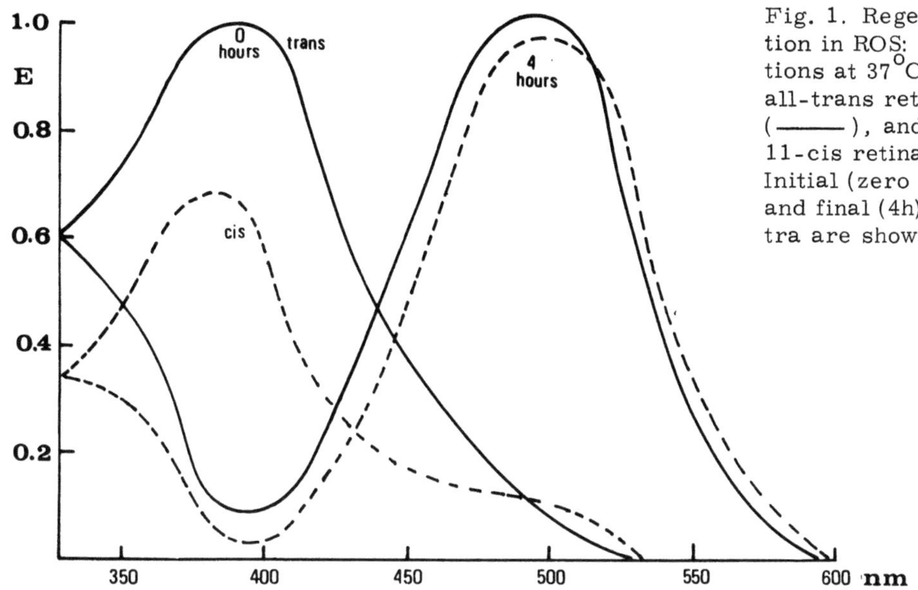

Fig. 1. Regeneration in ROS: Incubations at 37°C with all-trans retinal (———), and with 11-cis retinal (---). Initial (zero hours), and final (4h) spectra are shown

Scheme II Reduced rhodopsin derivative $\xrightarrow{NaOH}$ ε - N - retinyl - lysine $\xrightarrow{H_2}$ ε-perhydroretinyl-lysine $\xrightarrow{\text{dinitrofluorobenzene}}$ α, ε, bis-DNP-ε-perhydroretinyl-lysine.

The results in Table I show that at all stages the two derivatives behaved identically and gave about 70% of the original protein bound radioactivity in a fragment identified as ε-N-retinyl-lysine.

The work unambiguously shows that the retinyl moiety in both the rhodopsin derivatives is attached to an ε-amino group of lysine. In view of the fact that the retinylidene chromophore at the metarhodopsin II stage is stabilized under mild and efficient conditions it is justified in assuming that in this case an active site structure has been frozen. On the other hand, at the rhodopsin stage the linkage under discussion is reduced under somewhat more disruptive conditions which could induce an intramolecular rearrangement. In this case the work suggests but does not establish that the lysine residue to which the retinyl moiety is linked is in fact the lysine at the active site of native rhodopsin. Neither can one assert from the observations available to date that the same lysine is involved at the active site of native rhodopsin and metarhodopsin II. The results on the elucidation of rhodopsin active site from other laboratories has been presented by previous speakers (16,17).

The Formation of Rhodopsin from all-trans Retinal

The knowledge that rhodopsin is formed by the reaction of opsin with 11-cis retinal, and that on the absorption of light it is converted into opsin and all-trans retinal led to the theoretical deduction, that in order to complete the visual cycle there must be present *in vivo* an enzyme system that catalyses the conversion, all-trans retinal ⟶ 11-cis retinal. We describe below observations relevant to this aspect.

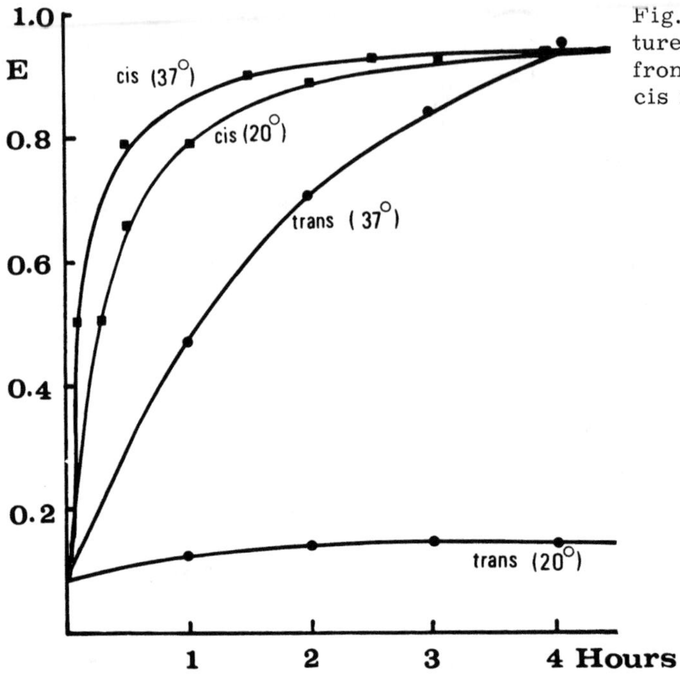

Fig. 2. The effect of temperature on pigment regeneration from all-trans retinal and 11-cis retinal

ROS prepared from fresh bovine retinae were separately incubated in the dark at 37°C with equivalent quantities of either 11-cis retinal or highly purified all-trans retinal. Fig. 1 shows that after 4 hours both samples regenerated rhodopsin. The amount of pigment regenerated (about 0.11 μMoles/ml) from all-trans retinal (about 0.1 μMoles/ml) was almost 100% of the theoretically expected amount. The stoiciometric conversion of all the added all-trans retinal into the pigment demonstrated that the regeneration of the pigment was exclusively due to the all-trans retinal rather than to any contaminants of 9-cis or 11-cis isomers.

The possibility that a contaminant may participate in the pigment formation in the above experiment was further ruled out by the fact that samples of all-trans retinal used did not regenerate rhodopsin when incubated at 37°C with a digitonin extract of opsin. This experiment also rules out the other possibility that a thermal isomerization of all-trans retinal into 11-cis retinal might have been responsible for the pigment formation in the rod preparation. The determination of the spectra of the regenerated pigment in ROS suspension showed that the species derived from all-trans retinal usually had λ_{max} at lower wavelengths compared to that observed for 11-cis retinal. A similar discrepancy was maintained when the regenerated pigments were extracted in digitonin and subjected to spectral measurement against a digitonin blank. In this connection attention is drawn to papers presented by Rotmans et al. (18) and by Futterman and Rollins (19) in this symposium.

The rate of regeneration of rhodopsin from equimolar quantities of 11-cis retinal and all-trans retinal was then studied over a period of 4 hours. 11-cis retinal gave rapid regeneration, reaching about 75% of the final amount within the first 15 min. Thereafter the rate slowed down to achieve about 100% regeneration in 4 hours (Fig. 2).

Regeneration from all-trans retinal was found to be a slower process achieving 75% regeneration in $2\frac{1}{2}$ hours (Fig. 2). This rate of pigment formation compares fa-

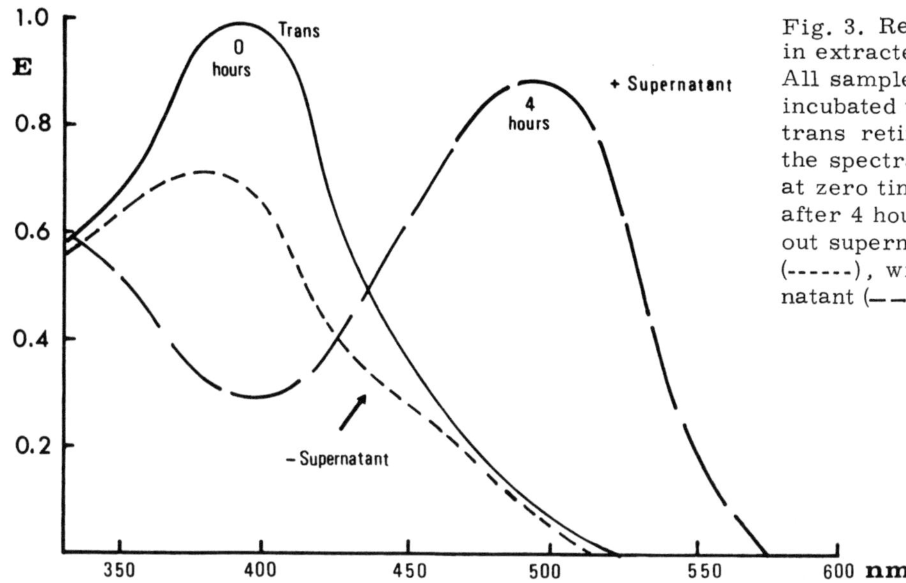

Fig. 3. Regeneration in extracted ROS: All samples were incubated with all-trans retinal and the spectra recorded at zero time (———); after 4 hours without supernatant (------), with supernatant (— — —)

vourably with in vivo studies of Dowling (20) in which it was shown that a completely light adapted rat took about 3 hours to regenerate 100% rhodopsin.

The rate of pigment formation in the ROS was dependent on the concentration of all-trans retinal and considerably higher rates than those observed in Fig. 2 were obtained with larger amounts of the substrate. However in order to ensure that contaminants of cis isomers (if present) did not contribute to the rhodopsin formation, the amount of all-trans retinal employed throughout the investigation was adjusted to give a quantitative utilization of the substrate.

Apart from its dependence on the substrate concentration, the rate of regeneration from all-trans retinal was shown to be dependent on the temperature of incubation. Thus samples incubated at 20°C failed to form pigment from all-trans retinal, while in a parallel experiment complete regeneration from 11-cis retinal occured though at a slightly reduced rate (Fig. 2). Preparations of ROS which had been allowed to age for a period of 2 weeks failed to exhibit regeneration when incubated with all-trans retinal, while an equal aliquot of the same sample gave complete regeneration with equimolar concentration of 11-cis retinal.

Some progress has also been made towards the solubilization of a component involved in the regeneration of the pigment from all-trans retinal. Thus freezing of rod outer segments in sodium phosphate buffer at -15°C for 24 hours followed by centrifugation gave a supernatant (referred to below as ROS-supernatant) and a ROS pellet which was washed twice. These washed and extracted ROS retained the ability to synthesize rhodopsin form 11-cis retinal but not from all-trans retinal. However the biosynthesis of the pigment from all-trans retinal in extracted ROS could be re-established on supplementation with ROS-supernatant (Fig. 3). The ROS-supernatant lost its activity when heated to 100°C for 5 min (21). Further studies involved ultrafiltration of the ROS-supernatant using UM 2oE membrane which gave two fractions, a) concentrate and b) filtrate; all the activity was present in

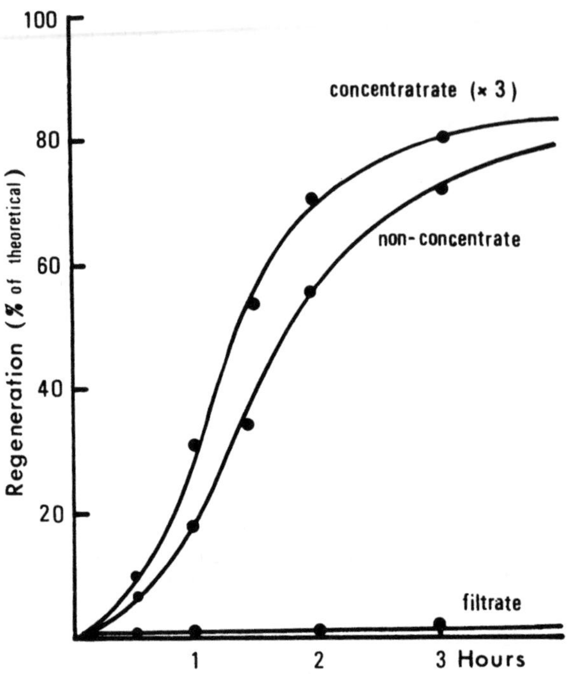

Fig. 4. The effectiveness of the concentrate and the filtrate obtained from the ROS-supernatant in the regeneration from all-trans retinal

the concentrate (Fig. 4). Several other properties of the system were also studied (22) and included the demonstration that the pH optimum for the pigment formation from all-trans retinal was at 7.0, whereas the pH optimum for rhodopsin synthesis from 11-cis retinal was at about 5.0.

In the experiments described above the biosynthesis of pigment from all-trans retinal was achieved within the ROS. Preliminary findings indicate that a system of all-trans retinal, ROS-supernatant, and opsin in digitonin solution fails to form the pigment. If however a small quantity of ROS fragments is introduced to this system, pigment formation takes place. Emphasis is laid on the fact that the simulnatant and ROS-fragments) is required for pigment formation in solution, the omission of any one of them abolishes regeneration.

Acknowledgement. The participation of Dr. P.T. Blosse, and Dr. P.B. Dewhurst in the initial phase of this work is gratefully acknowledged. - We are greatly indebted, also, to Professor K.A. Munday for his active support and encouragement. - A part of this work was supported by a grant from The Science Research Council.

References

1. WALD, G.: The molecular basis of visual excitation. Nature (Lond.) 219, 800-807 (1968).
2. PITT, G.A.J.: Carotenoides, ed. O. ISLER, pp. 217. Birkhäuser-Verlag, Basel (1972).
3. PITT, G.A.J., F.D. COLLINS, R.A. MORTON, P. STOK: Retinylidenemethylamine an indicator yellow analogue. Biochem.J. 59, 122-128 (1955).
4. AKHTAR, M., P.T. BLOSSE, P.B. DEWHURST: The reduction of a rhodopsin derivative. Life Sci. 4, 1221-1224 (1965).

5. BOWNDS, D., G. WALD: Reaction of the rhodopsin chromophore with sodium borohydride. Nature (Lond.) 205, 254-257 (1965).
6. AKHTAR, M., P.T. BLOSSE, P.B. DEWHURST: The active site of the visual protein, opsin. Chem.Commun. 631-632 (1967).
7. AKHTAR, M., P.T BLOSSE, P.B. DEWHURST: The nature of the retinal-opsin linkage. Biochem.J. 110, 693-702 (1968).
8. BOWNDS, D.: Site of attachement of retinal in rhodopsin. Nature (Lond.) 216, 1178-1181 (1967).
9. HIRTENSTEIN, M.D., AKHTAR, M.: A convenient synthesis of labelled rhodopsin and studies on its active site. Biochem.J. 119, 359-366 (1970).
10. ANDERSON, R.E., R.T. HOFFMAN, M.O. HALL: Linkage of retinal to opsin. Nature New Biol. 229, 249-250 (1971).
11. BORGGREVEN, J.M.P.M., J.P. ROTMANS, S.L. BONTING, F.J.M. DAEMEN: Biochemical aspects of the visual process. XIII. The role of phospholipids in cattle rhodopsin studied with Phospholipase C. Archives Biochem. Biophys. 145, 290-299 (1971).
12. DAEMEN, F.J.M., P.A.A. JANSEN, S.L. BONTING: Biochemical aspects of the visual process. XIV. The binding site of retinaldehyde in rhodopsin studied with model aldimines. Archives Biochem.Biophys. 145, 300-309 (1971).
13. HIRTENSTEIN, M.D., AKHTAR, M.: Chromophore-Environment interactions of visual pigments in model systems. Nature New Biol. 233, 94-95 (1971).
14. IRVING, C.S., G.W. BYERS, P.A. LEERMAKERS: Spectroscopic model for the visual pigments influence of microenvironmental polarizability. Biochemistry 9, 858-868 (1970).
15. WADDELL, W., R.S. BECKER: Hydrogen bonded (protonated) Schiff base of all-trans retinal. J.Am.Chem.Soc. 93, 3788-3789 (1971).
16. ABRAHAMSON, C.W.: The kinetics of early intermediate processes in the photolysis of visual pigments. This volume, pp. 47-55.
17. BONTING, S.L., J.P. ROTMANS, F.J.M. DAEMEN: Chromophore migration after illumination of rhodopsin. This volume, pp. 39-44
18. ROTMANS, J.P., F. J.M. DAEMEN, S.L. BONTING: Formation of isorhodopsin from photolyzed rhodopsin in darkness. This volume, pp. 139-145.
19. FUTTERMAN, S., M.H. ROLLINS: A catalytic role of dihydroriboflavin in the geometrical isomerization of all-trans retinal. This volume, pp. 123-130.
20. DOWLING, J.E.: Neural and photochemical mechanism of visual adaptation in the rat. J.Gen.Physiol 46, 1287-1301 (1963).
21. AMER, S., AKHTAR, M.: The regeneration of rhodopsin from all-trans-retinal; the solubilization of an enzyme system involved in the visual cycle. Biochem.J. 128, 987-989 (1972).
22. AMER, S., AKHTAR, M.: Studies on a missing reaction in the visual cycle. Nature New Biol. 237, 266-267 (1972).

Discussion

S.L. Bonting: Anticipating on the paper by Rotmans, may I ask: Have you identified the chromophore of the regenerated pigment by thin layer chromatography?

M. Akhtar: We have concentrated primarily on the identification of the regenerated rhodopsin.

S.L. Bonting: Have you excluded a role of bacteria in your regeneration experiments?

M. Akhtar: The cumulative results I have presented in the lecture, coupled with those already published in our recent papers, make the involvement of bacterial growth in contributing to the effect under discussion very unlikely. Unless of course the bacteria existed in the animal and their amount as normally present in the retina is sufficient to participate in the conversion of the trans isomer into the cis-form.

Note Added in Proof

Since the submission of this paper, we have shown that faster rates of regeneration than those reported here may be obtained when fresh ROS are used. These rates are unaffected when the incubations are carried out in the presence up to 300 μg/ml of one or more of the following antibiotics, benzylpenicillin, streptomycin sulphate, cephaloridine, and phenethiscillin.

On the question of the identification of the cis isomer formed in these experiments, the pigment was regenerated from labelled all-trans retinal, and the isolation of the chromophore and TLC analyses showed that the radioactivity was present in the regions corresponding to all-trans retinal, 9-cis and 11-cis retinal, though predominantly in the 9-cis compound.

Formation of Isorhodopsin from Photolyzed Rhodopsin in Darkness

J. P. Rotmans, F. J. M. Daemen, and S. L. Bonting
Department of Biochemistry, University of Nijmegen, Nijmegen/The Netherlands

In order to permit continued function of the rod photoreceptor rhodopsin must be continuously resynthesized from its bleaching products. The mechanism of this regeneration process, which can take place in the dark, is still largely unknown. The crucial step must involve isomerization of the all-trans form of retinaldehyde or retinol to the 11-cis isomer. It is generally accepted that no 11-cis retinaldehyde or retinol occurs outside the eye (1), but stores of 11-cis retinol are found in the eye (2,3). Hence, there must be a mechanism in the eye which converts all-trans retinaldehyde or retinol into the corresonding 11-cis isomer. Although in the frog this process seems to occur in the pigment epithelium, the rat itself clearly possesses an isomerizing mechanism in the retina (4,5).

We decided to look for a retinal mechanism in cattle retina, since we thought that retinaldehyde isomerization within the outer segment could be of considerable physiological importance. The only report of a retinal isomerizing enzyme is that of Hubbard (6), who showed that an ammoniumsulfate fraction of water-soluble proteins from cattle retina was able to isomerize all-trans retinaldehyde to 11-cis retinaldehyde. This isomerization occured only when the isomerizing enzyme was present during the illumination of the rhodopsin solution. Since the in vivo rate of regeneration of rhodopsin in rat and man (7,8) is equal in darkness and in light, we investigated whether isomerase activity could be demonstrated in the dark in cattle retina.

Experiments on Retinal Protein Extract

When repeating Hubbard's experiments on the retinal isomerase (6), we observed that in the presence of a water-soluble retinal protein extract a visual pigment was formed in the dark with illuminated rhodopsin (9). Analysis showed that the retinal protein solution contained no retinaldehyde. The synthesis of visual pigment can therefore not be ascribed to a reaction of photolyzed rhodopsin with 11-cis retinaldehyde, but only to reisomerization of the all-trans retinaldehyde present in the illuminated rhodopsin suspension and subsequent binding of the resulting isomerized retinaldehyde.

These results seemed to indicate the presence of an all-trans retinaldehyde isomerase in the retinal protein extract. Closer examination, however, revealed the following facts:

1. The absorption spectrum of the synthesized photopigment has an absorption maximum at 485 nm, suggesting that it is isorhodopsin, an analogue of rhodopsin not occuring in vivo. This was proven by extracting with 90 per cent ethanol and subjecting the extract to thin layer chromatography. The major spot proved to be 9-cis retinaldehyde (10).

2. Formation of isorhodopsin is rather slow requiring from 10 to 30 hrs at 20°C and 6 to 15 hrs at 37°C for maximal regeneration (Fig. 1). Often a lag-time of sev-

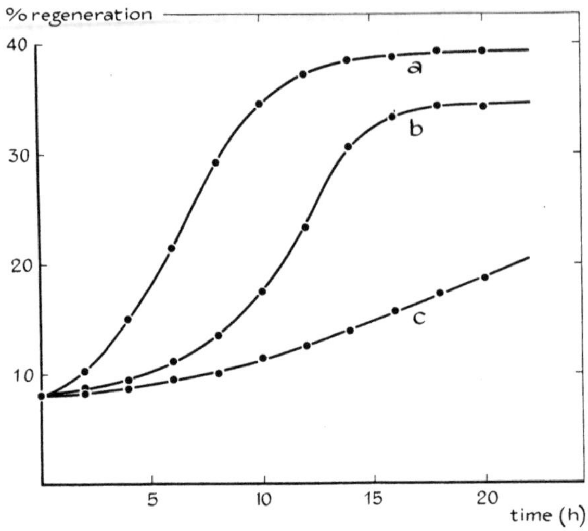

Fig. 1. Visual pigment synthesis stimulated by a retinal protein extract. 20 nmol photolyzed rhodopsin in 1 ml 0.067 M phosphate buffer (pH 6.5) was incubated at room temperature with 300 (a) and 150 (b) μl retinal protein extract in the same buffer. Curve (c) represents the synthesis of visual pigment in a blank containing no retinal protein extract

eral hrs was observed. The isorhodopsin yield varied between 30 and 80% of the amount of rhodopsin present, assuming a value of 1.06 for the ratio of the molar absorbance of isorhodopsin and rhodopsin (6). In high concentrations photolyzed rhodopsin, upon standing at 20°C in the dark in the absence of exogenous all-trans retinaldehyde and retinal protein extract, also yields isorhodopsin (Fig. 2). Maximal yields (up to 70%) are obtained at rhodopsin concentrations above 30 nmol/ml. The isomerizing activity was retained when the protein extract previously has been exposed to strong acid or base (final pH 1 and 10, respectively) or had been heated to 100° for 5 min. When the retinal protein extract was subjected to chromatography on Sephadex-G-75, all fractions showed isomerizing activity. These findings strongly suggest that the isomerizing activity of the extract is not due to an enzyme.

The long incubation times required for isomerization and the long time preceeding isomerization made us suspect a bacterial involvement. The possibility of bacterial involvement was checked by the addition of antibiotics to the incubation mixture. Both penicillin and streptomycin (100 mg/ml incubation mixture) inhibited completely the isomerizing effect of the retinal protein extract.

Further confirmation for bacterial involvement was obtained by sterilization of the reaction mixture. The rhodopsin was chemically sterilized with ethylene oxide. A very concentrated suspension of rhodopsin in 0.067 M phosphate buffer of pH 7 was treated in the dark with ethylene oxide (15% in CO_2) at room temperature for 15 hrs under a total pressure of 5 kp/cm^2. A control sample was kept for 15 hrs at room temperature in normal air. To free the sterilized rhodopsin from excess ethylene oxide, the sample was left for 3 hrs in a nitrogen atmosphere. The retinal protein solution was sterilized by heating at 100°C for 5 min. Incubations were carried out under sterile conditions. Under totally sterile conditions the formation of photolysable pigment was greatly inhibited (Fig. 3a), while the use of non-sterile protein extract gave slightly more isorhodopsin (Fig. 3b), and the normal amount of pigment was formed if both components were non-sterile (Fig. 3c). The ethylene oxide treatment of the rhodopsin preparation did not affect its spectral characteristics or its ability to form isorhodopsin under non-sterile conditions (Fig. 3d, e). Replacement of the retinal proteins by a typical bacterial growth medium, brain-heart infusion, led to a rapid formation of isorhodopsin in high yields. This suggests

Fig. 2. Isorhodopsin formation from photolyzed rhodopsin. Rhodopsin suspended in 0.067 M phosphate buffer was illuminated through orange and infrared filters for 10 min. The illuminated suspension was kept in the dark at room temperature. At certain times samples were withdrawn and the amount of isorhodopsin formed was measured and expressed as percentage of the rhodopsin originally present. Three concentrations of rhodopsin were used (a) 4 nmol/ml, (b) 8 nmol/ml, (c) 60 nmol/ml

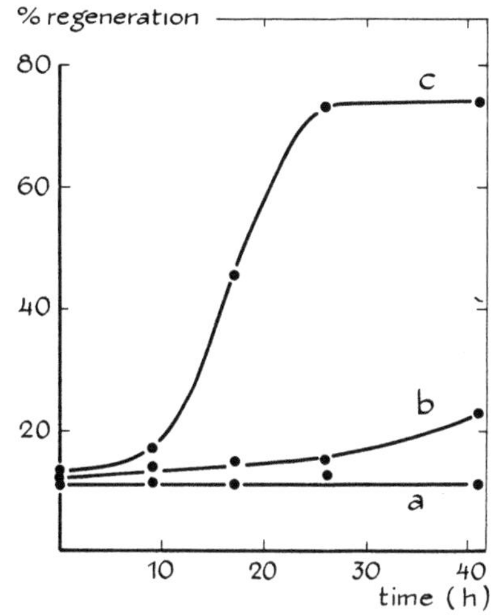

that the only function of the retinal protein extract is that of serving as a growth medium for the bacteria present in the rhodopsin preparation.

Isorhodopsin Formation by Bacterial Action

To confirm the bacterial influence in a more direct way, bacteria isolated from an isorhodopsin generating system were cultured anaerobically overnight at 37 ^{o}C in 300 ml brain-heart infusion and harvested by centrifugation (8,000· g, 30 min). The pellet was repeatedly washed with 0.067 M phosphate buffer of pH 6.5, resuspended

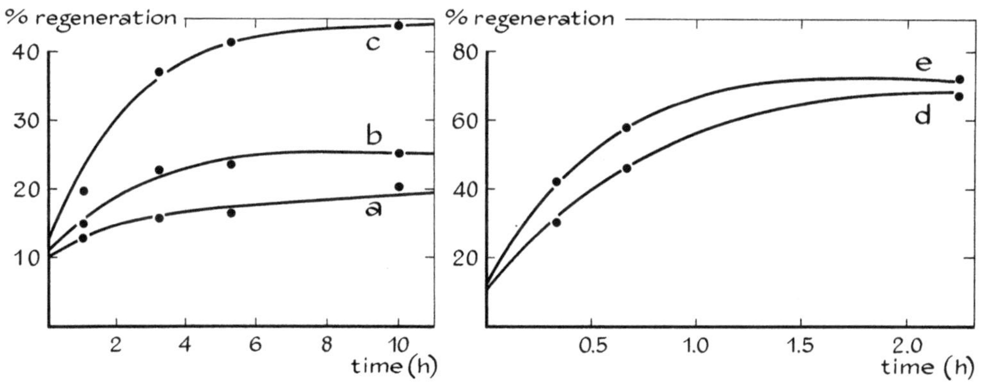

Fig. 3. Inhibition of isorhodopsin formation by sterilization with ethylene oxide. Sterilized photolyzed rhodopsin was incubated at room temperature with sterilized retinal protein extract (a), non-sterilized retinal protein extract (b). Curve (c) depicts the isorhodopsin formation from non-sterilized rhodopsin incubated with non-sterilized retinal protein extract. Sterilized photolyzed rhodopsin had retained its capacity to form isorhodopsin when incubated with bacteria (d). Curve (e) depicts the course of isorhodopsin formation from non-sterilized photolyzed rhodopsin when incubated with bacteria

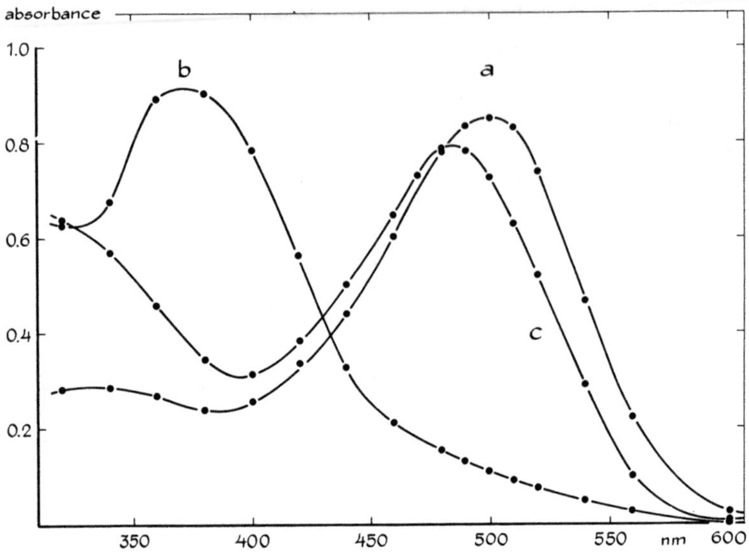

Fig. 4. Spectral evidence for the formation of isorhodopsin. Absorption spectra of rhodopsin (a), photolyzed rhodopsin (b), and isorhodopsin formed by bacterial action (c). The pigment suspensions were solubilized in a 1% Triton X-100 solution in 0.067 M phosphate buffer (pH 6.5) prior to spectral measurement

in 30 ml of the same buffer. Photolyzed rhodopsin was prepared by illuminating a rhodopsin suspension ($2.5 \cdot 10^{-5}$ M in 0.067 M phosphate buffer, pH 6.5) for 10 min through orange and infrared filters (OG 370 and KG 1 Schott-Jena) by a 75 W tungsten lamp at 15 cm distance. This leaves about 5 to 10 per cent of the visual pigment originally present as a mixture of photoregenerated rhodopsin and isorhodopsin. Equal volumes of the bacterial suspension and the illuminated rhodopsin suspension were incubated under nitrogen in the dark at 37°C. The amount of isorhodopsin formed reached a maximum after 3 to 4 hrs. At that time 65 to 95 per cent of the rhodopsin originally present had been converted to isorhodopsin, as shown by the absorption spectra in Fig. 4. Addition of antibiotics (penicillin or streptomycin, 100 mg/ml incubation mixture) resulted again in complete inhibition of pigment formation. Upon cultivation of the bacteria isolated from the rhodopsin preparations it was found that several species were present. Rather than isolating everyone of these and testing each species individually for isomerizing activity, we tested a number of organisms available to us in pure cultures. The results indicate that the isomerizing activity is not limited to a single species, although differences with respect to velocity and maximal level were observed. In subsequent experiments a facultative anaerobic organism, Proteus mirabilis, which showed the highest isomerizing activity, has been used. When isorhodopsin formation was complete, the reaction was first-order with respect to the opsin concentration and had a $t_{\frac{1}{2}}$ of about 35 min (Fig. 5). The initial rate and the extent of isorhodopsin formation are dependent on the number of bacteria present at the beginning of the incubation.

These observations demonstrate that experiments on the regeneration of rhodopsin are easily obscured by the action of bacteria. High concentrations of rhodopsin, the presence of non-rhodopsin protein, and an incubation temperature of 37°C are conditions, which lead to a considerable growth of bacteria. Action of these bacteria on photolyzed rod outer segments causes a rapid synthesis of isorhodopsin. This synthesis is accelerated by the presence of exogenous all-trans retinaldehyde.

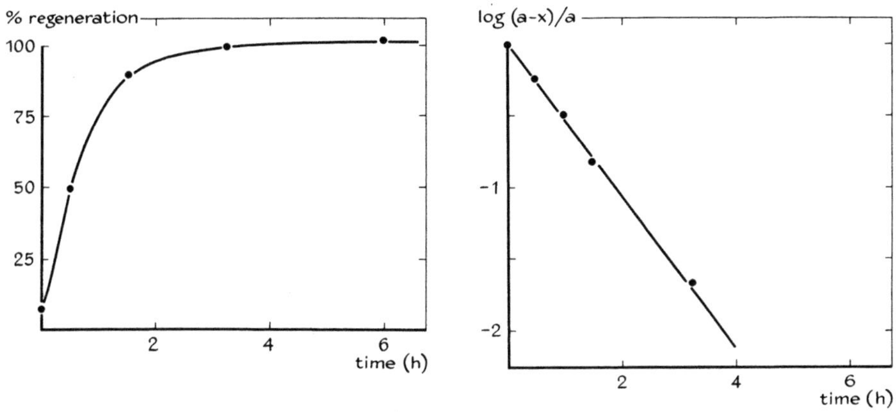

Fig. 5. Kinetics of isorhodopsin synthesis by bacterial action. 40 nmol photolyzed rhodopsin in 1 ml 0.067 M phosphate buffer (pH 6.5) was mixed with 1 ml bacterial suspension in the same buffer and incubated at 37 °C in the dark; 400 μl samples were taken and mixed with 50 μl 1 M NH_2OH and 50 μl Triton X-100 (10%). In the left figure the amount of isorhodopsin, calculated from ΔA_{485}, is expressed as a percentage of the rhodopsin originally present. In the right figure it is shown that the isorhodopsin formation is first-order for opsin

---

Oxygen Sensitivity of Isorhodopsin Formation

Immediately after the incubation of Proteus mirabilis with photolyzed rhodopsin, the redox potential of the suspension rapidly becomes more negative, until a value of about -100 mV is reached. Isorhodopsin synthesis starts only at this low redox potential, whereupon an increase in redox potential is observed. This increase is probably due to the utilisation of reduced substances by the isorhodopsin generating system. The bacterial isomerization process could easily be inhibited by the addition of oxygen (Fig. 6). This addition of oxygen immediately causes the redox potential to become positive. The introduction of a nitrogen atmosphere above the incubation mixture leads to a restoration of the low redox potential and a resumption of isorhodopsin synthesis.

Isorhodopsin Formation Catalyzed by an Isolated Bacterial Product

Ultrasonic treatment of a concentrated bacterial suspension usually caused complete disappearance of the isomerizing activity. However, most of the activity was retained, when ultrasonication was performed in a nitrogen atmosphere in the presence of 0.05 M 2-mercapto-ethanol. Incubation of photolyzed rhodopsin at 37 °C with the supernatant obtained by centrifugation (125,000· g, 30 min) yielded isorhodopsin. A maximal regeneration of 80 per cent was reached in 1½ hr. A control experiment, in which the bacterial supernatant was omitted but 0.05 M 2-mercapto-ethanol was present, showed no regeneration, indicating that this thiol compound had no isomerizing effect. Nearly all proteins could be removed from the bacterial supernatant by freeze-thawing without affecting the isomerizing activity, suggesting that the active factor is not a protein.

The degradation of free all-trans retinaldehyde upon incubation with the whole bacteria is not observed upon incubation with the bacterial supernatant. No detectable formation of 9-cis retinaldehyde is obtained when free retinaldehyde is incubated

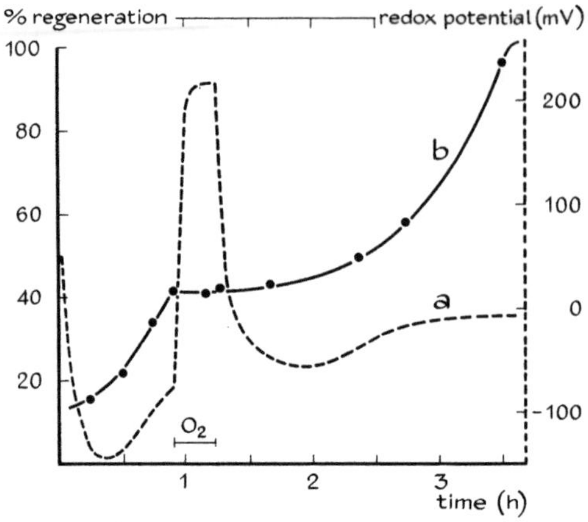

Fig. 6. Inhibition of isorhodopsin synthesis by oxygen. Photolyzed rhodopsin (25 nmol) in 1 ml 0.067 M phosphate buffer (pH 6.5) was incubated with 300 µ l bacterial suspension in the same buffer. The incubation was performed under a nitrogen atmosphere in the dark at 37°C. From the 55th to the 75th min oxygen was administered. The redox potential was continously recorded (curve a), while the regeneration percentage was determined periodically (curve b)

with bacterial supernatant, while addition of opsin immediately gives rapid formation of isorhodopsin.

Conclusions

Tentatively it is assumed that the bacteria form an isomerizing factor, which acts as a cofactor, while the rod membrane preparation acts as an enzyme, which isomerizes the substrate, all-trans retinaldehyde, and also binds the product, 9-cis retinaldehyde to form isorhodopsin.

Whether or not there is a relationship between dihydroriboflavin-catalyzed regeneration (11) and regeneration by bacterial action is not clear. We have not been able to demonstrate riboflavin nor its reduced form in the bacterial supernatant. Thus, the primary initiators of retinaldehyde isomerization in Futterman's and our experiments are probably different. This need not mean that also the actual mechanism is different. Both isomerizations are from the all-trans isomer to the 9-cis isomer and both occur in a strongly reducing medium.

For a further explanation of our findings, three observations are important. First, the synthesis of isorhodopsin occurs in darkness, excluding a light-activated isomerization of the chromophoric group. Secondly, the bacterial system only works in the presence of opsin, but not on free retinaldehyde alone, indicating a specific role of opsin in the isomerization process. Thirdly, the sensitivity of the pigment formation towards oxygen suggests that an easily oxidized substance produced by bacteria is able to change the opsin conformation in such a way that the adhering all-trans retinaldehyde is isomerized to 9-cis retinaldehyde, followed by formation of isorhodopsin.

References

1. AMES, S.R., W.J. SWANSON, P.L. HARRIS: Biochemical studies on vitamin A; XV. Biopotencies of geometric isomers of vitamin A aldehyde in the rat. J.Am.Chem.Soc. 77, 4136-4138 (1955).
2. HUBBARD, R., J.E. DOWLING: Formation and utilization of 11-cis vitamin A by the eye tissues during light and dark adaptation. Nature (Lond.) 193, 341-343 (1962).

3. KRINSKY, N.I.: The enzymatic esterification of vitamin A. J.Biol.Chem. 232, 881-894 (1958).
4. CONE, R.A., P.K. BROWN: Spontaneous regeneration of rhodopsin in the isolated rat retina. Nature (Lond.) 221, 818-820 (1969).
5. GOLDSTEIN, E.B.: Cone pigment regeneration in the isolated frog retina. Vision Res. 10, 1065-1068 (1970).
6. HUBBARD, R.: Retinene isomerase. J.Gen.Physiol. 39, 935-962 (1956).
7. LEWIS, D.M.: Regeneration of rhodopsin in the albino rat. J.Physiol. 136, 624-631 (1957).
8. RUSHTON, W.A.H.: Blue light and the regeneration of human rhodopsin in situ. J.Gen.Physiol. 41, 419-428 (1957).
9. ROTMANS, J.P., F.J.M. DAEMEN, S.L. BONTING: Retinaldehyde isomerization in rhodopsin regeneration. Fed.Proc. 30, 1197 (1971).
10. ROTMANS, J.P., F.J.M. DAEMEN, S.L. BONTING: Biochemical aspects of the visual process XIX. Formation of isorhodopsin from photolyzed rhodopsin by bacterial action. Biochem.Biophys.Acta 267, 583-587 (1972).
11. FUTTERMAN, S., M.H. ROLLINS: Role of dihydroflavin in regeneration of rhodopsin in vitro. Fed.Proc. 30, 1197 (1971).

[illegible]

Behavior of Rhodopsin and Metarhodopsin in Isolated Rhabdoms of Crabs and Lobster

Timothy H. Goldsmith and Merle S. Bruno*
Department of Biology, Yale University, New Haven, CT 06520/USA
Marine Biological Laboratory, Woods Hole, MA 02543/USA

The photoreceptor organelles - the rhabdoms - of decapod crustacea are elongate structures consisting of interleaved layers of microvilli from seven or eight retinular cells (1, 2, 3). When the eyes are broken open in an appropriate saline, the rhabdoms detach from the surrounding retinular cells, and individual rhabdoms can be identified under the microscope and studied spectrophotometrically with laterally-incident microbeams. All of the spectrophotometric measurements of rhabdoms described in this communication were performed on an instrument similar to that of Liebman and Entine (4). We have examined freshly-isolated rhabdoms as well as organelles fixed in 5% glutaraldehyde. The original reason for introducing glutaraldehyde fixation was to stabilize the structure of the isolated rhabdoms for periods commensurate with the measurements of pigment we wish to make (5). Although glutaraldehyde does not alter the absorption spectrum of either rhodopsin or metarhodopsin, it can hasten the destruction of both pigments, and therefore it is a useful tool in analyzing the pigment composition of rhabdoms by means of difference spectra.

A. The Lobster, Homarus americanus

Fresh Rhabdoms

The lobster of the western Atlantic coast contains a single visual pigment with λ_{max} at 515 nm. On exposure to light, this pigment is converted to a stable metarhodopsin with λ_{max} at 490 nm. Fig. 1 shows this conversion in a fresh rhabdom at room temperature. Control experiments (not shown) demonstrate that the act of measurement does not, in itself, produce any significant spectral change. Spectrum 2 was recorded after an exposure to a bright orange light sufficient to convert all of the rhodopsin to a thermally stable photoproduct. The same results are obtained at alkaline pH or in the presence of hydroxylamine.

The photoproduct is a metarhodopsin in the sense that it is apparently a single molecular species resulting from photoisomerization of the chromophore of the rhodopsin. Following spectrum 2, the rhabdom was next irradiated with blue light at 440 nm, and spectrum 3 was measured. Clearly blue light partially reverses the spectral changes brought about by orange light, with an isosbestic point at about 515 nm. This photosteady state will be discussed further below.

Bleaching in the Presence of Glutaraldehyde

In the presence of 5% glutaraldehyde at room temperature, absorption fades slowly in the dark with a half-time of 2-3 hours. Light causes a more rapid decay, for

*Present address: School of Natural Science and Mathematics, Hampshire College Amherst, Massachusetts 01002, U.S.A.

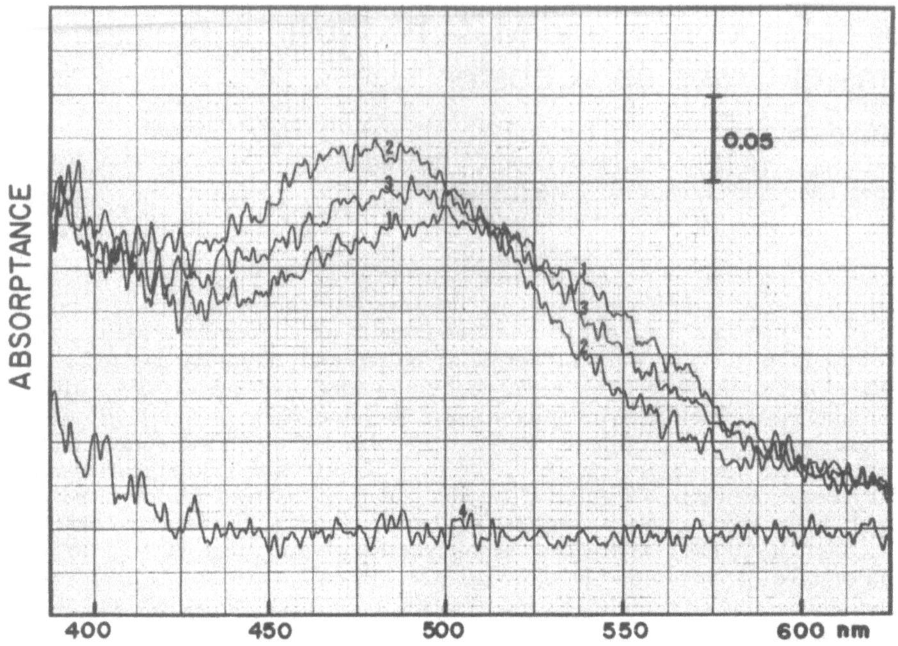

Fig. 1. Photoconversion of lobster rhodopsin and metarhodopsin in a freshly-isolated rhabdom suspended in sea water. Spectrum 1, original scan; 2 after a 30 sec irradiation with orange light (λ's > 550 nm). The shift in absorption to shorter wavelengths indicates the conversion of rhodopsin (λ_{max} 515 nm) to metarhodopsin (λ_{max} 490 nm). Spectrum 3 was recorded after 30 sec exposure to a bright blue light (440 nm interference filter). Note the isosbestic point at about 515 nm. Transverse irradiation with a 4 x 10 µm measuring beam, scanning rate 20 nm sec^{-1}, spectral band width 3.8 nm, 22 °C. Absorptance is fraction absorbed; 0.05 corresponds to an absorbance (optical density) of 0.022

---

glutaraldehyde increases both the thermal lability and the rate of photobleaching of metarhodopsin. This phenomenon has been observed in crayfish (5) and has been more extensively described for the crab Libinia (6).

Fig. 2 shows an experiment on a rhabdom fixed for 15 minutes in 5% glutaraldehyde and then resuspended in 1% glutaraldehyde-sea water. Under these conditions the metarhodopsin is stable enough to measure at room temperature and can be photobleached as well. Spectrum 1 is the original scan; 2 followed 20 sec of orange light; 3 was after 3 minutes of bright yellow light; and 4, after 6 more min in yellow light. Under these rather special conditions we have observed the accumulation of a photoproduct in the near UV, with absorption appropriate for retinaldehyde. Under most conditions that bleach the metarhodopsin, retinal seems to leave the rhabdom as fast as it is liberated from the opsin. The usual lack of near UV absorption on bleaching cannot be attributed to a change in molecular orientation within the microvillar membranes, for in crayfish we have studied bleaching with variably polarized measuring beams incident both parallel and at right angles to the microvillar axes. The relatively low amount of absorption of retinaldehyde relative to rhodopsin and metarhodopsin evident in Fig. 2 therefore should not be taken as evidence that the molar extinction coefficient of lobster rhodopsin is greatly in excess of 40,000; more likely some retinal has diffused out of the microvilli.

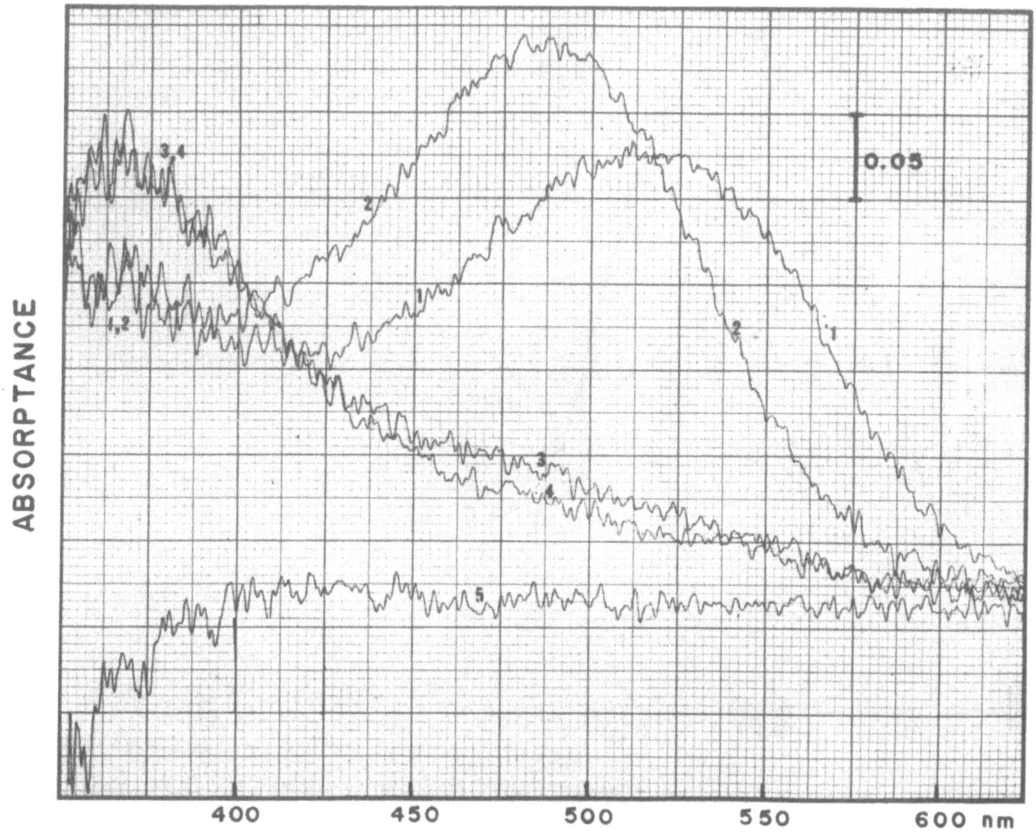

Fig. 2. Bleaching of a lobster rhabdom. 1, rhodopsin; 2, metarhodopsin; 3, retinal. See the text for experimental details and see the caption to Fig. 1 for recording conditions

Difference Spectra

Difference spectra for complete bleaching of lobster rhabdoms indicate that the λ_{max} of the rhodopsin lies at 515nm. Fig. 3 shows the close agreement between our average results on 22 rhabdoms (filled circles ± S.E., heavy solid curve); the results of Wald and Hubbard (7) on a digitonin extract of lobster rhodopsin (dashed curve) and the appropriate Dartnall (8) nomogram pigment (open circles).

Difference spectra for lobster metarhodopsin have λ_{max} at 490 nm. The ratio of the molar extinction coefficients of metarhodopsin and rhodopsin is about 1.2 both measured at the λ_{max}. In both these respects the present results are in close agreement with the in vitro data of Wald and Hubbard (7), a point that takes on interest in considering the visual pigment of the crab Callinectes (see below).

Photoreactions in Glutaraldehyde-Fixed and Washed Rhabdoms

If rhabdoms that have been suspended in 5% glutaraldehyde for 15 minutes and then washed in saline are exposed to bright lights, the metarhodopsin is not easily bleach-

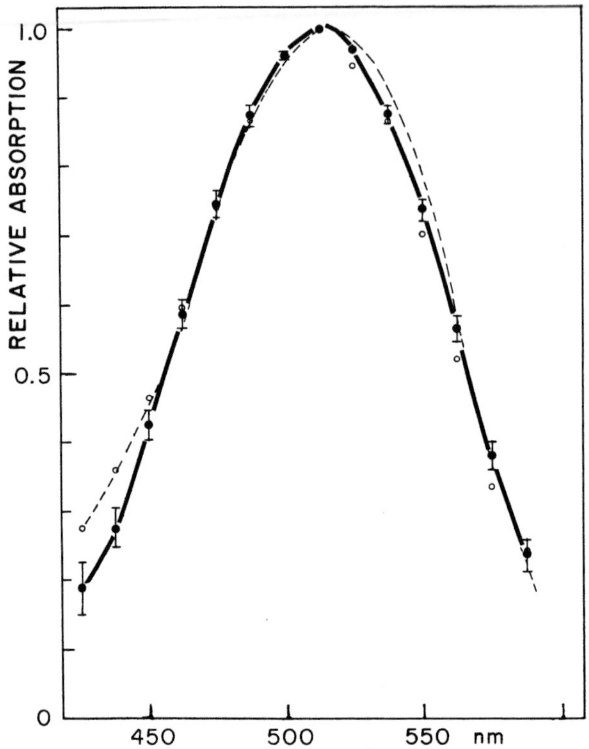

Fig. 3. Average bleaching difference spectrum for 22 lobster rhabdoms (filled circles ± S.E. and heavy curve) compared with lobster rhodopsin in digitonin extracts (broken curve, from Wald and Hubbard, 1957) and a hypothetical 515 nm visual pigment from Dartnall's (1953) nomogram (open circles)

ed. In this respect they are like unfixed organelles, but they can be studied for longer times without evidence of structural deterioration.

Even in such fixed material, photoregeneration of rhodopsin from metarhodopsin can be effected by blue light, but the photosteady state seems to favor metarhodopsin. From the relative molar absorption coefficients at 440 nm and the compositions of the pigment mixtures in the photosteady state it is possible to estimate the ratio of quantum efficiencies for the rhodopsin → meta : meta → rhodopsin conversions. This ratio is about 1.3 in fresh rhabdoms and about 3.8 in fixed material.

Behavior of Rhodopsin in vivo

We have attempted to alter the rhodopsin : metarhodopsin ratio in the living lobster by light adaptation. The assay involves rapid removal of the eye and preparation of fixed-and-washed rhabdoms at 1 °C. Spectra are recorded both before and after exposure to a bright orange light. The absorbance change at 552 nm is a measure of the amount of rhodopsin converted by the orange assay light, and the absorption at the isosbestic point is an estimate of the total amount of pigment present in the measuring beam. The ratio of these quantities is therefore an index of the amount of rhodopsin originally present in the rhabdom, normalized for path length (rhabdom thickness) and other variables. The ratio is in excess of 0.4 when all the pigment is originally present as rhodopsin, is zero when only metarhodopsin is present, and varies linearly between these values with the rhodopsin : metarhodopsin ratio.

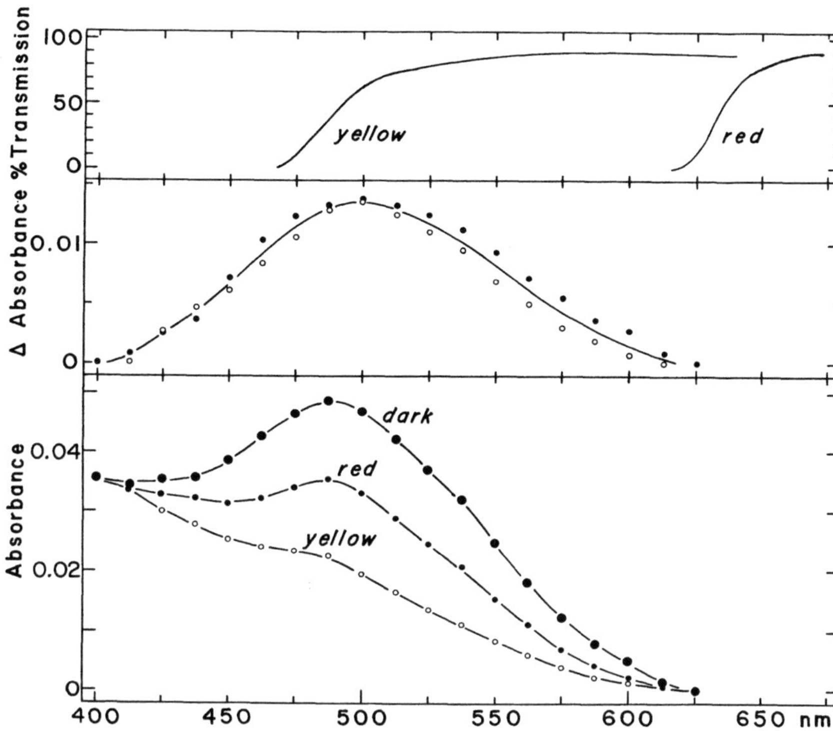

Fig. 4. Fractional bleach of a green crab (Carcinus) rhabdom in 5% glutaraldehyde. Transmission spectra of the filters used for bleaching are shown in the upper panel. Spectra of the rhabdom, obtained from records similar to Fig. 2 by subtracting the instrumental base line, are shown in the lower frame. "Absorbance" includes some light scattering by the optically inhomogeneous rhabdom. "Dark" is the initial spectrum; "red" after 14 min red light; "yellow" after 14 min yellow light. The middle panel shows the difference spectra for the two increments of bleaching. These spectra represent changes in molecular absorption and are undistorted by light scattering. There appears to be but a single visual pigment present

Lobsters were secured in a tank of sea water at room temperature with one eye stalk extended about 8 cm from a heat-filtered 6V microscope lamp and light-adapted for one hour. In some experiments the contralateral eye was kept covered and used as an internal control. The results were surprising, for with neither white light nor red light was it possible to demonstrate any accumulation of metarhodopsin. It therefore appeared that either the screening pigments effectively protected the rhabdoms from the adapting light, or in vivo the lobster possesses an efficient mechanism for regenerating rhodopsin, a mechanism that can keep pace with the light flux delivered by our adapting source.

The latter hypothesis is supported by experiments in which the lobsters were chilled to 0-5 °C during the time of light adaptation. Low temperature does not cause metarhodopsin to form in the dark, but in the presence of the adapting light as much as 85% of the pigment of the rhabdom is converted to metarhodopsin. These experiments mean that the accessory screening pigments do not block out the adapting light; and moreover, that the eye has a temperature-dependent process for regenerating rhodopsin.

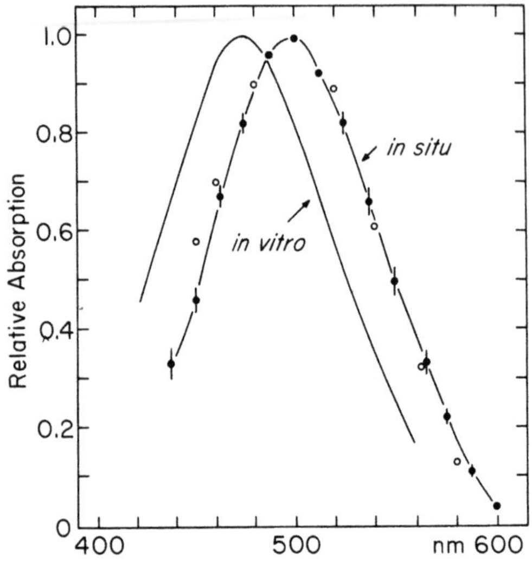

Fig. 5. Callinectes (blue crab) rhodopsin. The in vitro curve is an average difference spectrum for photobleaching in a digitonin extract in the presence of hydroxylamine or at pH 10. The λ_{max} is at about 477 nm. The in situ curve and filled circles (± S.E.) are average results for total bleaching of 18 isolated rhabdoms (cf. Fig. 4). Open circles are a 500 nm rhodopsin from Dartnall's (1953) nomogram. The in situ data provide a good fit to the spectral sensitivity function of Callinectes. See the text for a discussion of the 23 nm discrepancy between measurement in vitro and in situ

B. Crabs

We have examined three species of crab, Libinia, Carcinus, and Callinectes, and the results are sufficiently alike that they can be discussed together. Each seems to contain a single visual pigment with λ_{max} in bleaching difference spectra as follows:

Libinia (spider crab)	493 nm
Callinectes (blue crab)	500 nm
Carcinus (green crab)	504 nm

The metarhodopsins are stable in fresh rhabdoms, and have nearly identical absorption spectra to the parent visual pigments. Consequently, actinic irradiation is accompanied by negligible changes in spectral absorption. The formation of metarhodopsin can be inferred, however, from its greater lability in the presence of glutaraldehyde (6).

Efforts to bleach the rhabdoms in stages, starting with long wavelength irradiation, have provided no evidence for more than one visual pigment in each species. Such an experiment is shown in Fig. 4 for Carcinus. Moreover, in each case the absorption spectra of these pigments account adequately for the spectral sensitivity functions of the crabs (cf. 6, 9).

Digitonin extracts of Callinectes have been made, following floatation of macerated eyes on 45% sucrose following procedures that have been successful both with vertebrate retinas and with lobster (7). The difference spectra measured by photobleaching in the presence of hydroxylamine show the major loss of absorption at 480 ± 3 nm and the formation of a photoproduct at about 365 nm; appropriate for retinal oxime (Bruno and Goldsmith, unpublished observations; 11). The discrepancy of about 20 nm between absorption measured in the rhabdom and in digitonin extracts (Fig. 5) is, we believe, not due to distortion of difference spectra by photoproducts, but reflects a hypsochromic shift in absorption in vitro caused by removal of the visual pigment from the environment of the lipoprotein membranes. That such a displacement can occur does not seem remarkable; what is interesting is that it does not occur with most visual pigments (12).

Acknowledgement. This research was supported by USPHS Grant EY 00222.

References

1. EGUCHI, E.: Rhabdom structure and receptor potentials in single crayfish retinular cells. J.Cell.Comp.Physiol. 66, 411-430 (1965).
2. EGUCHI, E., T.H. WATERMAN: Fine structure patterns in crustacean rhabdoms. In: The Functional Organization of the Compound Eye, C.G. Bernhard, ed., Pergamon Press, Oxford-New York, 1966.
3. RUTHERFORD, D.J., G.A. HORRIDGE: The rhabdom of the lobster eye. Quart.J.Micr.Sci. 106, 119-130 (1965).
4. LIEBMAN, P.A., G. ENTINE: Sensitive low-light-level microspectrophotometer: detection of photosensitive pigments of retinal cones. J.Opt.Soc. Amer. 54, 1451-1459 (1964).
5. WATERMAN, T.H., H.R. FERNÁNDEZ, T.H. GOLDSMITH: Dichroism of photosensitive pigments in rhabdoms of the crayfish Orconectes. J.Gen.Physiol. 54, 415-432 (1969).
6. HAYS, D., T.H. GOLDSMITH: Microspectrophotometry of the visual pigment of the spider crab Libinia emarginata. Z.vergl.Physiol. 65, 218-232 (1969).
7. WALD, G., R. HUBBARD: Visual pigment of a decapod crustacean: the lobster. Nature 180, 278-280 (1957).
8. DARTNALL, H.J.A.: The interpretation of spectral sensitivity curves. Brit.Med.Bull. 9, 24-30 (1953).
9. BRUNO, M.S., M.I. MOTE, T.H. GOLDSMITH: Microspectrophotometry of visual pigment and spectral sensitivity of retinular cells in the crab Carcinus. Biol.Bull. 139, 416-417 (1970).
10. BRUNO, M.S., M.I. MOTE, T.H. GOLDSMITH: Spectral absorption and sensitivity measurements in single ommatidia of the green crab, Carcinus. J.Comp.Physiol. (in press).
11. FERNÁNDEZ, H.R.: A survey of the visual pigments of decapod crustacea of South Florida. Ph.D. Thesis, University of Miami, Coral Gables, Florida, 1965.
12. BOWMAKER, J.K.: Kundt's rule: the spectral absorbance of visual pigments in situ and in solution. Vis.Res. 12, 529-548 (1972).

Discussion

T.P. Williams: Have you warmed your light-adapted lobster and then checked to see if the rhodopsin comes back?

T.H. Goldsmith: We are currently attempting to measure the rate of regeneration.

[illegible]

References

1. [illegible] and receptive [illegible] (1968).
2. [illegible] structure [illegible] Neurological Organization of the [illegible]
3. [illegible] (1968).
4. [illegible] observations [illegible]
5. [illegible]
6. [illegible]
7. [illegible] spectral sensitivity [illegible] (1970).
8. [illegible] single [illegible]
9. [illegible] Thesis, University of [illegible]
10. [illegible]

Discussion

[illegible]

Photoregeneration and Sensitivity Control of Photoreceptors of Invertebrates

K. Hamdorf, R. Paulsen, and J. Schwemer
Institut für Tierphysiologie, Ruhr-Universität Bochum, Bochum-Querenburg/W. Germany

It is a well known fact that the metarhodopsins of cephalopods are thermostable and photoreconvertible into rhodopsin (Fig. 1). In the octopus species Eledone moschata the wavelength maxima of the rhodopsin (R) and acid metarhodopsin (M) are separated by about 50 nm. Due to the large distance between the wavelength maxima, this R-M-system is well suited for photokinetic studies (1-3). Monochromatic illumination causes a photoequilibrium between R (11-cis R, isoR) and its acid M. The equilibrium is determined solely by the ratio betweeen the absorption coefficients of the two pigments. Illumination at the isosbestic point yields 50% R and 50% M. Minimal R-concentration (38%) is obtained at 440 nm. At longer wavelength the M is quantitatively reconverted to R.

Thermostability and photoreconversion have been demonstrated for several photoreceptors of the rhabdomeric type. The present paper concerns the photoreceptors of three insects. The frontal eye of the neuropter Ascalaphus macaronius is sensitive to ultraviolet light only (Fig. 2) (4). Absorption of an UV quantum by the extracted pigment causes the formation of a photoproduct absorbing at long wavelengths (5-8). This pigment has a high molar absorbance. Long wavelength illumination

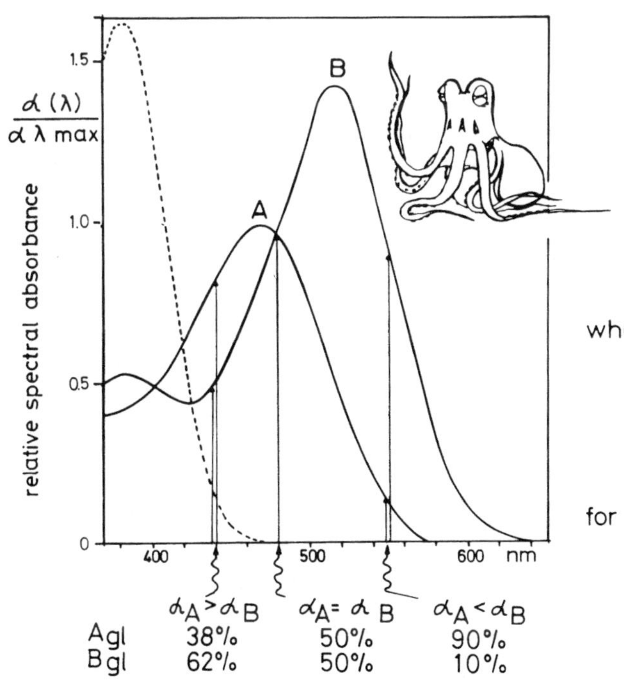

$$A \underset{k_2}{\overset{k_1}{\rightleftharpoons}} B$$

$$A_{gl} = A_0 \frac{k_2}{k_1 + k_2}, \quad B_{gl} = A_0 \frac{k_1}{k_1 + k_2},$$

where

$$k_1 = \alpha_A(\lambda) \cdot \gamma_A(\lambda) \cdot J,$$
$$k_2 = \alpha_B(\lambda) \cdot \gamma_B(\lambda) \cdot J;$$

$$\frac{A_{gl}}{B_{gl}} = \frac{k_2}{k_1} = \frac{\alpha_B(\lambda) \cdot \gamma_B(\lambda) \cdot J}{\alpha_A(\lambda) \cdot \gamma_A(\lambda) \cdot J}$$

for the relative quantum efficiency:

$$\frac{\gamma_B(\lambda)}{\gamma_A(\lambda)} = \frac{A_{gl} \cdot \alpha_A(\lambda)}{B_{gl} \cdot \alpha_B(\lambda)}$$

Fig. 1. Photoequilibrium in rhodopsin system of Eledone. The ratio between rhodopsin (A) and acid metarhodopsin (B) in photoequilibrium is given by equation A_{gl}/B_{gl} (right half)

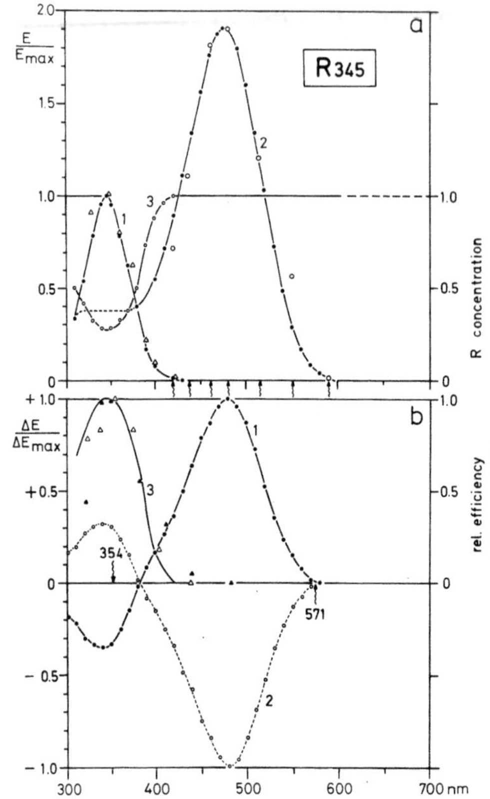

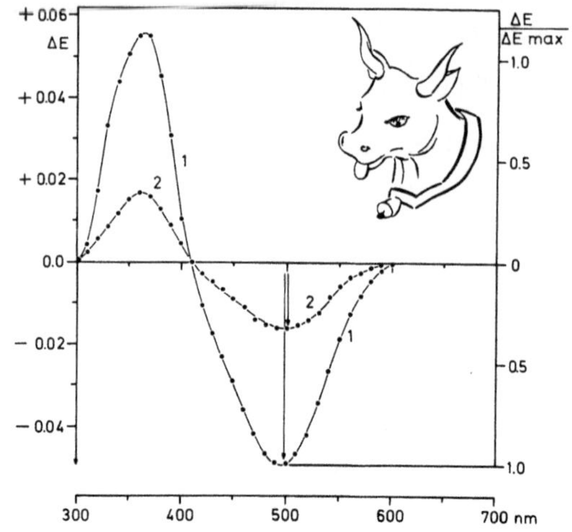

Fig. 3. Cattle rhodopsin synthesized from chromophore of UV pigment in Ascalaphus. Difference spectra show bleaching of synthesized cattle rhodopsin in presence of NH_2OH. Cattle rhodopsin formed from denatured R 345 (1). - Cattle rhodopsin formed from UV (354 nm) irradiated R 345 (2). In the second experiment only one third of the previously formed cattle rhodopsin is formed

Fig. 2. Visual pigment system of Ascalaphus.
a) Spectral absorbance of R 345 (1) and its acid M 475 (2), and concentration of R 345 in photoequilibrium in relation to wavelength of illumination (3). - Open triangles: Relative spectral sensitivity of the frontal eye. Open circles: Spectral efficiency for photoreconversion based on electrophysiological data like that in Fig. 6.
b) Change in absorbance in R 345 extract after UV (1) and after subsequent blue irradiation (2). Filled triangles: Experimental data from extract. Open triangles: Experimental data from the retina

completely reconverts this thermostable pigment into the UV visual pigment. In contrast, UV light leads to a photochemical equilibrium between the UV pigment and its photoproduct. Like in Eledone, the equilibrium is directly proportional to the ratio between the absorption coefficients for the two pigments. The content of the UV pigment and its photoproduct after monochromatic illumination can be calculated from this ratio (Fig. 2a). There is good agreement between the calculated contents, and the contents found in measurements on pigment extract and in microspectrophotometric measurements of the pigments in situ (Fig. 2b).

The UV visual pigment is an 11-cis retinal proteid (9). The chromophore can be separated from the pigment molecule and combined with cattle opsin. Bleaching of this "synthesized" cattle rhodopsin in the presence of NH_2OH leads to the difference spectrum of high amplitude shown in Fig. 3. When the same experiment is carried out on a UV irradiated pigment, only one third of the amount of cattle rhodopsin is formed. This value corresponds to the content of 11-cis retinal which is to be expected at the photochemical equilibrium between R 345 and its thermostable M 475.

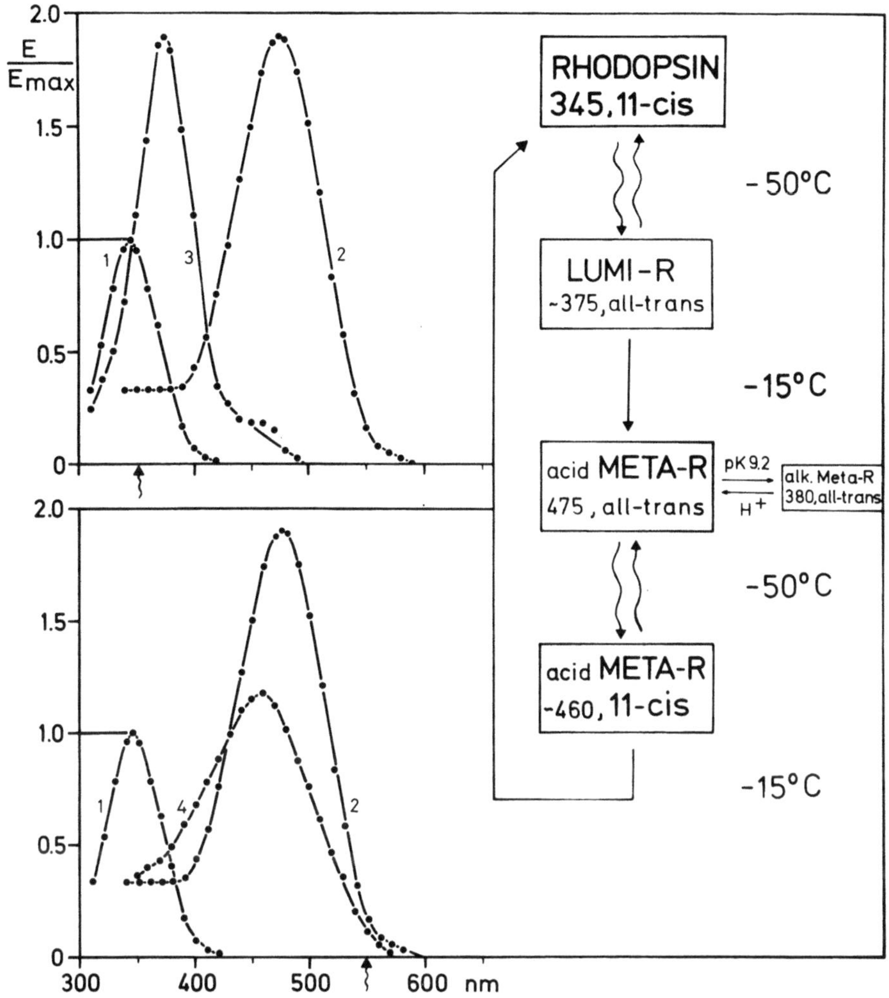

Fig. 4. Light and dark reactions of the R 345-system of Ascalaphus.
Left: Spectral absorbance of rhodopsin 345 (1), acid metarhodopsin 475 (2), lumirhodopsin (3), and 11-cis metarhodopsin (4). Spectra of intermediates (3 and 4) are calculated from difference spectra.
Right: Scheme of photoreactions (wavy lines) and dark reactions (straight lines). The alkaline metarhodopsin does not exist under physiological pH conditions, its spectral absorbance is not shown

---

In UV light at -50°C, R 345 converts into an all-trans pigment of high absorbance in the UV region (Fig. 4). This intermediate, probably lumirhodopsin, when warmed converts into acid M 475. Long wavelength illumination of acid M 475 at -50°C leads to another intermediate. The protein configuration of this compound should be similar to that of the all-trans M 475. However, the chromophore is 11-cis retinal, because upon warming in darkness the intermediate converts into R 345. The intermediate is therefore designated as acid 11-cis metarhodopsin. The same intermediate occurs in Eledone during photoreconversion. (In this species also the alkaline all-trans M converts into a corresponding alkaline 11-cis M).

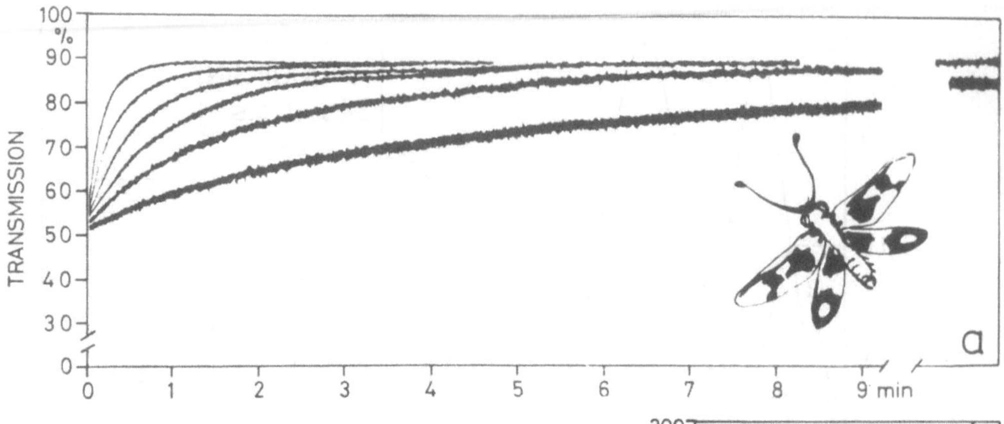

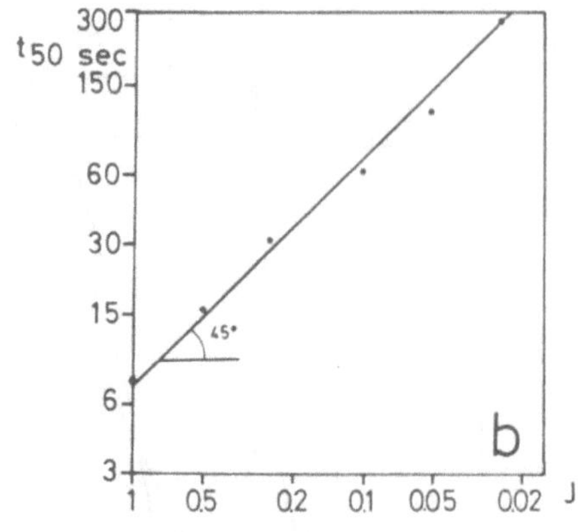

Fig. 5. Photoreconversion of acid M475 in retina of Ascalaphus by blue light of various intensities.
a) Time course of transmission change during illumination by 475 nm.
b) Demonstration of direct proportionality between the rate of photoreconversion and intensity of regenerating light

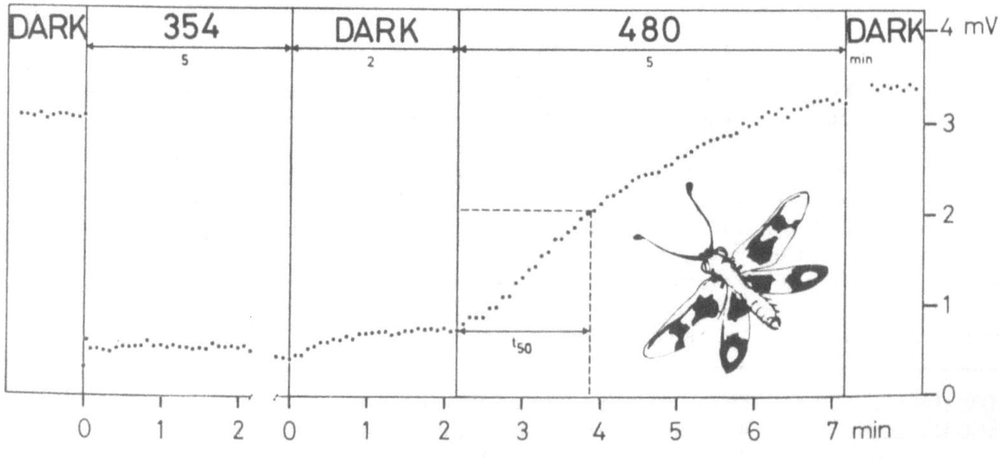

Fig. 6. Regeneration of retina sensitivity by blue light (Ascalaphus): Amplitude of extracellularly recorded receptor potential (dots) to low intensity UV test-flash in darkness, during bright UV adaptation (354 nm), and during illumination with blue light (480 nm). The rate of sensitivity increase (t_{50}) depends on wavelength, using light of equal quanta

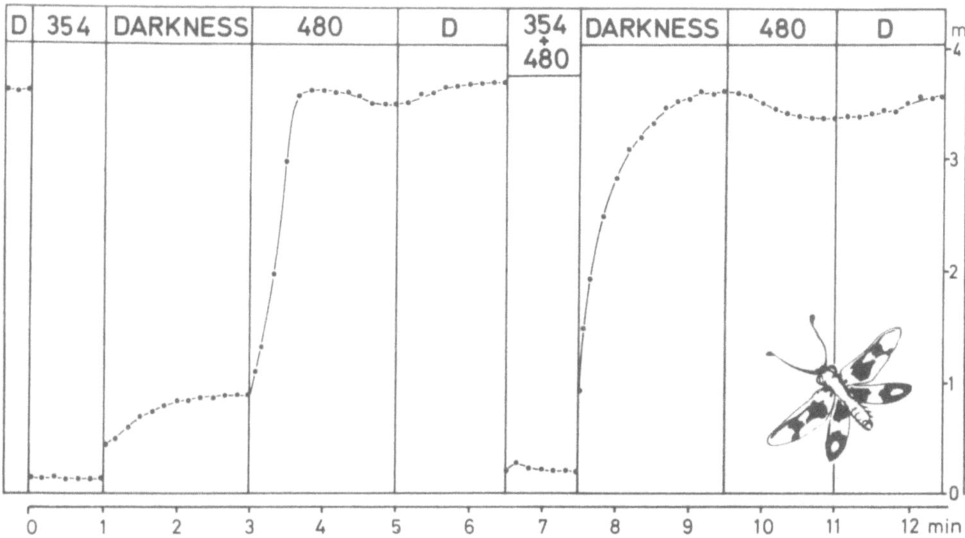

Fig. 7. Regeneration of retina sensitivity by blue light (Ascalaphus): Amplitude of extracellularly recorded receptor potential (dots) to low intensity UV test-flash in darkness, during bright UV adaptation (354 nm), during illumination by blue light (480 nm), and by a combination of UV light (354 nm) and blue light (480 nm). - Right ordinate: Relative sensitivity at each response amplitude

---

The rate of photoregeneration from acid M to R 345 depends on the intensity of illumination, as shown by microphotometric measurements on the retina (Fig. 5). The rate of transmission change at 475 nm was determined after saturating UV illumination. The rate increases with the intensity of the regenerating light. Fig. 5b shows a direct proportionality between the intensity of the regenerating light and the rate of photoreconversion.

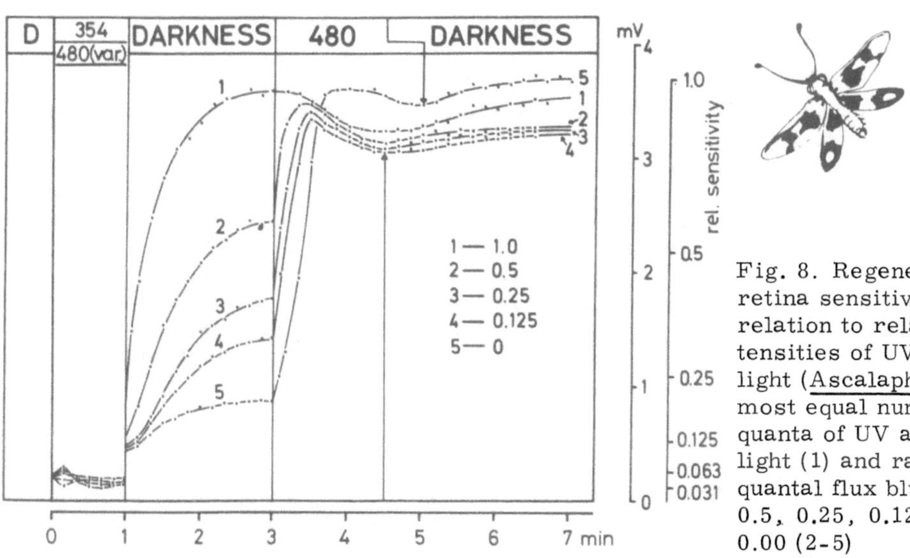

Fig. 8. Regeneration of retina sensitivity in relation to relative intensities of UV and blue light (Ascalaphus): Almost equal number of quanta of UV and blue light (1) and ratios of quantal flux blue/UV: 0.5, 0.25, 0.125, and 0.00 (2-5)

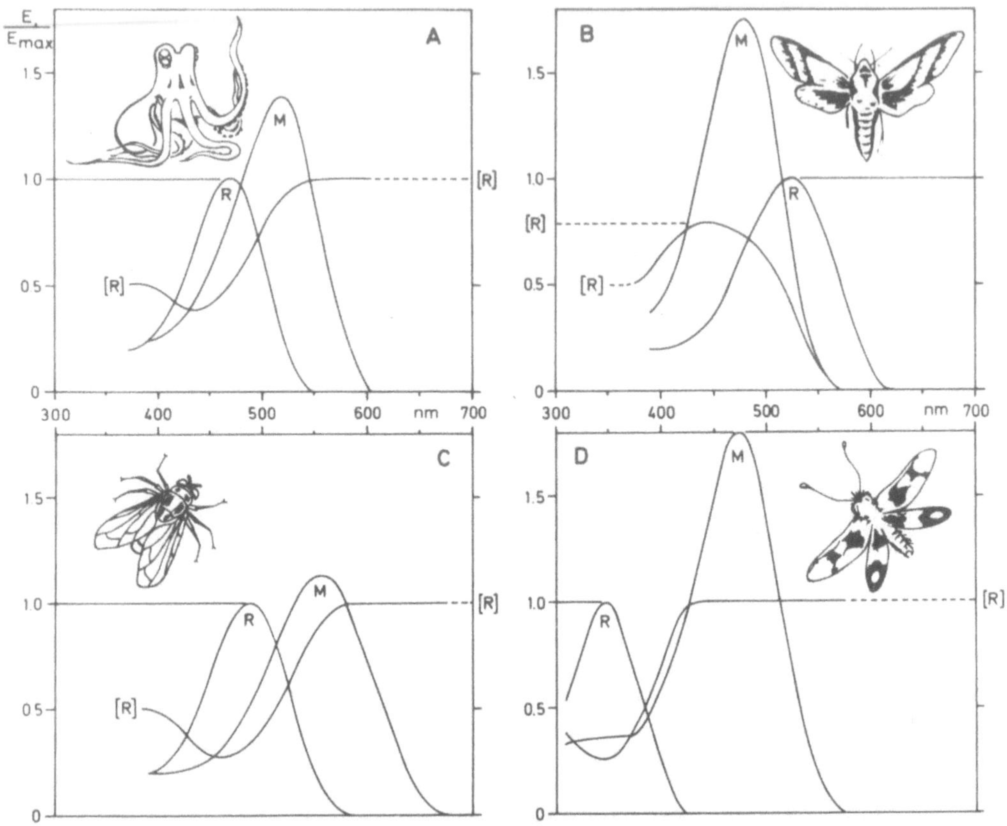

Fig. 9. Photoconvertible visual pigment systems of four invertebrates: A) Eledone moschata, B) Deilephila elpenor, C) Calliphora erythrocephala, D) Ascalaphus macaronius - R: Rhodopsin; M: Acid metarhodopsin. R : Content of rhodopsin in receptor in relation to wavelength of monochromatic adapting light, calculated for relative quantum efficiency of one

---

The proportionality between light intensity and rate of regeneration can also be demonstrated electrophysiologically (8). Exposure of the eye in Ascalaphus to UV light causes a considerable decrease in the response amplitude (Fig. 6). The amplitude increases only slightly after cessation of the adapting light. During exposure to blue light, the amplitude increases to the original value. The rate of this sensitivity increase is directly proportional to the intensity of the blue light. - The rate of the sensitivity increase varies with the wavelength of the regenerating light (8). Using irradiation of equal quanta in the wavelength range 400 nm to 589 nm, the regeneration rate was found to be highest at 475 nm (Fig. 2a, open circles). The spectral efficiency for the electrophysiologically recorded sensitivity increase is directly proportional to the absorbance spectrum of the acid M 475.

Information about the sensitivity regulation for Ascalaphus in its natural environment was obtained by simultaneously exposing the eye to about an equal number of quanta of UV and blue light (Fig. 7). The maximal sensitivity was recorded immediately after cessation of the adapting light. The sensitivity increase is determined by the intensity of the blue light relative to that of the UV light (Fig. 8) (10).

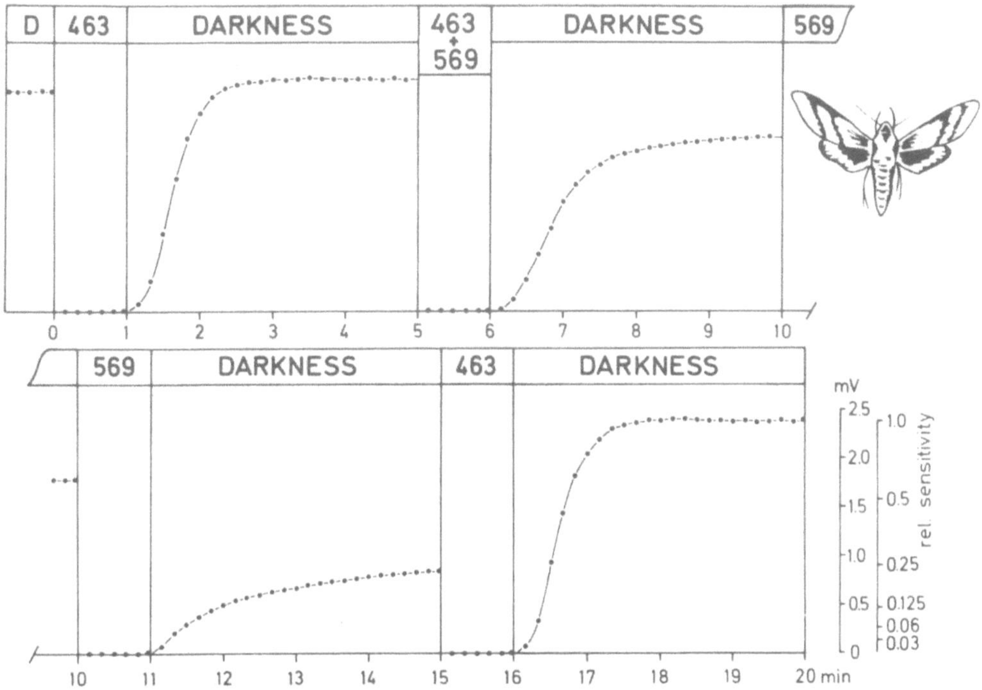

Fig. 10. Effect of blue (463 nm) and yellow (569 nm) adapting light, and combination of blue and yellow light on retina sensitivity of Deilephila. After cessation of blue adapting light the electrical response of the retina to a short stimulus (40 msec, 608 nm) increases within 2 min to a maximal value. In contrast, after yellow adapting light the sensitivity increases only slightly. After the combination of blue and yellow the response increases nearly to the maximal value

---

The spectral content of the light from the sky has a maximum in the blue region, and another in the UV (8). The high content of blue light, and the high absorption of the acid M 475 in this region, causes almost all M to be reconverted into R 345. Calculations show that the R content is about 90%. The sensitivity to UV light is therefore almost nearly maximal.

Rhodopsins absorbing maximally in the green region of the spectrum occur in the moth Deilephila elpenor (11,12) and the fly Calliphora erythrocephala (13). In Deilephila the M absorbs at wavelengths shorter than the absorption maximum of R (11,12) while in Calliphora the maxima are reversed (Fig. 9C). Both metarhodopsins are reconverted by light. The variation in R concentration with wavelength is shown in Fig. 9. It is to be expected that in Deilephila blue light reconverts the M into R, and that long wavelengths completely convert the R into its M. In the fly, on the other hand, long wavelengths should regenerate, and blue light should cause a maximal conversion into M. Experimental results confirm this hypothesis: In Deilephila after cessation of a blue adapting light (Fig. 10), the sensitivity increases to the original value in less than two minutes. In contrast, the sensitivity increases much less after adapting to long wavelengths. After subsequent exposure to blue light the sensitivity increases again to its maximal value. After a combination of blue and long wavelength adapting light, the sensitivity is almost equal to that after blue light alone, in spite of the higher photon flux during the light adaptation (14).

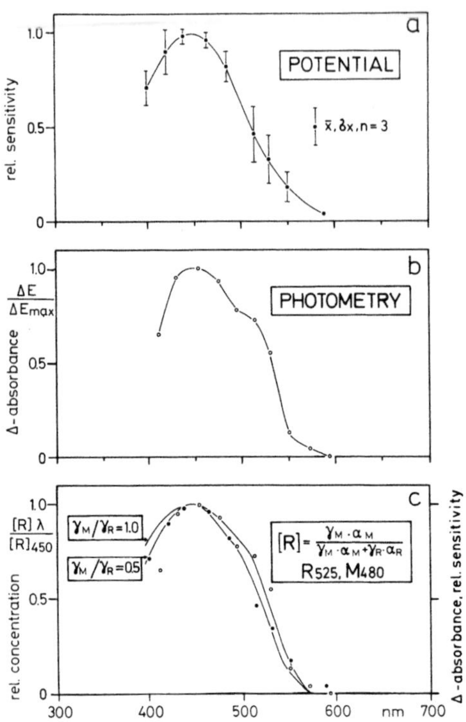

Fig. 11. Photoequilibrium and sensitivity in Deilephila.
a) Relative sensitivity of retina 2 min after monochromatic light adaptation (equal quantal flux). Abscissa: Wavelength of irradiation.
b) Change in absorbance during monochromatic light adaptation obtained from measurements on isolated retina. Values are proportional to R concentration.
c) Comparison between theoretical function of R content in photoequilibrium (calculated for relative quantum efficiencies, $\gamma M/\gamma R$, 1.0 and 0.5), and empirical data from a) and b)

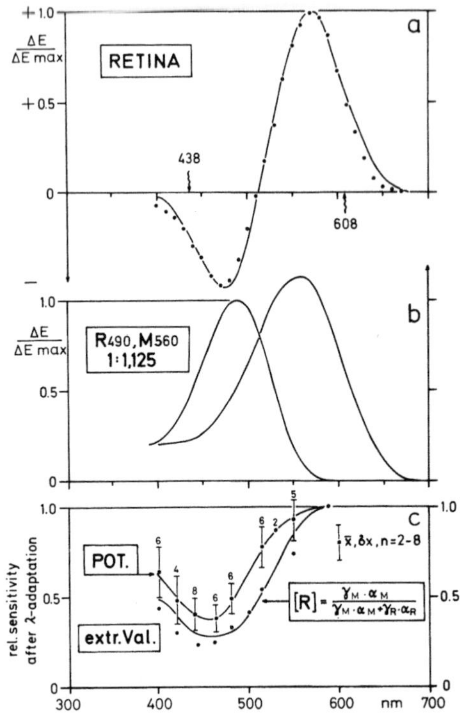

Fig. 12. Photoequilibrium and sensitivity in Calliphora.
a) Difference in absorbance caused by alternating illumination (608 nm and 438 nm) of retina.
b) Calculated spectra of R 490 and M 560 (from a).
c) Comparison between calculated R content in photoequilibrium (relative quantum efficiency 1.0), and sensitivity of retina 2 min after monochromatic light adaptation. Dots with bars: mean values of the electrophysiological data. Dots without bars: extreme values of a single electrophysiological experiment showing good fit with calculations of R content

The sensitivity increase after light adaptation varies with the wavelength of the adapting light (Fig. 11 a). The increase is maximal at 450 nm. The spectral efficiency for the sensitivity increase is very similar to the photometrically measured reconversion of M into R. Maximal reconversion into R 525 is caused by light of 450 nm wavelength (Fig. 11 b). The R concentration at various wavelengths can be calculated from the absorbance spectra (11). Fig. 11 c shows the R concentration at a relative quantum efficiency (M/R) of 1.0 and 0.5. There is a good agreement with the data from the photometrical and electrophysiological measurements. Therefore, it can be calculated that during 2 min dark-adaptation the sensitivity depends only on the probability for quantum absorption by the visual pigment.

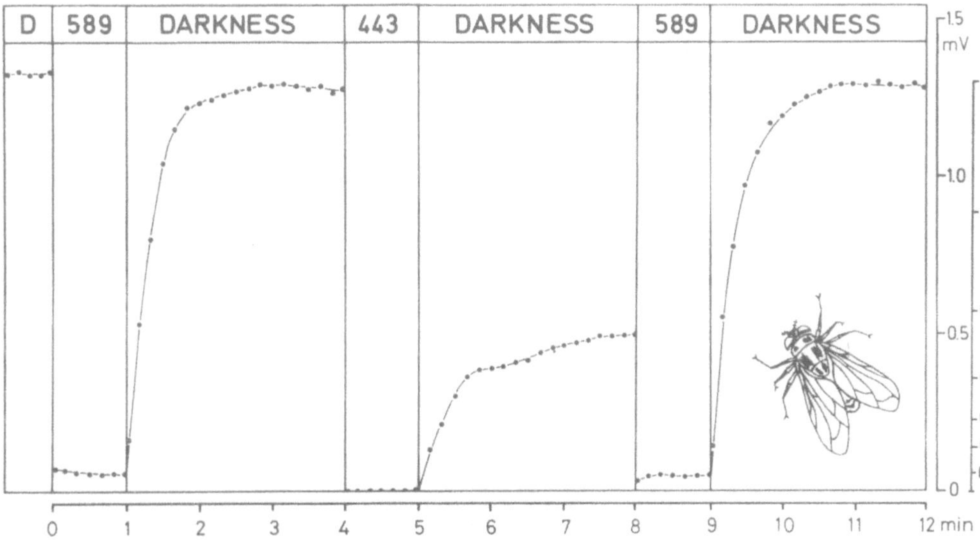

Fig. 13. Effect of blue (443 nm) and yellow (589 nm) adapting light on retina sensitivity of Calliphora. After cessation of yellow, the electrical response to a test flash increases to a maximal value within 2 min. After blue adaptation the sensitivity is strongly reduced. Subsequent yellow illumination causes maximal sensitivity

---

Exposure of the retina in Calliphora to light of short wavelength, followed by exposure to light of long wavelength, gives the difference spectrum shown in Fig. 12 a. Calliphora is maximally sensitive to light of wavelength 490 nm. The visual pigment therefore absorbs maximally at this wavelength. The wavelength maximum of the acid M, and its absorbance relative to that of R, can be calculated from the difference spectrum and the spectral sensitivity (Fig. 12 b). Assuming that the acid M is completely reisomerized by light, the variation in R concentration with wavelength can be calculated (Fig. 12 c). The regulation of the sensitivity by light was demonstrated similarly for Calliphora as for Ascalaphus and Deilephila. The eye was exposed to light of wavelength 443 nm (Fig. 13). After illumination, the sensitivity increased only slightly. Following subsequent exposure to long wavelength illumination the sensitivity rapidly increased by about 2 log units. The spectral efficiency of sensitivity increase is similar to the calculated spectral variation in the photoequilibrium between R 490 and M 560 (Fig. 12c) (15).

It can be concluded that in all three species the sensitivity after a short dark period depends only on the concentration of R in the receptor membrane. The absolute sensitivity is nearly directly proportional to the R concentration and is determined only by the probability for light absorption by rhodopsin. The concentration of the acid M thus does not influence the sensitivity. In all three receptors the light from the sky keeps the R concentration above 50%. Thereby an absolute sensitivity of 50% or more is attained after a very short period in darkness. It is questionable if there is chemical regeneration from M to R. Such a chemical regeneration seems unnecessary in diurnal insects because a chemical regeneration to 100% R only doubles the absolute sensitivity.

A chemical regeneration may be of importance to animals which, like Deilephila, are active at low light intensities. This reconversion can be very slow. Fig. 14 il-

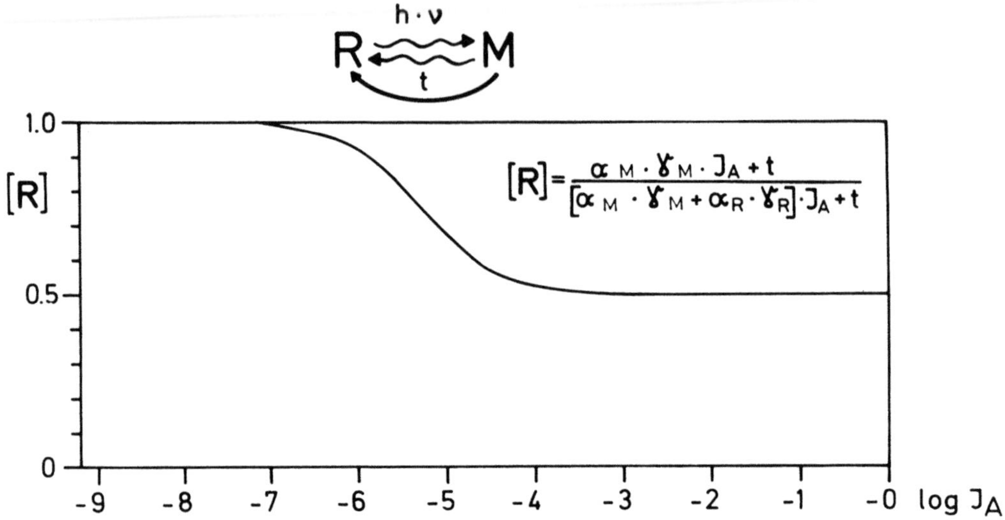

Fig. 14. R concentration in a photoreceptor with low rate of chemical regeneration in relation to intensity of adapting light. - The photoequilibrium becomes 0.5 R by illumination with monochromatic light at the isosbestic point

---

lustrates the variation in R concentration with the quantal flux of the adapting light. It is assumed that the probability for chemical regeneration of a M molecule is independent of the light intensity. The R concentration in the receptor is nearly 100% at light intensities that are so low that the quantal absorption by the pigment is less than the probability for chemical reconversion of the photoproducts. A photochemical equilibrium between R and M occurs at the intensity range at which the photochemical reconversion becomes almost equal to the chemical regeneration. The equilibrium concentrations remain constant when the adapting light is further increased.

Acknowledgement. We are grateful to Mrs. Ellen Dresel for drawing the graphs. - This work was supported by the Deutsche Forschungsgemeinschaft, Sonderforschungsbereich 114, "Bionach".

References

1. HAMDORF, K., J. SCHWEMER, U. TÄUBER: Der Sehfarbstoff, die Absorption der Rezeptoren und die spektrale Empfindlichkeit der Retina von Eledone moschata. Z. vergl. Physiol. 60, 375-415 (1968).
2. SCHWEMER, J.: Der Sehfarbstoff von Eledone moschata und seine Umsetzung in der lebenden Netzhaut. Z. vergl. Physiol. 62, 121-152 (1969).
3. HAMDORF, K.: Korrelation zwischen Sehfarbstoffgehalt und Empfindlichkeit bei Photorezeptoren. Verh. Dtsch. Zool. Ges., Köln 1970, 64, 148-158 (1970).
4. GOGALA, M.: Die spektrale Empfindlichkeit der Doppelaugen von Ascalaphus macaronius Scop. (Neuroptera, Ascalaphidae). Z. vergl. Physiol. 57, 232-243 (1967).
5. GOGALA, M., K. HAMDORF, J. SCHWEMER: UV-Sehfarbstoff bei Insekten. Z. vergl. Physiol. 70, 410-413 (1970).

6. HAMDORF, K., J. SCHWEMER, M. GOGALA: Insect visual pigment sensitive to ultraviolet light. Nature (Lond.) 231, 458-459 (1971).
7. SCHWEMER, J., M. GOGALA, K. HAMDORF: Der UV-Sehfarbstoff der Insekten: Photochemie in vitro und in vivo. Z. vergl. Physiol. 75, 174-188 (1971).
8. HAMDORF, K., M. GOGALA, J. SCHWEMER: Beschleunigung der "Dunkeladaptation" eines UV-Rezeptors durch sichtbare Strahlung. Z. vergl. Physiol. 75, 189-199 (1971).
9. PAULSEN, R., J. SCHWEMER: Studies on the insect visual pigment sensitive to ultraviolet light: Retinal as the chromophoric group. Biochim. Biophys. Acta 283, 520-529 (1972).
10. HAMDORF, K., M. GOGALA: (unpublished data).
11. HAMDORF, K., G. HÖGLUND, H. LANGER: Mikrophotometrische Untersuchungen an der Retinula des Nachtschmetterlings Deilephila elpenor. Verh. Dtsch. Zool. Ges., Helgoland 1971, 65, 275-280 (1972).
12. HÖGLUND, G., K. HAMDORF, H. LANGER, R. PAULSEN, J. SCHWEMER: The photopigments in an insect retina. This volume, pp. 167-174.
13. LANGER, H., B. THORELL: Microspectrophotometry of single rhabdomeres in the insect eye. Exp. Cell Res. 41, 673-677 (1966).
14. HAMDORF, K., G. HÖGLUND, G. ROSNER: (unpublished data).

Discussion

T.P. Williams: First, I should like to say that your spectrum of the 11-cis metarhodopsin is excellent. Much better than the one I obtained. Next, I would like to ask if you measured the rate at which it is converted to rhodopsin? Is this experiment possible with your system?

K. Hamdorf: Between 11-cis metarhodopsin and all-trans metarhodopsin there exists a photoequilibrium depending on the wavelength of irradiation. Our present measurements indicated that all the 11-cis metarhodopsin formed changes into rhodopsin. Kinetic data for this thermal transition have not yet been determined. Technically these measurements can be carried out with our equipment.

T. Ebrey: You stated that irradiating metarhodopsin at $-50^{\circ}C$ can, at least partially, produce a species, "11-cis" metarhodopsin, which can, upon warming, yield rhodopsin. Isn't it also possible that other isomers of retinal may be present on the illuminated metarhodopsin? Some of these could be stable (9-cis?) but others probably would not be stable and thus difficult to identify with your techniques.

K. Hamdorf: It may be possible that several retinal isomers are produced by illumination. Yet, the experimental results on the UV-rhodopsin extracted from Ascalaphus show that in this case only the 11-cis configuration is formed in provable quantities, not only at $-50^{\circ}C$, but also in the physiological temperature range: You can see this best in the experiment described in Fig. 3: If other isomers were produced in considerable amounts at physiological temperature, a shift of λ_{max} of the synthesized cattle rhodopsin were to be expected. Furthermore, we might have been able to detect these other isomers by our TLC-experiments (in any case the 9-cis isomer).

G. Wald: In an eye that depends upon photoregeneration, would it not be an advantage for the visual pigment and its photoproduct to overlap maximally - as they nearly do in the American squid Loligo pealii - so that there would be no "bleaching" without regeneration?

K. Hamdorf: It is impossible to give a generalizing answer to this question. In the case of total overlapping of R and M, the photoequilibrium between the two pigments is constant, independent from the spectral distribution of the light source. The concentration of R and M in this equilibrium is determined only by the ratio of the molar absorption coefficients at λ_{max}. It depends on the behaviour and environment of Loligo, whether this system is an advantage; this should be examined.
In the case of different λ_{max} of R and M, the photoequilibrium is determined by the spectral emission as well as by the molar absorption coefficients. As shown for Ascalaphus, the strong separation of the two pigments provides a very high R concentration under natural light conditions. This system would allow a change in R content according to the particular conditions of illumination. Furthermore, in the Deilephila-system, which will be described by Dr. Höglund (12), the separation of R and M surely is an advantage for the constancy of color discrimination in trichromatic vision: The blue light of the sky absorbed equally by all the three metarhodopsins. Therefore, it is to be concluded that the ratio of absolute sensitivities is constant for the three receptor types.

W. de Grip: Have you ever performed similar measurements on night insects?

K. Hamdorf: No, so far experiments have been carried out only on the moth Deilephila elpenor, which is active in twilight.

The Photopigments in an Insect Retina

G. Höglund*, K. Hamdorf, H. Langer, R. Paulsen, and J. Schwemer
Institut für Tierphysiologie, Ruhr-Universität Bochum, Bochum-Querenburg/W. Germany

Colour vision is not an exclusive property of vertebrates. Also insects can discriminate wavelengths. The best known example is the honeybee, as shown by training experiments (1) and electrophysiological recordings (2,3). The peripheral wavelength discrimination is accomplished by at least three receptor types. The spectral sensitivity of the receptors fairly well agrees with resonance spectra for rhodopsins (3), and bee heads contain retinol and retinal (4). These results suggest that the visual pigments in insects are rhodopsins, i.e. they consist of retinal bound to a protein.

Although these results were obtained on the bee, the eyes of other species may be technically better suited for further studies of the visual pigments in insects. Behavioural studies (5,6,7) indicate that some Lepidoptera have colour vision, and electrophysiological mass recordings (8) have shown that the spectral sensitivity of a sphingid moth (Manduca sexta) is similar to that of the bee. This result suggests that the eye contains at least two photopigments. The photoreceptors form a large part of the eye in sphingid moths, and most of the screening pigment can be dissected away from the receptors. For these reasons the compound eye of the sphingid moth Deilephila elpenor was chosen for the present study of the visual pigments in insects. The results are based on electrophysiological and spectrophotometric methods.

A. Electrophysiological Measurements

Selective adaptation was made to obtain information on the number of photoreceptors in the retina, and the number of pigments in each receptor.

Methods. Light from a Xenon arc (2 kW) passed through interference filters (20 filters spaced between 338 and 608 nm). An equal number of photons at each wavelength was obtained by neutral density filters. A 30 msec test flash was given at 2 sec intervals. The maximum intensity was $1 \cdot 10^{15}$ quanta$\cdot$cm$^{-2}\cdot$sec^{-1}. The responses were recorded by an Ag-AgCl electrode. In order to illuminate all ommatidia a large part of the cornea and distal screening pigment were dissected away from the dark-adapted eye. The preparation was covered by a quartz glass, and kept in a moist chamber at + 25°C during the experiment.

Results. Fig. 1 shows the spectral sensitivity of the retina in Deilephila. The responses to short, middle and long wavelength stimulation could be selectively decreased by adapting illumination (Fig. 2). The measurements showed that the sensitivity to ultraviolet light decreased only slightly when the eye was exposed to a green adapting light. When the eye was exposed to a combination of green and violet adapting illumination, the sensitivity decreased less than one third. The decrease probably corresponds to the absorption of the ultraviolet test light by the pigments absorbing maximally at about 440 and 525 nm.

---

*Department of Physiology, Karolinska Institutet, S-104 01 Stockholm, Sweden.

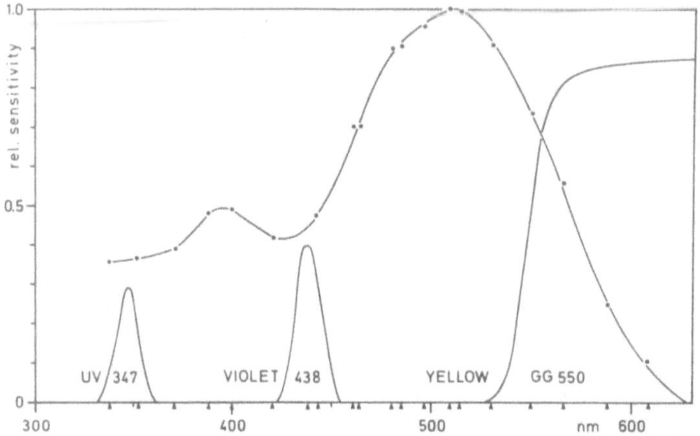

Fig. 1. Spectral sensitivity (mass recording) of dark adapted retina in Deilephila elpenor, and light transmission of filters used for selective adaptation. Transmission of the filters is maximal at 347 nm (ultraviolet), 438 nm (violet), and at wavelengths longer than 550 nm (yellow, Schott GG 550)

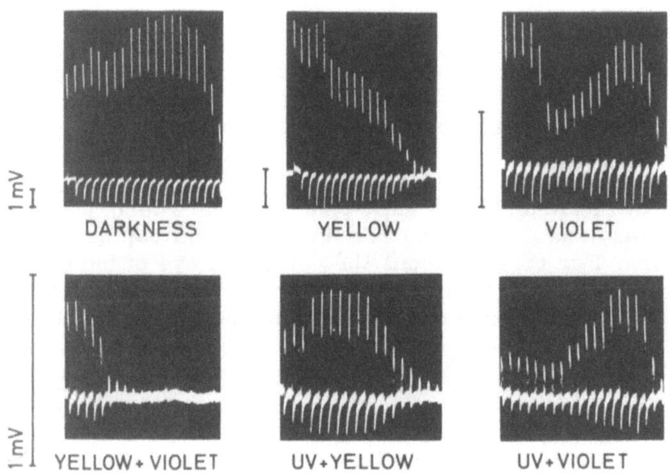

Fig. 2. Amplitude of responses (mass recording) to test flashes (wavelengths 338, 352, 371, 388, 402, 420, 438, 443, 461, 464, 480, 485, 498, 509, 514, 530, 550, 566, 589, 608 nm) in darkness, and during exposure to adapting light (ultraviolet, violet, and yellow)

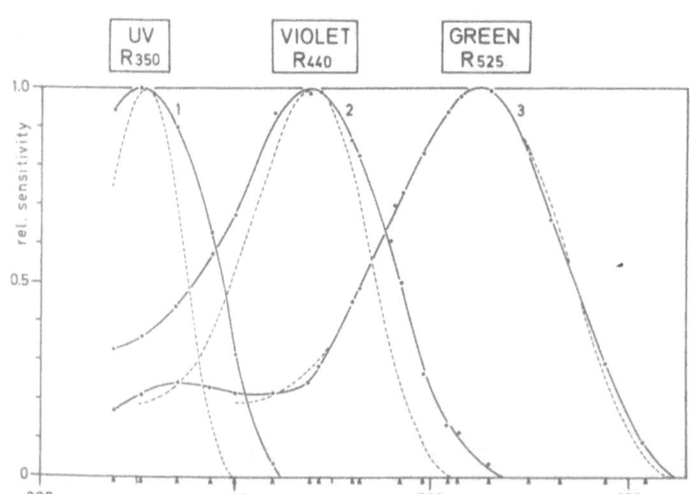

Fig. 3. Spectral sensitivity (mass recording) of retina during selective adaptation (continuous line) by violet and yellow light (1), ultraviolet and yellow light (2), and ultraviolet and violet light (3). Broken lines: Resonance spectra for rhodopsins absorbing maximally at 350 nm (R 350), 440 nm (R 440), and 525 nm (R 525)

Fig. 3 shows that the spectral sensitivity rather well agrees with calculated resonance spectra (9) for rhodopsins absorbing maximally at 350, 440, and 525 nm. The results show the presence of three receptor types with spectral sensitivity maxima in these regions. The results suggest that each pigment is located in a separate receptor. Therefore the pigments should be similar to those of the bee.

B. Extraction

An attempt was made to extract the photopigments and to identify the compounds formed by illumination.

Methods. Isolated retinae were washed in phosphate buffer and extracted by 2% aqueous digitonin at pH 6.1. The spectral absorbance of the extract was measured in a recording two-wavelength double beam spectrophotometer (Hitachi 356). An actinic light source of high intensity permitted continuous monochromatic illumination (Schott IL interference filters) of the extract.

Results. The change in absorbance caused by the wavelengths 628 and 498 nm at +20°C is shown in Fig. 4. A maximal decrease in absorbance is seen at about 520 and 450 nm, respectively, and a maximal increase for both at about 380 nm. Adding hydroxylamine to the irradiated extract caused a shift to about 360 nm and an increase in absorbance. This result strongly suggests that retinal is the chromophore of the pigments. From this experiment can be concluded that the extract contained at least two rhodopsins with maximal absorbance at about 520 (R 520) and 450 nm (R 450), respectively. The pigments are bleached to free retinal and protein. The metarhodopsins of these pigments were demonstrated at -15°C. Illumination by 628 nm caused a maximal decrease in absorbance at 550 nm, and a maximal increase at about 460 nm (Fig. 5). On warming to +20°C in darkness, this photoproduct hydrolyzes to yield free retinal and protein. The re-cooled extract was illuminated by 435 nm which converted R 450 into a photoproduct absorbing maximally at longer wavelength. From the difference spectra it is concluded that the maximal absorbance of the metarhodopsin from R 520 is at about 470 nm, and that from R 450 at about 490 nm.

The ultraviolet absorbing pigment could be demonstrated at -15°C (Fig. 5). Exposure to 353 nm caused a decrease in absorbance in this wavelength region. A simultaneous considerable increase at about 460 nm showed the formation of metarhodopsin. At +20°C, the maximum at 460 nm disappeared and a maximum at about 380 nm was obtained. This result shows that the metarhodopsin hydrolyzes to free retinal and protein part. The presence of R 350 can thus not be demonstrated at +20°C, since the decrease in concentration of R 350 during illumination is compensated for by the simultaneous formation of retinal. While free retinal is formed in solution, all metarhodopsins are thermostable in the receptor membrane.

C. Spectrophotometry

The thermostability of R 525 in the membrane was demonstrated by spectrophotometry.

Method. The cornea and distal screening pigments were removed from the dark-adapted eyes. The isolated retinae were mounted in the spectrophotometer. The reference beam passed through one retina, and the test beam through the other.

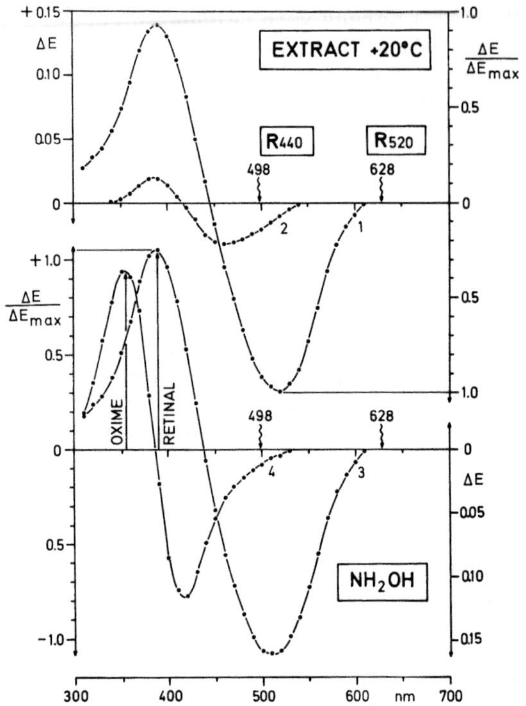

Fig. 4. Selective illumination of visual pigment extract at +20°C. Upper half: Change in spectral absorbance after exposure to 628 nm (1) and 498 nm (2). Lower half: Change in spectral absorbance after illumination by 628 and 498 nm (3), and change in absorbance after subsequent adding of NH_2OH (4)

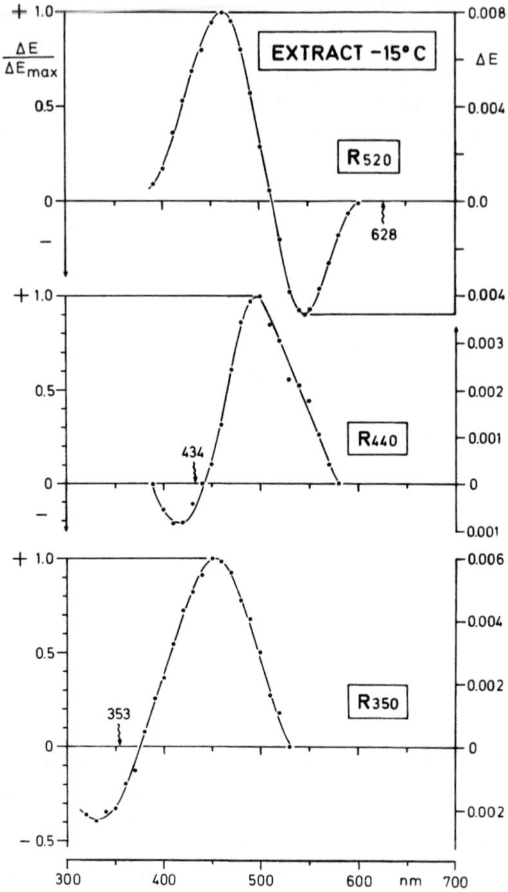

Fig. 5. Selective illumination of visual pigment extract at -15°C. Change in spectral absorbance after exposure to 628 nm (upper part), 434 nm (middle part), and 353 nm (lower part)

The latter retina could be illuminated with monochromatic light by the actinic light source.

Results. Measurements of the spectral absorbance by the retina before and after illumination by 594 nm resulted in a difference spectrum with a minimum at about 540 nm, and a maximum at about 460 nm (Fig. 6). The difference spectrum agrees with that obtained from an extract illuminated by 628 nm at -15°C. Subsequent irradiation by 434 nm causes a reversal of the difference spectrum. It is concluded that the metarhodopsin of R 525 is thermostable in the membrane, and that it is reconverted by light.

D. Microspectrophotometry

R 350 and R 525 and their metarhodopsins were analyzed microphotometrically.

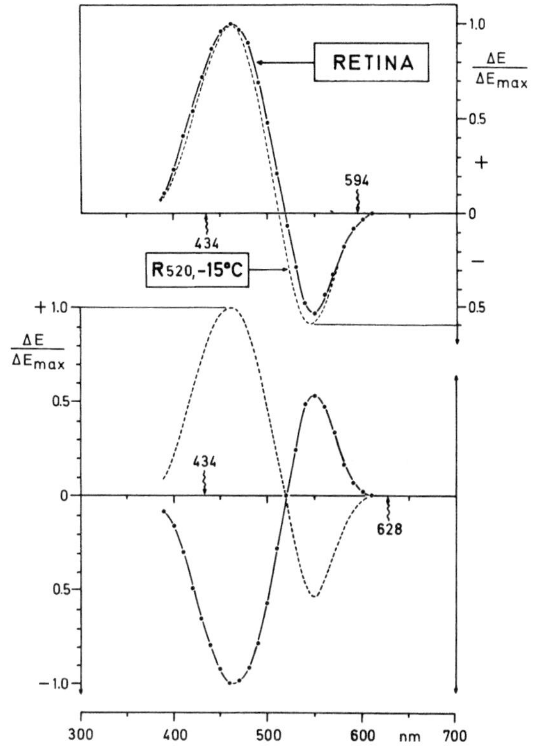

Fig. 6. Photoconversion of rhodopsin in retina. Upper half: Change in spectral absorbance after exposure to 594 nm (continuous line) and absorbance by pigment extract (-15°C) after exposure to 628 nm (broken line, from Fig. 5). Lower half: Change in spectral absorbance after subsequent exposure of the same retina to 434 nm (continuous line) and after exposure to 594 nm (broken line)

Methods. A slice from a fresh or deep-frozen (-20°C) retina was excised and suspended in saline (10). The change in absorbance of the receptors caused by illumination was measured in a one beam microphotometer (Zeiss MPM, Ultra-Fluar optics, Schott IL interference filters, diameter of measuring area 100 µm) at various wavelengths, usually at right angle to the optical axis of the receptors. The equipment permits measurements at wavelengths exceeding 370 nm.

Results. Ultraviolet illumination (372 nm) caused changes in absorbance in the wavelength region 400 to 500 nm. These changes (Fig. 7) agree with those for the formation of metarhodopsin in Ascalaphus (11). In this insect the metarhodopsin absorbs maximally at about 475 nm. In Deilephila illumination by different wavelengths in the ultraviolet region (Fig. 7) caused changes in absorbance, which were almost identical to those found for the rhodopsin in Ascalaphus. The results show that R 350 in Deilephila is converted to a thermostable metarhodopsin, which is completely reconverted to R 350 by wavelengths of about 450 nm.

R 525 is completely converted to its metarhodopsin by exposure to a long wavelength that is not absorbed by the metarhodopsin. The spectral absorbance after exposure to such a wavelength (608 nm) was measured. The receptors were then exposed to a wavelength (453 nm) that is primarily absorbed by the metarhodopsin. Thereby the metarhodopsin was reconverted to R 525. The spectral absorbance was again measured. The difference spectrum (Fig. 8) between the spectral absorbance after illumination by 608 and 453 nm agrees with that for the R 520 extract at -15°C (Fig. 5). Difference spectra calculated from resonance spectra based on cattle rhodopsin (12) well agree with the present results (Fig. 8).

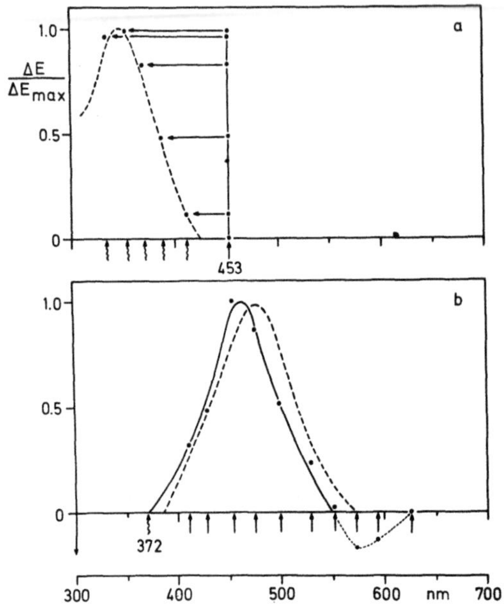

Fig. 7. Ultraviolet visual pigment in receptor. a) Absorbance change at 453 nm after exposure to wavelengths in ultraviolet region (dots). b) Absorbance change at 10 wavelengths after exposure to 372 nm (continuous line). Corresponding values for ultraviolet pigment in Ascalaphus (broken lines)

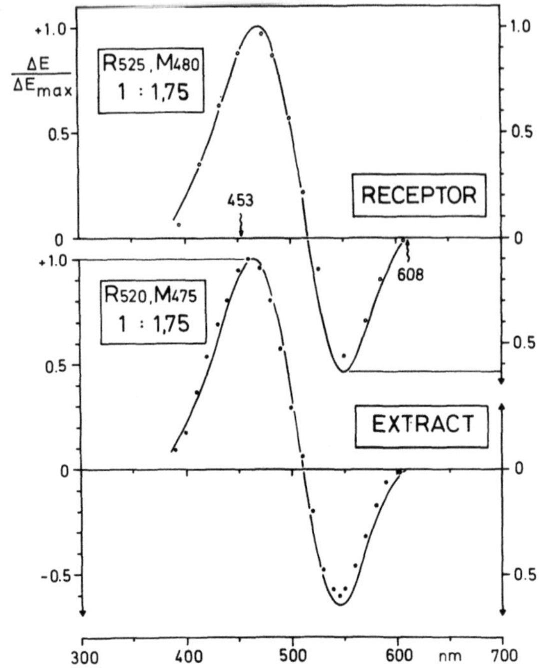

Fig. 8. Green visual pigment in receptor in comparison with extract. Upper half: Difference between spectral absorbances by receptors after exposure to 453 nm (abscissa) and subsequent exposure to 608 nm (circles). Calculated difference spectrum for a mixture of rhodopsin (R 525) and metarhodopsin (M 480) with a relative absorbance of 1:1.75 (continuous line). Lower half: Change in spectral absorbance in extract (-15°C) after exposure to 628 nm (dots, from Fig. 5). Calculated difference spectrum for a mixture of rhodopsin (R 520) and metarhodopsin (M 475) with relative absorbance 1:1.75 (continuous line)

The spectral efficiency for the regeneration of metarhodopsin to rhodopsin (R 525) was determined by measuring the change in spectral absorbance, caused by exposure to a constant test wavelength after exposure to another wavelength. Figs. 6 and 8 show that this change is best measured at about 450 nm. The absorbance change at the test wavelength varies with the wavelength of the preceeding illumination (Fig. 9). The change is proportional to the amount of rhodopsin formed.

E. Conclusions

The compound eye of Deilephila contains three rhodopsins absorbing maximally at 350, 440, and 525 nm. The metarhodopsins of all three pigments have a higher ab-

Fig. 9. Spectral efficiency for photoregeneration of R 525, and relative amount of rhodopsin formed. Regeneration measured as change in absorbance at 453 nm after exposure to 10 wavelengths between 412 and 628 nm

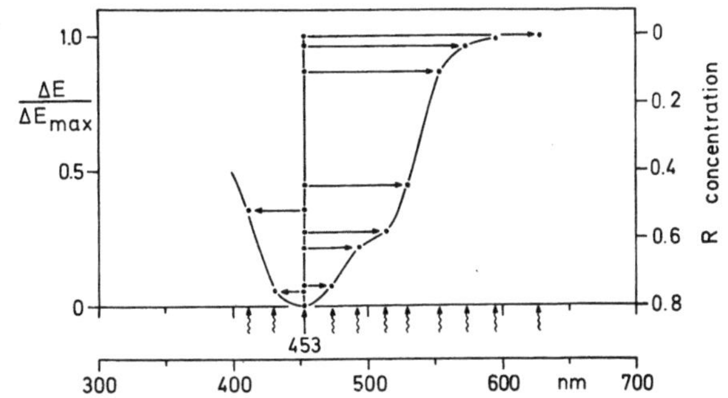

sorbance than their rhodopsins. They absorb maximally at wavelengths in the region at about 480 nm. The present results show that in the receptor membrane light reconverts two of the metarhodopsins (those of R 350 and R 525). It seems probable that the metarhodopsin of R 440 is also reconverted by light.

From the present results can be concluded that in Deilephila, during the day, the blue light from the sky keeps the concentration of rhodopsin constant, whereby also the sensitivity of the receptors is maintained constant. The concentration of R 350 is about 90 per cent (as shown for Ascalaphus; 13) and that of R 525 about 60 to 70 per cent. R 440 and its metarhodopsin both absorb maximally in the same wavelength region. The concentration of R 440 therefore probably exceeds 50 per cent. Behavioural experiments (14) have shown that the bee particularly well discriminates mixtures of ultraviolet and green wavelengths. This ability is probably explained by the high and constant concentration of R 350 and R 525 in the receptor at daylight. Deilephila is active also at twilight. It is therefore probable that at low light intensities the rhodopsins are regenerated biochemically to a concentration up to 100 per cent. The rate of dark-adaptation is low (15) which suggests that the chemical regeneration occurs slowly. At daylight the amount of chemically regenerated rhodopsin molecules per unit time may therefore be low in comparison with the amount of photochemically regenerated molecules.

References

1. FRISCH, K. v.: Tanzsprache und Orientierung der Bienen. Springer, Berlin-Heidelberg-New York (1965).
2. AUTRUM, H., V. von ZWEHL: Zur spektralen Empfindlichkeit einzelner Sehzellen der Drohne (Apis mellifica). Z. vergl. Physiol. 48, 8-12 (1962).
3. AUTRUM, H., V. von ZWEHL: Die spektrale Empfindlichkeit einzelner Sehzellen des Bienenauges. Z. vergl. Physiol. 48, 357-384 (1964).
4. GOLDSMITH, T. H., L. T. WARNER: Vitamin A in the vision of insects. J. Gen. Physiol. 47, 433-441 (1964).
5. KNOLL, F.: Lichtsinn und Blütenbesuch des Falters von Deilephila livornica. Z. vergl. Physiol. 2, 329-380 (1925).
6. ILSE, D.: Über den Farbensinn der Tagfalter. Z. vergl. Physiol. 8, 658-692 (1928).
7. CRANE, J.: Imaginal behaviour of a Trinidad butterfly, Heliconius erato hydara Hewitson, with special reference to the social use of colour. Zoologica 40, 167-196 (1955).
8. HÖGLUND, G., G. STRUWE: Pigment migration and spectral sensitivity in the compound eye of moths. Z. vergl. Physiol. 67, 229-237 (1970).

9. DARTNALL, H.J.A.: The interpretation of spectral sensitivity curves. Brit. Med. Bull. 9, 24-30 (1953).
10. EPHRUSSI, B., G.W. BEADLE: A technique of transplantation for Drosophila. Amer. Nat. 70, 218-225 (1936).
11. SCHWEMER, J., M. GOGALA, K. HAMDORF: Der UV-Sehfarbstoff der Insekten: Photochemie in vitro und in vivo. Z. vergl. Physiol. 75, 174-188 (1971).
12. WALD, G., J. DURELL, R.C.C. ST. GEORGE: The light reaction in the bleaching of rhodopsin. Science 111, 179-181 (1950).
13. HAMDORF, K., M. GOGALA, J. SCHWEMER: Beschleunigung der "Dunkeladaptation" eines UV-Rezeptors durch sichtbare Strahlung. Z. vergl. Physiol. 75, 189-199 (1971).
14. DAUMER, K.: Reizmetrische Untersuchungen des Farbensehens der Bienen. Z. vergl. Physiol. 38, 413-478 (1956).
15. HÖGLUND, G.: Pigment migration, light screening and receptor sensitivity in the compound eye of nocturnal Lepidoptera. Acta physiol. scand. 69, suppl. 282 (1966).

Rhodopsin Processes and the Function of the Pupil Mechanism in Flies

D. G. Stavenga, A. Zantema, and J. W. Kuiper
Department of Biophysics, University of Groningen, Groningen/The Netherlands

In the photoreceptor cells of the compound eye of the fly pigment granules migrate under the influence of a change in light intensity (1). The light flux in the cell's rhabdomere, which contains the visual pigment molecules and functions as a light guide, depends on the number of pigment granules present near the rhabdomere. This number of absorbing and scattering granules is small in the dark adapted state and increases with light adaptation. To describe this pupil mechanism we have proposed an electrophoresis model (2) in which the force exerted on the pigment granules is a consequence of the change in membrane potential following rhodopsin conversion. We present here results of more extensive studies on rhodopsin conversion processes and closely related properties of the pigment granules.

Methods

The tip of a quartz rod is placed in the back of the head of the wild type Calliphora erythrocephala. The deep-pseudopupil phenomenon (3) is used and the antidromic transmission T_a of rhabdomeres of receptor cells R_3 or R_{7+8}, situated in the lower part of the right eye, is measured. The measurements are performed 1 min after saturating orthodromic illumination of wavelength λ_o in order to avoid contamination by the pupil mechanism, which has a time constant of 2-5 sec. The calculated absorbance change is

$$\Delta E = {}^{10}\log \frac{T_a\,(635)}{T_a\,(\lambda_o)} \;.$$

Results

Fig. 1 shows the difference spectra of R_3 after illumination of 352, 470, and 514 nm with respect to 635 nm illumination. These spectra represent photochemical equilibria which can be converted into each other. The spectra are proportional and have extrema at about 580 nm and 470 nm; the isosbestic point is at 510 nm.

As can be deduced from Fig. 2, which represents antidromic transmission measured with sub-threshold light intensity of wavelengths 470 nm and 583 nm, respectively, rhodopsin is regenerated from the thermostable photoproduct metarhodopsin in the dark with a time constant of about 25 min. The metarhodopsin absorbs at longer wavelengths than the native rhodopsin.

The degree to which rhodopsin is converted by irradiation with a particular wavelength can be seen in Fig. 3. The metarhodopsin concentration after a saturating flash of white light is shown to be much greater than after a period of illumination during which the pupil has been closed. This shift in photochemical equilibrium is due to the blue peaking absorbance spectrum of the pupil given in Fig. 4 by the continuous line through the open circles. We conclude that the pupil shifts the photo

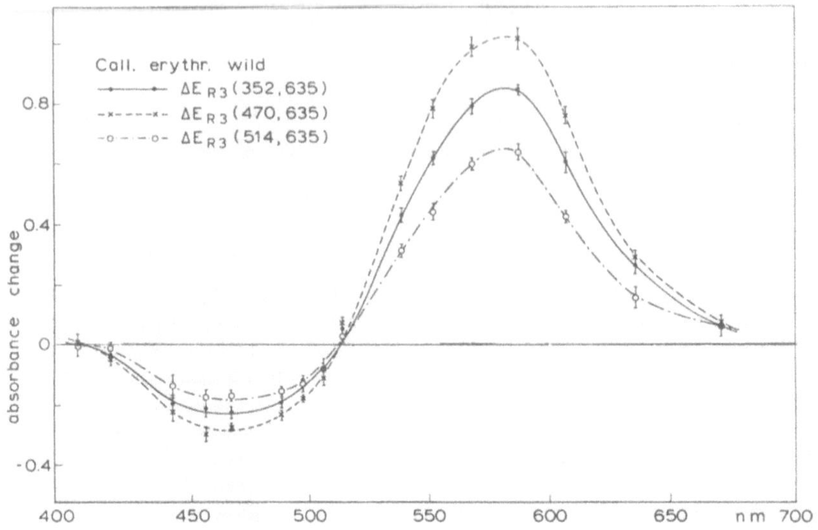

Fig. 1. Difference spectra of rhabdomeres of sense cells R_3

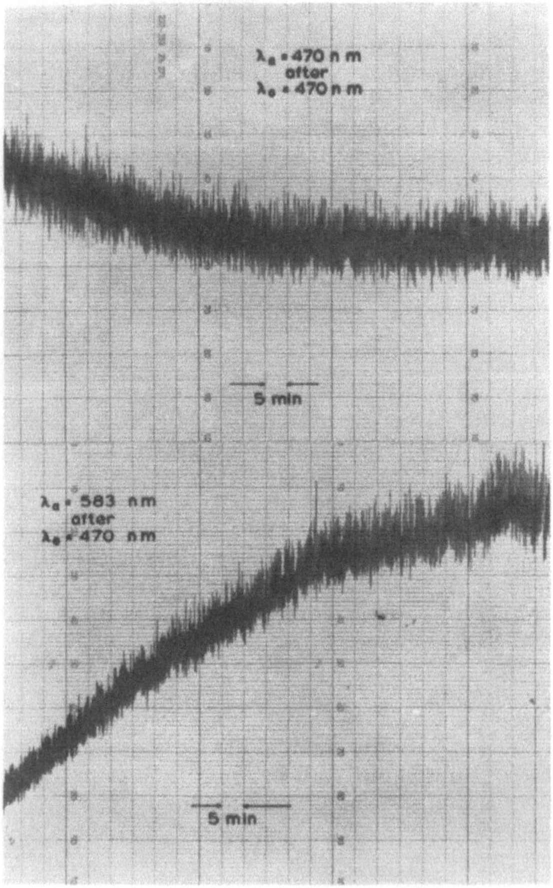

Fig. 2. Change of antidromic transmission measured with light of very low intensity after blue saturating orthodromic illumination with wavelengths 470 nm and 583 nm, respectively, showing the regeneration of rhodopsin in the dark

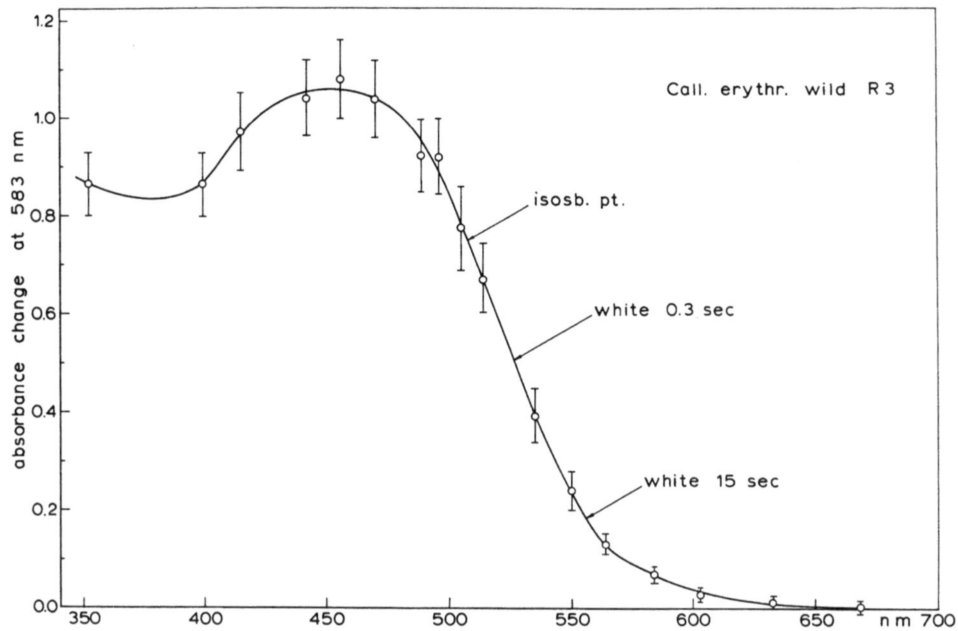

Fig. 3. The difference in absorbance from the dark value, after a photo-equilibrium is reached at wavelength λ_o, measured at 583 nm

equilibrium between rhodopsin and metarhodopsin by changing the effective spectral composition of the incident light from white to yellow.

The triangles in Fig. 4 represent the absorbance spectrum of the pigments in the pigment cells. Two curves measured by Langer (4) are compared with it. The small absorbance in the red has a function in the rhodopsin conversion: Illuminating the eye dorsally with white light causes a photochemical conversion of metarhodopsin into rhodopsin in the ventral part of the eye as a result of the red stray light.

The light flux in the rhabdomeres of retinula cells R_{1-6} (all containing the same type of rhodopsin) is influenced profoundly by the pupil, whereas in R_{7+8} the effect is negligible. The difference spectra of R_{7+8} (Fig. 5) behave in a special way also. The spectra of 514 nm and 456 nm with respect to 635 nm approximate those of R_{1-6} but the spectrum of 352 nm deviates predominately in the blue. This points to the presence of a UV-rhodopsin in R_7 and/or R_8, in addition to the rhodopsin present in R_{1-6}.

Discussion

The visual system of the fly contains a green and an ultraviolet absorbing rhodopsin. The thermostable metarhodopsins (at physiological pH) absorb in the yellow and the blue respectively. The fly's photochemistry is similar to that of other invertebrates (for references, see Hamdorf et al., 5).

The green absorbing rhodopsin is located in R_{1-6}. We put forward the hypothesis that this rhodopsin is also present in R_8 and that the UV-rhodopsin occurs in R_7.

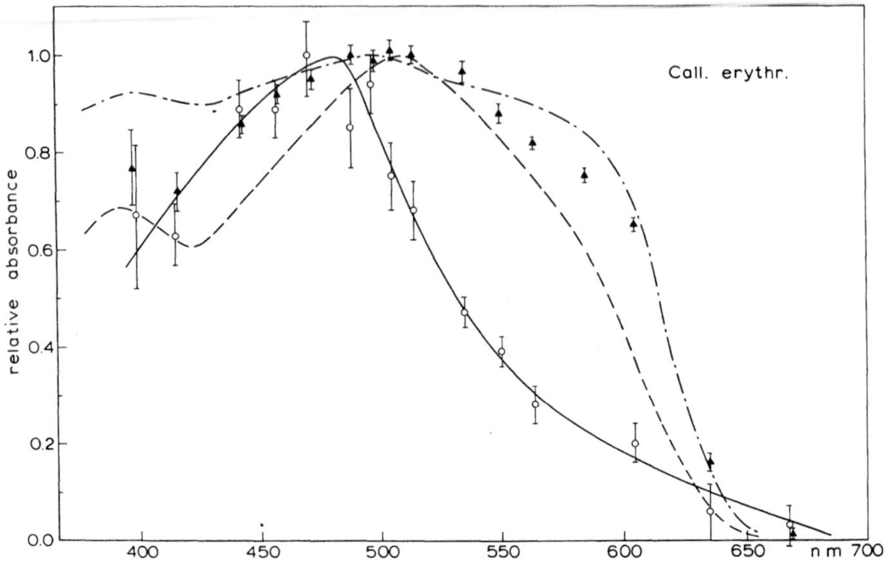

Fig. 4. Screening pigment spectra. Circles with continuous curve represent photoreceptor granules. Triangles represent pigment cells compared with two experimental curves of Langer (4)

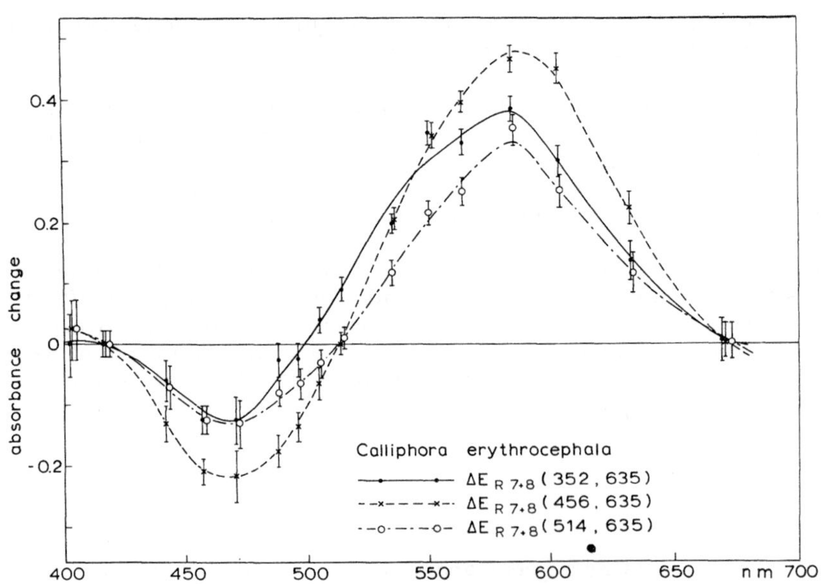

Fig. 5. Difference spectra of rhabdomeres of sense cells R_{7+8}. The curve representing the situation after UV-illumination deviates substantially in the blue; compare with Fig. 1. A metarhodopsin absorbing in the blue is known from the UV-rhodopsin of Ascalaphus (5)

The difference in spectral sensitivity between R_{1-6} and R_8 is about 20 nm at threshold (6) and must be caused by the difference in rhabdomere diameter (7). In our view, however, the described pupil mechanism is essential for colour vision. R_{1-6} functions at low intensities exclusively (8), which explains the absence of colour vision there (9). Increasing intensity causes a ("Purkinje") shift in sensitivity of R_{1-6} (10) which enables the fly to see colour by the co-operation of R_{1-6}, R_7, and R_8. The question whether the colour vision of flies is dichromatic or trichromatic has still to be solved.

References

1. KIRSCHFELD, K., N. FRANCESCHINI: Ein Mechanismus zur Steuerung des Lichtflusses in den Rhabdomeren des Komplexauges von Musca. Kybernetik 6, 13-22 (1969).
2. STAVENGA, D.G.: Adaptation in the compound eye. Proc. Int. Union Physiol. Sc. IX, 532, XXV Int. Congr. Munich (1971).
3. FRANCESCHINI, N., K. KIRSCHFELD: Les phénomènes de la pseudopupille dans l'oeil composé de Drosophila. Kybernetik 9, 159-182 (1971).
4. LANGER, H.: Über die Pigmentgranula im Facettenauge von Calliphora erythrocephala. Z. vergl. Physiol. 55, 354-377 (1967).
5. HAMDORF, K., R. PAULSEN, J. SCHWEMER: Photoregeneration and sensitivity control of photoreceptors of invertebrates. This volume, pp. 155-166.
6. ECKERT, H.: Die spektrale Empfindlichkeit des Komplexauges von Musca (Bestimmung aus Messungen der optomotorischen Reaktion). Kybernetik 9, 145-156 (1971).
7. SNYDER, A.W., W.H. MILLER: Fly Colour Vision. Vision Res. 12, 1389-1396 (1971).
8. KIRSCHFELD, K.: Aufnahme und Verarbeitung optischer Daten im Komplexauge von Insekten. Naturwissenschaften 58, 201-209 (1971).
9. FINGERMAN, M., F.A. BROWN JR.: Color discrimination and physiological duplicity of Drosophila vision. Physiol. Zool. 26, 59-67 (1953).
10. FINGERMAN, M., F.A. BROWN JR.: A "Purkinje shift" in insect vision. Science 116, 171-172 (1952).
11. BURKHARDT, D.: Spectral sensitivity and other response characteristics of single visual cells. Symp. Soc. exp. Biol. 16, 86-109 (1962).
12. LANGER, H.: Spektrometrische Untersuchung der Absorptionseigenschaften einzelner Rhabdomere im Facettenauge. Verh. Dtsch. Zool. Gesellsch. Jena 1965. Zoolog. Anz., Suppl. 29, 329-338 (1966).
13. LANGER, H.: Grundlagen der Wahrnehmung von Wellenlänge und Schwingungsebene des Lichtes. Verh. Dtsch. Zoolog. Gesellsch. Göttingen 1966. Zoolog. Anz., Suppl. 30, 195-233 (1967).

Discussion

H. Langer: Your investigation points again to an old discrepancy: The spectral sensitivity curves of these eyes - as measured by ERG - and also of most single receptor cells peak between 490 and 500 nm (for review see Burkhardt, 11). My own microspectrophotometric measurements on single rhabdomeres of the outer cells in the ommatidium showed maxima in a large range, between 490 and 540 nm, most around 515 nm (12,13). The measurements were run from short to long wavelengths, and therefore a partial conversion of rhodopsin to metarhodopsin by the measuring light must be brought into consideration. According to our new knowledge on the stability of metarhodopsins in insects, it is now possible to obtain a better understanding of these results. - In model calculations, we summed up spectral resonance curves of rhodopsins, λ_{max} 490 or 500 nm, and 550 or 560 nm, respectively, in varying relative heights. By summation of "rhodopsin" and only a small part (10-30%) of "metarhodopsin" we obtained curves, which peak between 510 and 520 nm and fit quite well with the microspectrometric measuring curves. - In my earliest publication on spectrophotometry on rhabdomeres, I gave the mean value of λ_{max} = 494 nm for a group of sensory cell no. 2 rhabdomeres (12, Table I). But I didn't believe this result because of these curves having very small signal-to-noise ratios; for the cross section of these rhabdomeres is small and their illumination was very poor. But therefore, the photoconversion may have been slow and the portion of metarhodopsin quite small, so that this figure may be the most probable one. The curves with maxima at longer wavelengths may come from mixtures between rhodopsin and a smaller or larger part of thermostable acid metarhodopsin. For metarhodopsin does not influence the spectral sensitivity, the discrepancy in results can be understood now.

Biochemical Properties of Retinochrome

Tomiyuki Hara and Reiko Hara
Department of Biology, Nara Medical University, Kashihara, Nara/Japan

The cephalopod retina has two kinds of photosensitive pigments. These pigments have been examined in various squids and octopii (1-4), and are called rhodopsin and retinochrome(5). Rhodopsin is present in the outer segments of the visual cells, and catches the light for visual excitation. On the other hand, retinochrome is abundantly contained in the inner segments of the visual cells. When the outer and inner portions of the retina are collected separately, each photopigment can be extracted by digitonin. The absorption spectrum of squid retinochrome is shown in Fig. 1. The shapes of the spectra of retinochrome and rhodopsin are very similar. Retinochrome is a chromoprotein that includes retinaldehyde as prosthetic group. However, retinochrome is different from rhodopsin not only in its location in the retina but also in photochemical behaviour, molecular architecture, and other chemical properties (5-7). The most essential difference is in the stereoisomeric configuration of the prosthetic group. Rhodopsin has the 11-cis isomer of retinal, whereas retinochrome has the all-trans. The outline of our earlier research on retinochrome is presented in several reviews (8-10).

Linkage between Retinal and Protein

When retinochrome is irradiated with orange light (>530 nm) at weakly acid or neutral pH, the absorbance falls at about 490 nm, and it is converted into a photoproduct with two low peaks at wavelengths near 465 nm and 380 nm. When the pH of the sample is increased, these peaks disappear, and the photoproduct is changed to a

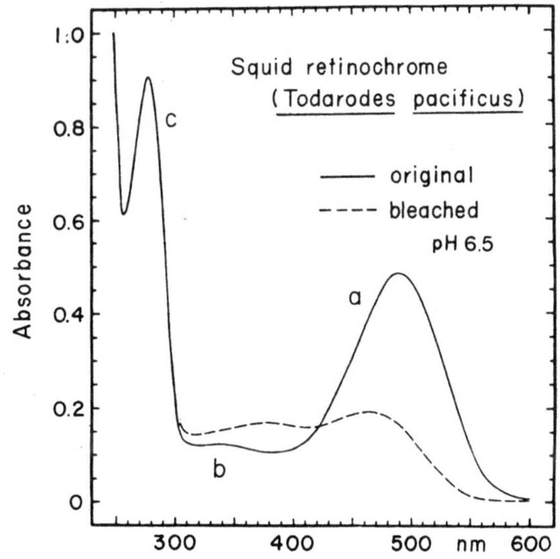

Fig. 1. Absorbance spectra of an extract of squid retinochrome and of the extract bleached by orange light

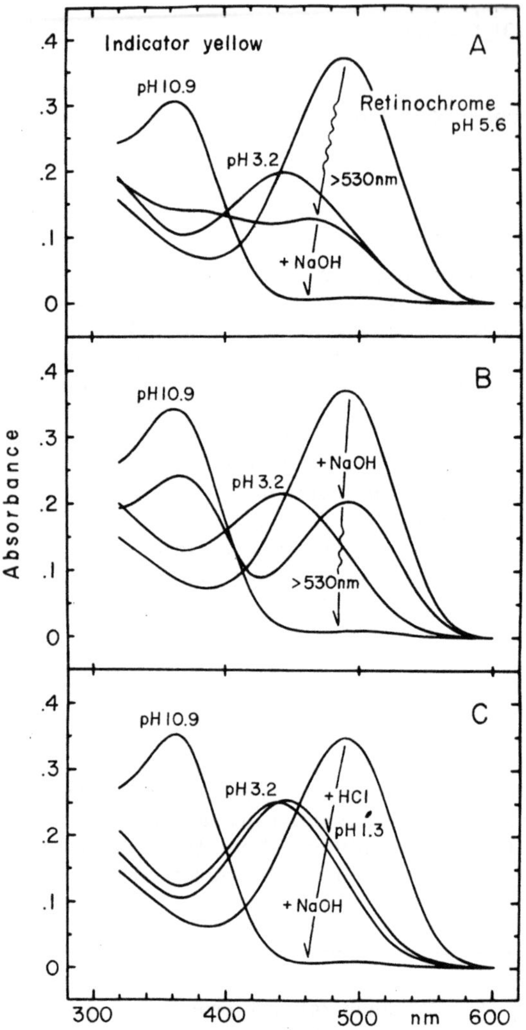

Fig. 2. Formation of indicator yellow from retinochrome bleached by orange light at acid (A) and alkaline (B) pH and from retinochrome denatured by hydrochloric acid (C)

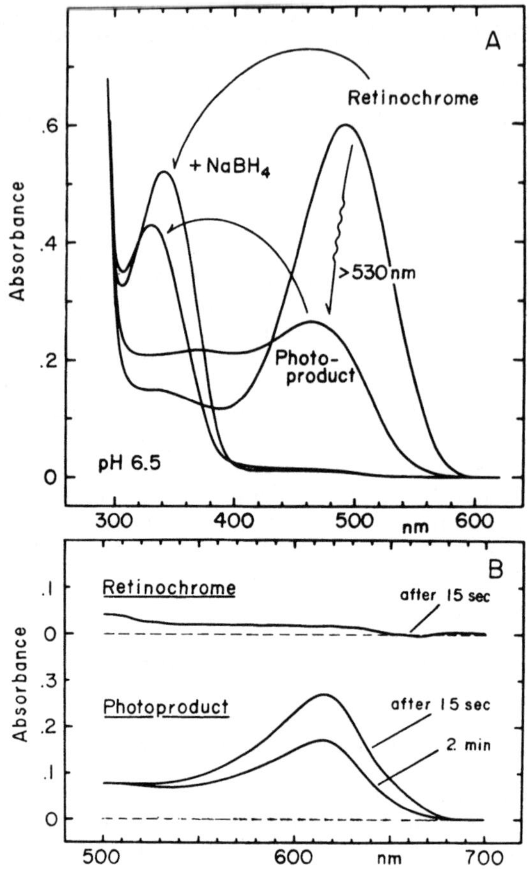

Fig. 3. Reduction of retinochrome and its photoprocuct with sodium borohydride. A. They are respectively reduced at once into 340-nm and 330-nm compounds. B. Antimony trichloride tests with these reduced compounds. The "reduced photoproduct" liberates retinol as shown by an absorbance band of 620 nm

compound with a peak at 365 nm (Fig. 2A). The absorbance spectrum of retinochrome is originally dependent on the hydrogen ion concentration. When the pH of the acid extract is increased, the absorbance falls near 490 nm and rises near 370 nm. Such alkaline retinochrome is also photosensitive. When irradiated with orange light, it is directly converted into a 365-nm compound (Fig. 2B). These compounds with absorbance peaks at 365 nm are alkaline indicator yellow. When the pH is decreased, they are converted into an acid form with an absorbance peak near 445 nm. Retinochrome itself becomes denatured at extremely acid pH. The acid-denatured retinochrome is also converted into indicator yellow by changing pH (Fig. 2C). Such

Fig. 4. Reduction of alkaline indicator yellow with sodium borohydride. A compound with λ_{max} at 330 nm is formed, which never liberates retinol

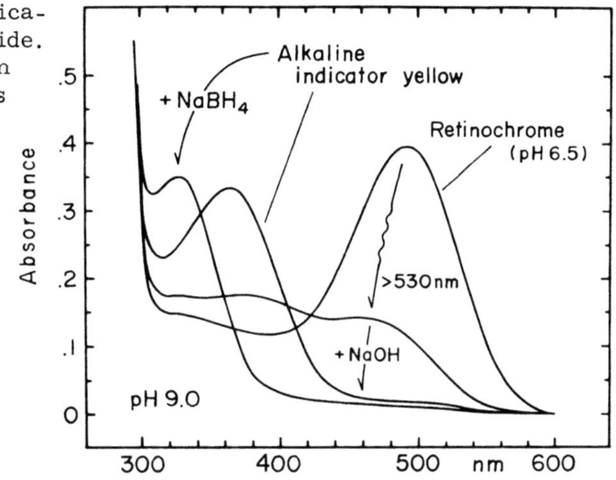

formation of indicator yellow suggests that, in the retinochrome molecule, the prosthetic group is bound to protein primarily in a Schiff-base linkage, which is formed by condensation of the aldehyde group of retinal with an amino group of the protein moiety.

In the retinochrome molecule, however, the secondary interaction between retinal and protein is far weaker than in rhodopsin. For instance, rhodopsin is hardly affected by addition of sodium borohydride (11), whereas retinochrome and its photoproduct easily react with this reagent (Fig 3A) (12). Retinochrome is at once reduced to a compound with λ_{max} at 340 nm. This compound seems to be a retinyl-protein. The photoproduct is also reduced to a different compound with λ_{max} at 330 nm. When treated with Carr-Price reagent in chloroform, the "reduced retinochrome" never shows any colour, but the "reduced photoproduct" shows a well-defined band around 620 nm characteristic of retinol. This means that the photoproduct liberates the retinol derived from reduction of prosthetic retinal (Fig. 3B). Alkaline indicator yellow is changed to a reduced compound with λ_{max} at 330 nm by addition of sodium borohydride (Fig. 4), but this compound cannot release retinol. Acid indicator yellow and acid-denatured retinochrome give the same results as well. Retinochrome is, unlike rhodopsin, affected by treatment with formaldehyde and some organic solvents, and its absorbance spectrum is modified to varying degrees. It is also broken by cetyltrimethylammonium bromide (CTAB), so that we have been unable to use this detergent for the extraction of retinochrome.

Rhodopsin is stable when treated with hydroxylamine. Retinaldehyde oxime is not formed until the retinal has been released by hydrolysis of metarhodopsin. In contrast, retinochrome is readily destroyed by hydroxylamine, and the oxime is formed at once, indicating again that the prosthetic retinal is bound to the protein less tightly than in the rhodopsin molecule. Fig. 5 presents the comparison between the retinaldehyde oximes formed from retinochrome and from its photoproduct. The "retinochrome oxime" shows higher absorbance in the near ultraviolet range than the "photoproduct oxime", and further shows a slightly longer λ_{max} (5). This fact suggests that, when retinochrome is exposed to light, the prosthetic group of retinochrome is changed from the all-trans to the cis configuration.

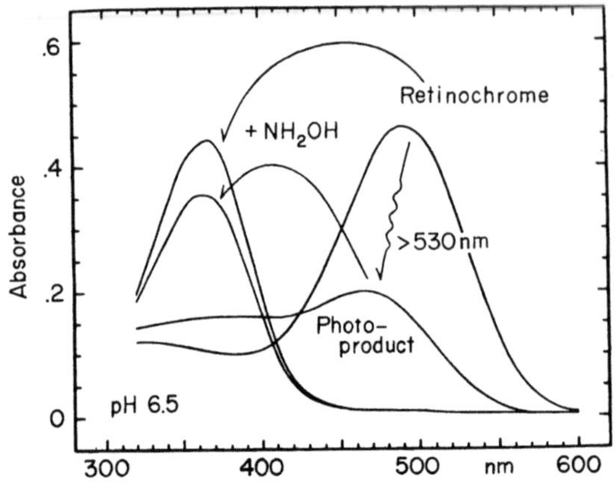

Fig. 5. Reaction of retinochrome and its photoproduct with hydroxylamine. Differences in absorbance and λ_{max} are seen between the two retinaldehyde oximes. This indicates that the prosthetic group of retinochrome is isomerized by light from all-trans to cis form

Affinity of Retinochrome-Protein for Isomers of Retinal

One of the resulting problems was to determine the origin of the 465-nm and 380-nm peaks in the absorbance spectrum of the photoproduct of retinochrome. We then went further to test the affinity of the protein moiety of retinochrome for various isomers of retinal (13). In order to free the protein moiety of retinochrome, the extract of retinochrome was irradiated with intense near ultraviolet light. This irradiation could gradually destroy the prosthetic group of retinochrome without seriously damaging the protein moiety. The near ultraviolet light used here was the 360-nm beam of a mercury lamp passed through a black glass filter which cut off light of wavelength longer than 405 nm and shorter than 285 nm. In practice, retinochrome was often irradiated first for 3 min with orange light and then with near ultraviolet light (Fig. 6A). In this case, at the beginning of the ultraviolet irradiation, retinochrome is slightly regenerated by photoreversal, and then gradually bleached by the destruction of the prosthetic group. The same result is achieved even if the first irradiation with orange light is omitted. In any case, only the protein moiety of retinochrome is left in digitonin solution after about half an hour. We used for experiments those solutions of retinochrome-protein which had an optical purity (E_{390}/E_{490}) of about 0.20 for retinochrome.

A few years ago, we demonstrated that retinochrome could be regenerated by addition of all-trans retinal to its photoproduct - retinochrome bleached by irradiation with orange light (7). Fig. 6A further shows that a red pigment is also formed when the all-trans retinal is added to the colourless retinochrome-protein obtained after further irradiation of the bleached retinochrome with near ultraviolet light. The pigment synthesized here is retinochrome, which shows the usual light-sensitivity and dependence of the spectrum on pH. This fact demonstrates that the prosthetic group of retinochrome is completely in the all-trans configuration. As shown in Fig. 6B, this synthesis of retinochrome with the freed protein is very quick, compared with that with the bleached retinochrome (cf., Fig. 3C in 7). It is half completed within 5 sec, almost fully completed after about 10 min, and finished after about 1 hour.

To synthesize retinochrome to the extent seen in Fig. 6A, it is necessary to add all-trans retinal in excess to the protein. However, in the following syntheses, we mixed an excess of protein with small quantities of retinal. Otherwise, the spectrum of synthesized pigment might be distorted by the retinal remaining in excess in the mixture. Experiments were then designed to produce about 50 per cent regeneration.

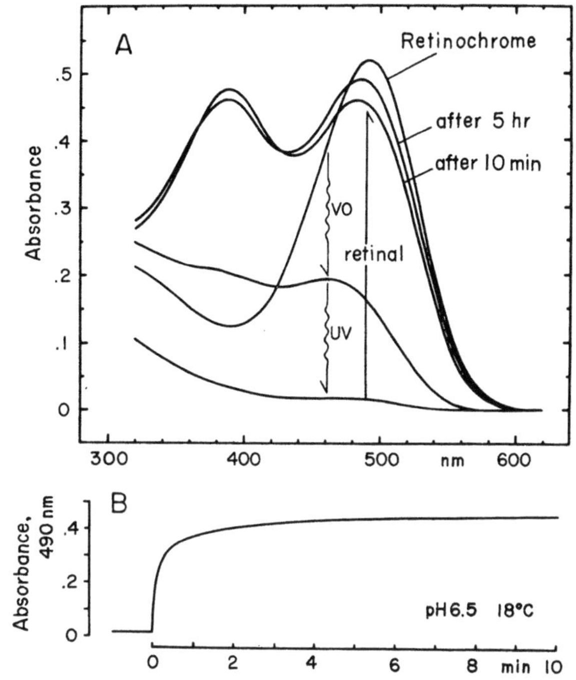

Fig. 6. Synthesis of retinochrome. A. Retinochrome (curve 1) was irradiated first for three min with orange light (curve 2) and further with near ultraviolet light (curve 3). The freed retinochrome-protein thus obtained was mixed with excess all-trans retinal. B. Retinochrome is synthesized very rapidly. The half-time of the reaction is about 5 sec at 18°C

Fig. 7 shows the results of addition of all-trans and 11-cis retinal to retinochrome-protein. The vertical bar inserted on the right from the spectra denotes the final concentration of each retinal added, in form of absorbance at its λ_{max}. Both retinal solutions, all-trans and 11-cis, are equivalent in molar concentration, though they are optically different. To the protein solution (curves 1 and 1') were added equivalent amounts of all-trans and 11-cis retinal, respectively, and the mixtures were incubated in the dark for 20 min. Straight lines denote the addition of retinal, wavy lines the irradiation with orange light, and broken lines the addition of sodium carbonate for changing pH. The addition of all-trans retinal produces a good yield of retinochrome (curve 2). When irradiated with orange light, it is bleached to the photoproduct characterized by two low peaks of absorbance (curve 3), and then changed to indicator yellow (curve 4). On the other hand, the addition of 11-cis retinal results in a product (curve 2'), which has the same spectrum as the light-bleached retinochrome. This product shows no further change in spectrum, however long it is irradiated with orange light (curve 3'). These facts clearly indicate that the 465-nm and 380-nm peaks of the light-bleached retinochrome originate from the 11-cis retinal, and its prosthetic group is therefore all in the 11-cis configuration.

Similarly, Fig. 8 shows the results of coupling experiments with various isomers of retinal, all-trans, 13-cis and 9-cis. Each isomer was mixed with aliquots of retinochrome-protein (curves 1, 1', 1'') to yield an absorbance at its λ_{max} of about 0.25. All the mixtures were then incubated in the dark for 20 min. During the incubation all-trans retinal forms retinochrome (curve 2). The addition of 13-cis retinal yields a retinochrome-like photopigment (curve 2'). The absorbance maximum also lies near 490 nm. Compared with retinochrome, the absorbance is relatively low in the visible, but high in the near ultraviolet. However, this pigment behaves like retinochrome with respect to light and pH-change, and could be called "iso-retinochrome". The addition of 9-cis retinal yields a product (curve 2''), the

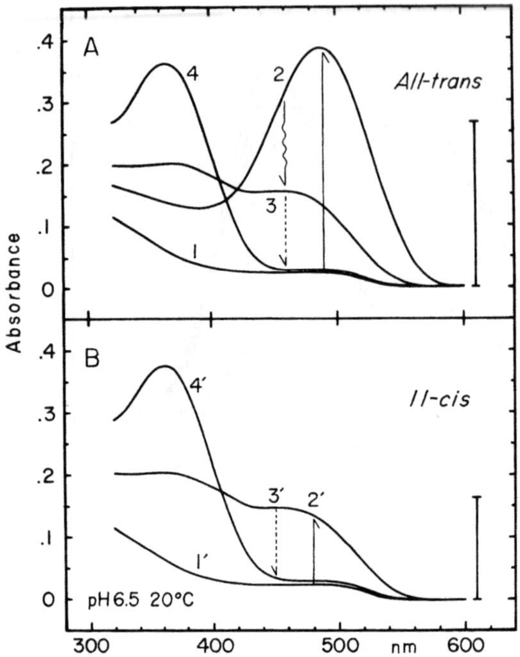

Fig. 7. Addition of all-trans and 11-cis retinal to retinochrome-protein. A. All-trans retinal yields retinochrome (curve 2). B. 11-cis retinal yields a product with the same spectrum as the retinochrome bleached by light (curve 2')

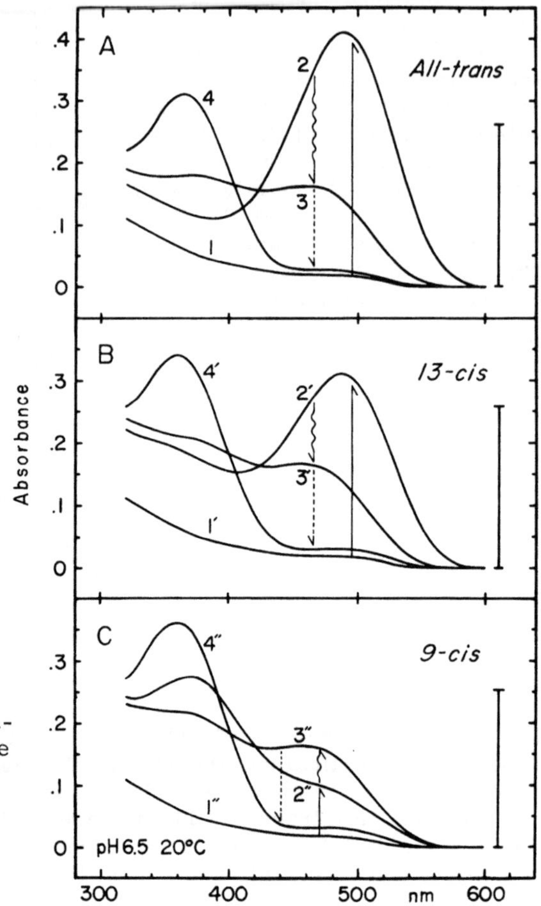

Fig. 8. Addition of all-trans, 13-cis and 9-cis retinal to retinochrome-protein. A. All-trans retinal forms retinochrome (curve 2). B. 13-cis retinal forms a retinochrome-like photopigment, iso-retinochrome (curve 2'). C. 9-cis forms a product with low absorbance in the visible (curve 2''). By irradiation with orange light, all the products (curves 2, 2', 2'') are converted into only one type of photoproduct (curves 3, 3', 3'') (cf. curve 3' in Fig. 7B)

---

absorbance of which is very low in the visible and very high in the near ultraviolet. When irradiated with orange light, it is slowly changed into the photoproduct of curve 3''. The final spectrum in the visible range matches closely those of the photoproducts of retinochrome (curve 3) and of iso-retinochrome (curve 3'). As already seen, the retinochrome-protein can yield different products depending upon the isomer of retinal used in the synthesis. However, these products are changed by irradiation with orange light eventually to only one type of photoproduct. This fact suggests that the protein moiety of retinochrome may be able to act as a catalyst in the light for converting various isomers of retinal into the 11-cis form.

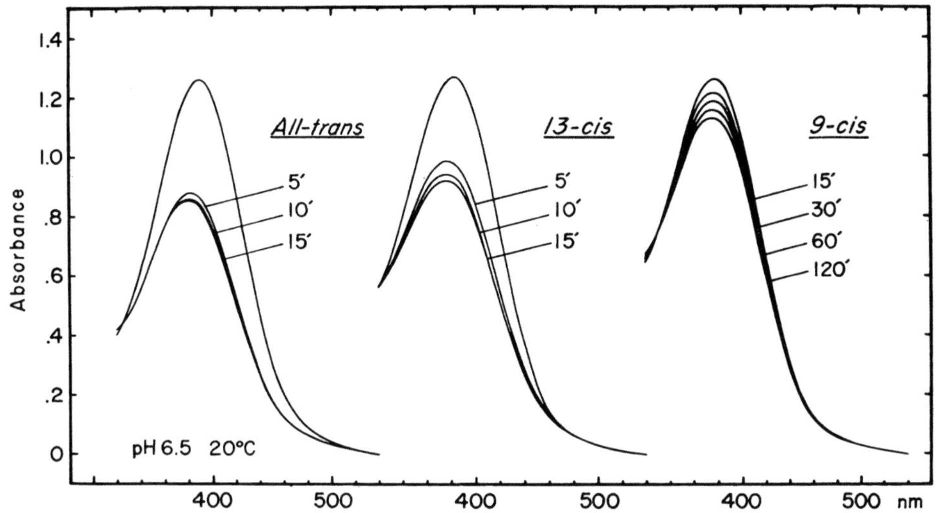

Fig. 9. Retinochrome-catalysed light isomerization of retinals, all-trans, 13-cis and 9-cis. When each isomer of retinal is steadily irradiated with orange light in the presence of retinochrome, the absorbance decreases with the progress of isomerization

---

Retinochrome-catalysed Light isomerisation of Retinals

The next experiments were aimed to elucidate a specific function of retinochrome in the light isomerization of retinals. A trace of retinochrome (absorbance at λ_{max} 0.02) was added to digitonin solutions of all-trans, 13-cis and 9-cis retinal, respectively. Each mixture was steadily irradiated with orange light, and spectra were measured from time to time (Fig. 9). In all cases the absorbance of retinal decreases to various degrees with time during the irradiation, showing the progress of isomerization. Because orange light can be absorbed by retinochrome alone, the isomerization is caused by the photochemical reaction of retinochrome. The change is over after about 10 min in all-trans retinal, and about 15 min in 13-cis, but it continues slowly in the 9-cis. What kind of retinal is accumulated in each of the mixtures? This problem was examined as follows.

Each of the three kinds of retinal and its isomerate obtained in the previous experiments was mixed with cattle opsin to test for the formation of photosensitive pigment. An example of a series of such syntheses is shown in Fig. 10. All the six mixtures were kept in the dark for 2 hours at 20 $^{\circ}$C, until the broken-line curves were achieved. Each retinal or isomerate was present in excess, and the synthesis was limited by the activity of opsin. Following the synthesis, hydroxylamine was added to all the mixtures, and their spectra were measured before and after irradiation with orange light. All-trans and 13-cis yielded no photopigment (cf. 14), but all the isomerates yielded large amounts of photopigment. Also, when opsin was mixed with the isomerate produced solely by irradiation of retinal with near ultraviolet light, a small amount of a rhodopsin-isorhodopsin mixture was formed, of less than 0.1 absorbance around 500 nm.

Difference spectra of the photopigments synthesized previously are shown in Fig.11. Of the three retinals, only 9-cis retinal gives a photopigment with λ_{max} at 485 nm

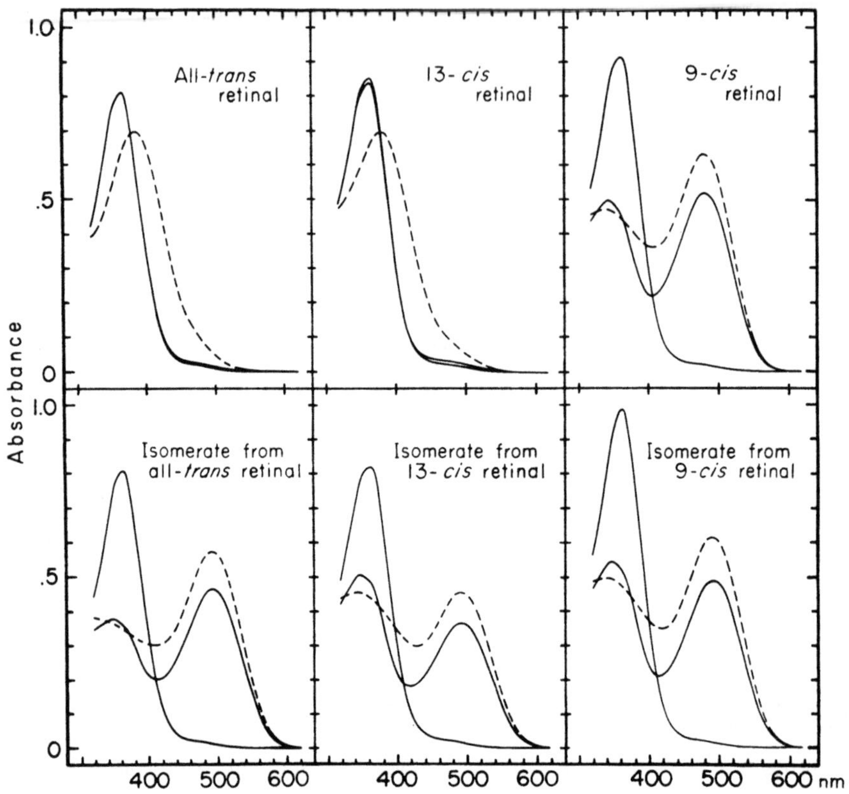

Fig. 10. Reactions of cattle opsin with the three isomers of retinal and with their isomerates obtained in the experiments shown in Fig. 9. After the reaction finished (broken-line curves), hydroxylamine was added to the mixture and spectra before and after irradiation with orange light were measured

(Fig. 11A). This is isorhodopsin. On the other hand, after the retinochrome-catalysed isomerization in visible light, each isomer of retinal forms a large amount of rhodopsin with λ_{max} at 497 nm (Fig. 11B). In the case of the isomerate from 9-cis retinal, the λ_{max} appears to be somewhat shifted towards shorter wavelengths, indicative of slight contamination with isorhodopsin. In Fig. 11C, the time courses of these photopigment syntheses are presented for comparison. In the synthesis of isorhodopsin with 9-cis retinal, the half time of the reaction is about 15 min. With all three isomerates, the half times are about 10 min and the curves are all the same, indicating again the same photopigment - rhodospin - has been formed.

As described above, the protein moiety of retinochrome has an ability to combine with various geometrical isomers of retinal to form chromoproteins which absorb visible light to varying degrees. When these are irradiated with visible light, their prosthetic groups in various isomeric forms are converted almost entirely into the 11-cis configuration, which is then readily released from the binding site. If this synthesis and decomposition are quickly repeated in a concentrated retinal solution under steady illumination, most of the retinal may be changed to the 11-cis form and accumulated in the medium. This is perhaps an important mechanism of light isomerization of retinal catalysed by retinochrome-protein.

Fig. 11. Difference spectra of photopigments synthesized with cattle opsin in the experiments shown in Fig. 10. A. Among the three isomers used, only 9-cis retinal gives a photopigment, isorhodopsin (λ_{max}: 485 nm). B. All the isomerates are capable of forming a photopigment, rhodopsin (λ_{max}: 497 nm). C. The time course of these photopigment syntheses is the same with all three isomerates, showing that the same photopigment, rhodopsin, has been formed

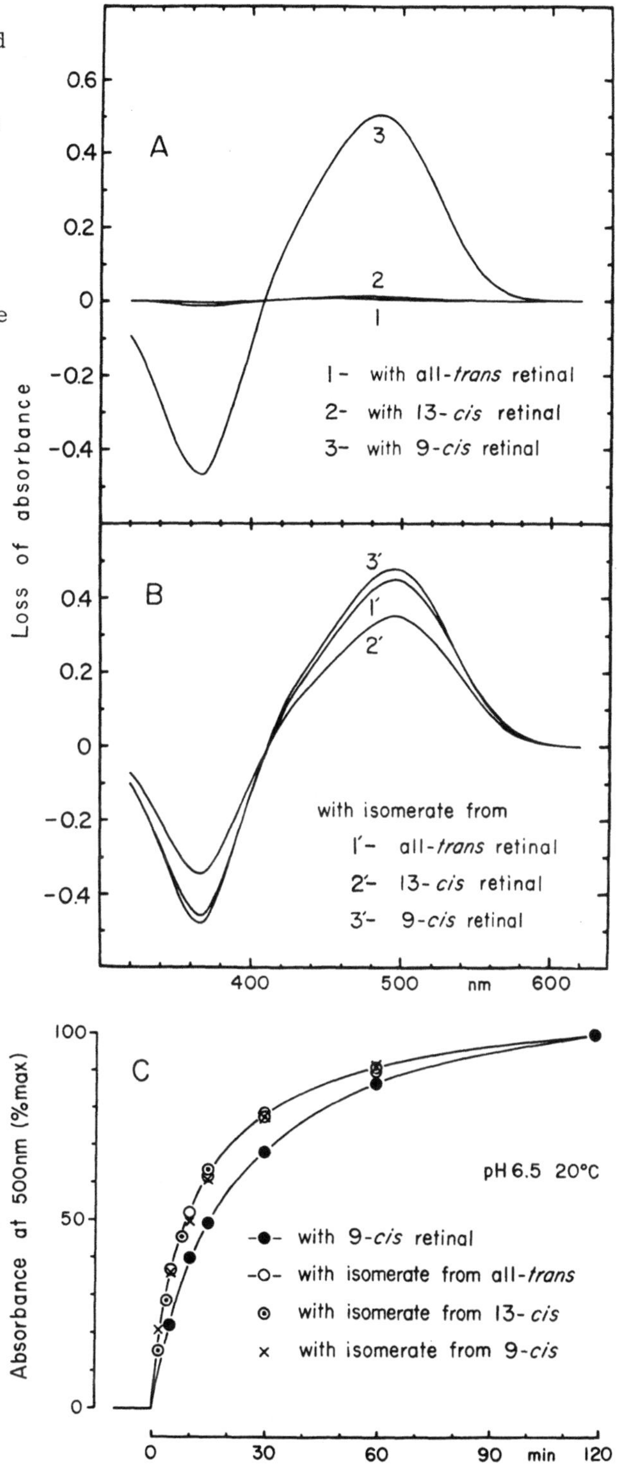

Some Remarks on Retinochrome Function

When irradiated with near ultraviolet light, retinochrome is decomposed far more easily than rhodopsin and metarhodopsin, though the mechanism is not yet clear. The retinochrome-protein, unlike opsin, can combine with various isomers of retinal, and moreover very rapidly. It may be assumed therefore that the prosthetic group of retinochrome is in contact with the surface structure of the protein moiety more loosely than in rhodopsin. Further examination of the architecture of the retinochrome molecule could yield information concerning the molecular structure and the photoreceptive mechanism of visual pigments.

As discussed elsewhere (7,10), the function of retinochrome would contribute to the reconstitution of active retinal so as to maintain the visual pigment in the retina. From this point of view, it is curious why in the retina the retinochrome is located far away from the rhodopsin and metarhodopsin, across the basement membrane and the layer of black pigment. In fact, retinochrome is present largely in the inner portions of the visual cells, and we used to extract it from rhabdome-free retinas. However, according to our recent studies on the retinal location of retinochrome, it could be also extracted from the outer portions of the visual cells only. We therefore believe there is a possibility that a small amount of retinochrome is distributed in the layer of rhabdomes, perhaps in the stems of the outer segments. If so, the retinochrome may be exposed directly to the incident light like rhodopsin. Such a fraction of retinochrome may diffuse forward into the layer of rhabdomes together with migrating granules of black pigment, leaving most of the retinochrome contained probably in the laminated structures (15,16) in the inner segments.

References

1. HARA, T., R. HARA: New photosensitive pigment found in the retina of the squid Ommastrephes. Nature (Lond.) 206, 1331-1334 (1965).
2. HARA, T., R. HARA: Photosensitive pigments found in the cephalopod retinas. Zool. Mag. (Tokyo) 75, 264-269 (1966).
3. TAKEUCHI, J.: Photosensitive pigments in the cephalopod retinas. J. Nara Med. Ass. 17, 433-448 (1966).
4. HARA, T., R. HARA, J. TAKEUCHI: Vision in octopus and squid; Rhodopsin and retinochrome in the octopus retina. Nature (Lond.) 214, 572-573 (1967).
5. HARA, T., R. HARA: Vision in octopus and squid; rhodopsin and retinochrome in the squid retina. Nature (Lond.) 214, 573-575 (1967).
6. FUJISEKI, Y.: Studies on rhodopsin and retinochrome in the cuttlefish retina. J. Nara Med. Ass. 19, 161-173 (1968).
7. HARA, T., R. HARA: Regeneration of squid retinochrome. Nature (Lond.) 219, 450-454 (1968).
8. HARA, T.: Photosensitive pigments in the cephalopod retina. Zool. Mag. (Tokyo) 77, 99-108 (1968).
9. HARA, T.: Visual photoreception. In: Seibutsu no butsuri (Biophysics), ed. by Physical Society of Japan, pp. 230-252. Maruzen, Tokyo (1971).
10. HARA, T., R. HARA: Cephalopod retinochrome. In: Handbook of Sensory Physiology, Vol. VII/1, Photochemistry of Vision, ed. H. J. A. DARTNALL, pp. 720-746. Springer, Berlin-Heidelberg-New York (1972).
11. BOWNDS, D., G. WALD: Reaction of the rhodopsin chromophore with sodium borohydride. Nature (Lond.) 205, 254-257 (1965).
12. HARA, T., R. HARA, T. NEMOTO: Squid retinochrome. Zool. Mag. (Tokyo) 79, 327 (1970).

13. HARA, T., R. HARA: Isomerization of retinal catalysed by retinochrome in the light. Nature (Lond.), in press (1972).
14. HUBBARD, R., G. WALD: Cis-trans isomers of vitamin A and retinene in the rhodopsin system. J.Gen.Physiol. 36, 269-315 (1952).
15. ZONANA, H.V.: Fine structure of the squid retina. Bull.Johns Hopkins Hosp. 109, 185-205 (1961).
16. YAMAMOTO, T., K. TASAKI, T. SUGAWARA, A. TONOSAKI: Fine structure of the octopus retina. J.Cell Biol. 25, 345-360 (1965).

Discussion

T.G. Ebrey: What is the quantum efficiency of the retinochrome catalyzed isomerization?

T. Hara: The quantum efficiency for the bleaching of retinochrome has not yet been determined. It is supposed to be a value similar to that of rhodopsin. When irradiated at alkaline pH with orange light, retinochrome is bleached about 1.6 times as fast as rhodopsin. Such a high bleaching rate of retinochrome seems to be only due to its larger molar extinction.

T.H. Goldsmith: What fraction of the total retinochrome is found in the outer segments of the squid retina?

T. Hara: The fraction extractable from the outer segments may be about one-tenth of the total retinochrome contained in the visual cells.

T.H. Goldsmith: In the outer segments, what is the molar ratio of rhodopsin: retinochrome?

T. Hara: Though we cannot say it definitely, it seems to be about ten.

A.I. Cohen: Since the membranes containing retinochrome are fairly well shielded from light, what do you believe their function to be?

T. Hara: Though little light may reach the inner portions of the visual cells through the lens of the eye, the amount of retinochrome *in situ* does vary with the level of illumination, possibly influenced by the light diffusing inwards through the wide outer surface of the eyeball. Thus the physiological function of retinochrome must be based on its photosensitivity, and then on the 11-cis retinal formation through photoisomerization.

G. Wald: The big trouble with trying to understand what retinochrome is doing for the squid is that we don't know what to do in squid with 11-cis retinal. No one has yet demonstrated any regeneration of squid rhodopsin from 11-cis retinal. At this moment it would be easier to see what retinochrome could do for a cow than for a squid. I should add my great admiration for the Hara's work. Ruth Hubbard and Linda Sperling have by now confirmed virtually all their observations.

T. Hara: Thank you very much.

IV. Excitation and Adaptation of Photoreceptor Cells

Energy in Vertebrate Photoreceptor Function*

W. Sickel
Institut für normale und pathologische Physiologie der Universität Köln, Köln-Lindenthal/W. Germany

Usefulness and Feasability of Energy Measurements

In addition to a morphological description on a cellular and molecular level and to the demonstration of physiological consequences of the quantum catch measurements of the energy involved in the process of light transformation would seem desirable, particularly so for two reasons: First, in spite of detailed knowledge on the intermediate steps the rhodopsin molecule undergoes when hit by a quantum, little is known where the energy comes in to bring it back in the responsive state, and how much of it is needed. The energy content of the incident quanta appears too high for any of the conserving reactions to be stored over any length of time. Secondly, after a quantum has been caught considerable time elapses before the receptor output signal is generated, time enough for several processes to bring about amplification, which by definition implies the utilization of extraneous energy. But how is it liberated? Is it produced ad hoc, i.e. is the metabolic process itself "electrogenic", or does it serve to refill batteries?

With energy-dissipating and energy-restoring reactions distributed among different members of a coupled reaction sequence it is mandatory for the energy measurements to be meaningful to keep this sequence intact. It is, therefore, the design of this approach to preserve receptor function as close to normal as possible. The perfused retina technique seems to come close to this requirement. The isolated retina - detached from the pigment epithelium - exhibits activity of higher-order neurons in addition. For this reason it will be necessary to single out their share, which can be achieved by a suitable exploratory procedure. But an advantage is that the second-order neuron signal may serve as a "synaptic probe" comparable with the junction of an inserted electrode. Thus the retina furnishes its own built-in amplifier. It is by virtue of this fact that small effects, resulting from experimental manipulations within physiological limits, are detectable, and that instantaneously.

Method

Simultaneous data on the processing of photic stimuli and the energy involved were obtained from frog retinas positioned between layers of supporting mesh in a compartment (Fig. 1) which had its inflow from and return to a magnetically driven centrifugal pump. The recirculating system was in turn connected to an external reservoir supplying room-air saturated nutrient medium as required. Transretinal electrical activity was led off from a pair of silver electrodes close to the retina, and a polarized platinum cathode served to monitor the partial pressure of oxygen of the medium which continuously washed the retina. While during fast perfusion of the chamber the pO_2 seen by the oxygen electrode would be essentially that of the reservoir, it would drop on cutting the external flow, when the oxygen available to the

* Supported by the Deutsche Forschungsgemeinschaft

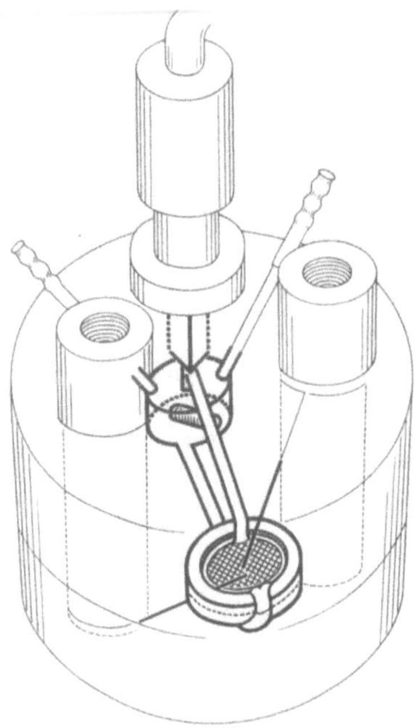

Fig. 1. Recirculating system to measure polarographically the oxygen uptake and light responses of perfused retina. See text.
Perfusate (mM/l): NaCl 80, KCl 2, $CaCl_2$ 0.15, $MgCl_2$ 0.1, phosphate buffer (pH 7.7) 15, glucose 5, O_2 0.26

retina was confined to the quantitiy dissolved in the one milliliter of recirculating solution. Some 5% of this is expended within 3 min, so that a periodic flow-stop procedure gives a conveniently measurable modulation of the pO_2 reflecting retinal oxygen uptake in a quasi-continuous fashion.

Alternatively, a constant slow flow of nutrient medium through the chamber could be established, while fast circulation within the chamber was maintained. The pO_2, as a result, would eventually reach a stationary level which is determined by the supply from the reservoir and the extraction by the retina, respectively, of oxygen. Thus the meter reading can directly be calibrated in terms of oxygen uptake, increasing downward, in the recordings.

Phenomena and Correlations

This latter procedure was adopted in the experiment of Fig. 2. The sample recording is to demonstrate three distinct effects of light on retinal energy metabolism, side-by-side. It shows above the stimulus track electroretinograms and in the upper trace the oxygen uptake. The recording represents a 3-hour experimental period but the time base is discontinuous: the recorder was advanced for a few sec (=millimeters) only, to protocol the ERG amplitudes together with the instantaneous PO_2, but was stopped thereafter for a 3 min interval, each.

1) Following an equilibration period with no stimulation light flashes, subthreshold for cones, of one second duration were applied regularly at 3 min intervals. This resulted in a slight downward movement of the upper trace, indicating additional oxygen extraction from the medium caused by the responding to the widely spaced stimuli. From the displacement of the upper trace the amount of extra-oxygen uptake per flash can be determined as $5 \cdot 10^{-10}$ mol.

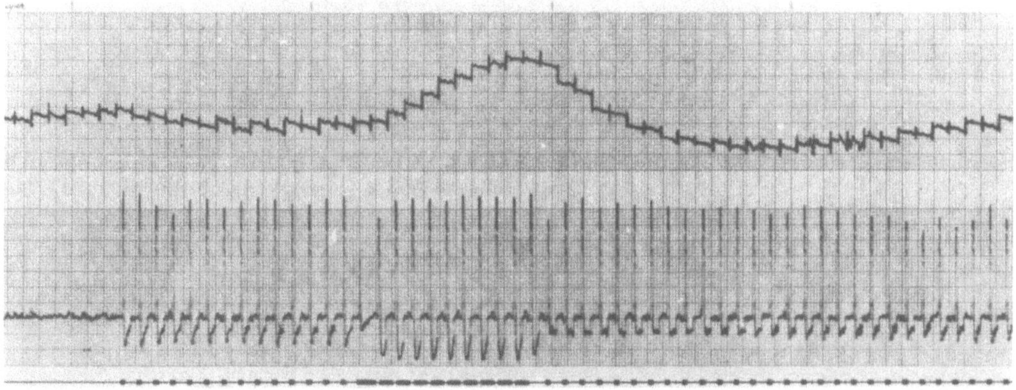

Fig. 2. Oxygen uptake and electrical responses of perfused retina as affected by brief stimuli and steady light. - Upper trace: change of oxygen uptake (increase downward); full scale 75 nM/h. - Lower trace: ERG-responses; 200 µV/cm; RC 1 sec. - Stimulus track showing light and "dark" flashes (1 sec duration in steady light, 10^{-2} lx)
The time base is discontinuous: the recorder was advanced for 5 sec (5 mm) only, with intervals of 3 min each; altogether a 3-hour experimental period is displayed. - Temp. 20°C; perfusion of chamber at 6 ml/hr; dry weight of retina 1.75 mg.
Notice (see text):

Stimuli (on or off)	increase O_2-uptake	from neural processing
Light (*per se*)	reduces O_2-uptake	at reduced receptor current
Pigment bleach	results in O_2-debt	for pigment regeneration

This quantity of extra-energy expenditure is not caused by receptor activity, because optical techniques had disclosed increments of cellular respiration both at on and off the stimulus, which the technique of oxygen measurement is not adequate to resolve. But extra-oxygen uptake was found in response to dark flashes in continuous light exposure. Therefore, it is changes of light - that trigger neural activity - which give rise to the increased energy demand, regardless whether the light goes on or off.

2) After a stationary state of oxygen uptake was reached under regular stimulation in dark adaptation the operation of the shutter was reversed: the light was now on continuously but for short 1 sec interruptions every 3 min; stimulation continued in light adaptation. The PO_2-trace climbed slowly towards a new stationary level indicating that the continuous irradiation had resulted in a decreased O_2-uptake; from the area under the curve by 12 nM O_2 for the half-hour period.
The time course of the pO_2 change is artefactually distorted but the onset and end of the effect is prompt, and the level reached would be, and has been found to be, maintained, for more than 5 hrs. Primarily on the basis of this behaviour the depressing effect of light on oxygen uptake is attributed to receptor cells, because it is a property of receptors to respond -electrically- to light with a sustained hyperpolarization, whereas higher-order neurons do not share this property but are activated transitorily by light changes. Supporting evidence comes from a chemical "slicing". Aspartate, which completely eradicates the above mentioned stimulated increments of oxygen uptake and the P II component of the electrical response - which is generated beyond the first synapse -, fails to do so with the light-induced depression of oxygen uptake and the P III component of the ERG, which reflects receptor activity. Furthermore, light intensities too strong to allow electrical res-

ponses to recover in dark adaptation, also prevented the preexisting higher rate of oxygen uptake to be resumed. In spite of the irreversible damage from the excessive light to the scotopic receptor population the rest of the retina proved intact, when activated through the other input, the cones, at higher light levels.

3) On removing the background light, and returning to the previous light stimulation schedule, the former - higher - rate of oxygen uptake is not resumed along a simple exponential time course, but with an overshoot. This overshoot would be more pronounced after longer light exposure, with stationarily reduced oxygen uptake. Thus a "debt" has accumulated, equivalent to 7.5 nM of oxygen in the present case, from the previous quantum catch, to be paid for not until the light has been removed. From the prolonged period of accumulation, again, one is led to attribute the effect to early processes in the chain, prior to a synapse. Indeed, the system to provide an energetic advance that size must have an extraordinary capacity, hardly imaginable other than in the large number of visual pigment molecules, ready to be bleached. It would, therefore, seem reasonable - and is tentatively suggested - to attribute this portion of extra-O_2 uptake that follows the exposure to light to the powering of regeneration of rhodopsin. However, the evidence is less compelling and the implications perhaps not expected.

Regeneration of Rhodopsin

In spite of the ease with which regeneration can come about in solution the process seems much more complicated in situ, and has been denied to occur in a retina detached from the underlying pigment epithelium. Numerous experiments have been designed to determine limiting steps, focussing on prosthetic groups, isomerizing enzymes, and energy sources.

For the contention arrived at above, that it is regeneration that takes up the energy liberated after light exposure, it is of course decisive whether regeneration occurs. This claim is more easily stated than proven, because the range within which regeneration takes place is small and not easily delt with employing conventional spectrophotometric instrumentation. In fact, it is the limitations within which regeneration was observed subjectively in the preparation and reversible extra-oxygen uptake measured that from their coincidence would add credit to the correlation. Purple color would not reappear after bleaching when the perfusion of the retina was inadequate, or cyanide added to the perfusate. Therefrom it is taken that of the "cofactors" enumerated above the energy supply attaches the utmost importance, because when the viability of the outer segments is impeded from energy shortage - or excessive light - the other components might leak out. They may then be found in the adjacent pigment epithelium, well in proportion to the preceding bleach, but not necessarily undergoing an integral step of regeneration there.

Another means of assessing bleaching and regeneration of rhodopsin with a high degree of resolution, i.e. for small changes, is offered from electrical measurements of a special kind: Employing DC-amplification one observes in transretinal recordings an off-effect which is triggered by the return of the rod receptor potential and occurs, after the latter has reached ceiling amplitude, the later the more intense the stimulus, and the earlier the more of the rhodopsin has been bleached previously. Other than its amplitude, its time of occurence is invariant with most parameters tested, except for 1) the numbers of pigment molecules present and 2) of the light quanta incident. In the experiment of Fig. 3 six responses to identical stimuli are superimposed; response 1 - having one of the high amplitude

Fig. 3. Electroretinograms before and after bleaching of approximately 10% of the rhodopsin. - Test stimuli (10^{-2} lx, 1 sec) were applied before (1) and 1 (2), 10 (3), 19 (4), 28 (5), and 37 (6) min after exposure for 1 min to 30 lx; DC-amplification; deflection of stimulus mark - 200 μV.
Notice: different recovery rates for b-wave amplitude and e-wave delay

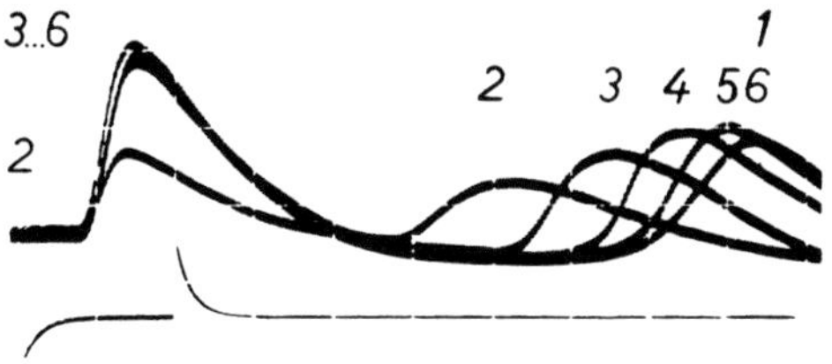

on-effects (b-wave) and the latest of the off-effects (e-wave) - was obtained before, and the others after a light exposure that bleached not more than 10% of the rhodopsin present. It will be seen that the response immediately following the bleach has a smaller b-wave and the early e-wave. But while the b-wave recovery is complete within 10 min, it takes more than 30 min to restore the pigment content, as seen from the coincidence of the e-waves.

The energy-utilizing step may well be the reduction of the retinaldehyde, which in the re-oxydation of the vitamin A is put to use in the isomerization of the prosthetic group. For reduced nicotin amid-adenine nucleotides seem an essential requirement. The reduced form of these nucleotides represents the higher amount of energy. The reduction is brought about metabolically at several sites, in the glycolytic chain, from reversed electron flow in the mitochondria, or at the expense of ATP. The involvement of an oxydation-reduction mechanism in the pigment regeneration may at the same time explain the high susceptibility of the process - and rod viability - to oxygen at higher partial pressure. The quantity of oxygen consumed for the amount of rhodopsin restored would appear high on a molecule-for-molecule basis, but the exact stoichiometry has to be worked out.

Dark Current

One point should perhaps be stressed, because it may bear on the difficulty to obtain regeneration in a retina after detachment from the pigment epithelium (which may result in more than just the physical separation of the two layers): The energy to be spent in the regeneration has to be available when and where needed. It is a most attractive hypothesis advanced by the authors when commenting on the significance of the receptor dark current, that at least in part it may serve to transport material electrophoretically from the inner to the outer segment of the receptors. If that is so, it would also account for the observation, that thermal regeneration - as determined metabolically - does not take place while the light shines. If there was regeneration in that period, perhaps it was photic.

The existence of the dark current as well as the hyperpolarizing nature of the vertebrate photoreceptor response, for which it is an explanation, is no longer doubted, and it is perhaps satisfying to see it energetically accounted for by the second of the effects described above. The dark current may carry the information to be transmitted across the synapse. Considerable work is presently being spent in investigations of its mode of generation and control. Probably more details will emerge from such work to fill the gap between the quantum catch and the generation of the synaptic signal. It may, however, be of intersest at this point, and eliminate ad hoc hypotheses, to state that up to and including the second-order neuron light res-

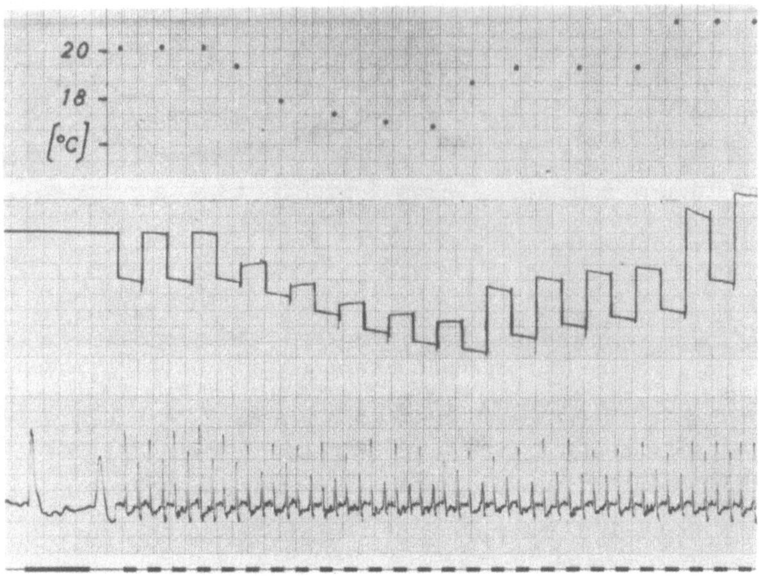

Fig. 4. Oxygen uptake and light responses in varied temperature. - Uppermost: temperature reading from air-thermostate housing the chamber of Fig. 1. - Upper trace: pO_2 recording; span: 0.208...0.260 $\mu M\ O_2/ml$. - Lower trace: ERG responses; as in Fig. 2. Stimulus duration 4 sec. - Time base: interrupted recordings as in Fig. 2 (first response recorded at 5 mm/sec).
Alternatingly the chamber had been rapidly perfused from the reservoir and the external flow been stopped, for 3 min each. Thus 1 mm downward displacement of the upper trace corresponds to an uptake of oxygen of $2 \cdot 10^{-8}$ M/hr.
Notice: high temperature influence upon oxygen uptake and off-responses, but not upon on-responses

---

ponse no processes other than of the diffusion type are involved. This would follow from experiments such as that of Fig. 4, aimed at determining Q_{10}.

Oxygen uptake (upper trace) and light responses (below) were recorded at varied temperature (top). The light stimuli, applied every 3 min, were longer in duration to bring out more clearly the on- and off-component of the response. The oxygen uptake was determined by the flow stop technique: the modulation depth of the pO_2-trace gives the amount of oxygen taken up per 3 min. Due to temperature dependencies of the measuring technique the baseline (upper level) was not constant, but this would affect the results but little, as the baseline (zero-pO_2) was suppressed by 5 chart spans.

It is immediately seen that the oxygen uptake is severely depressed in the lowered temperature, but the on-responses are hardly so. What does change is the off-responses. But this is not due to a trigger of opposite sign, because they would be equally resistant to temperature changes, if they preceded the on-response, i.e. on "dark flashes". In the experiment the off-responses behave as the second of a double-stimulus response. It is the recovery process that parallels in its temperature coefficient that of the oxygen uptake, whereas that of the immediate response does not exceed Q_{10} = 1.4.

Fig. 5. Electroretinograms in different Ca^{++}-concentrations. - Technique and display as in Fig. 3.
Notice: prompter on-reaction and more delayed off-reaction in the higher concentration of calcium, as though more pigment molecules had been hit

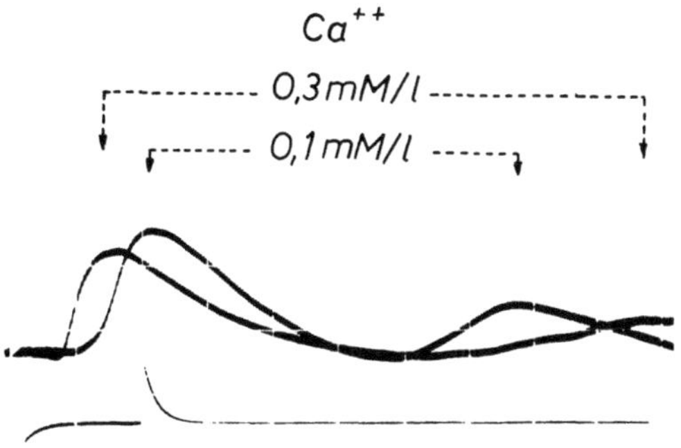

If, therefore, for the initiation of the receptor output signal passive ion fluxes are responsible, a gating process operating on membrane permeability is to be demonstrated. Calcium would seem a likely candidate to be involved. In fact, calcium proved to be an exceptionally effective parameter to take influence on the shape of the electrical light response - in the sense of adding to the effect of light (Fig. 5): shortening the on-latency and lengthening the delay of the off-reaction. Translated in terms of receptor activity it is a more sudden onset and a delayed decay with decreased slope of the receptor potential. The effect of calcium resembles that of hydrogen ions, which had previously been shown to mimic several aspects of light adaptation.

Conclusions

A substantial fraction of retinal energy metabolism is blocked in the light. It covers most of the energy expended by the photoreceptors, which is spent in the production of a dark current.

The control of the dark current involves a light activated Ca^{++}-mechanism which acts on membrane permeability, with the resulting fluxes of univalent ions activating ATPase and oxygen uptake.

Onset and decay of the receptor potential coincide over a wide range with the triggering of second-order neuron activity. Therefore, the dark current, which is responsible for the receptor potential, carries the information to be transmitted.

The dark current may also serve to transport material necessary in the regeneration of bleached pigment to the site where it is needed. Regeneration in the absence of a dark current, in the light, if it occurs, may be photic.

The energy required in thermal regeneration of photopigments is supplied probably in the form of reduced pyridine nucleotides, which are produced under extra-oxygen uptake, and utilized for re-isomerization in the re-oxidation of the retinol.

The involvement in the photochemical cycle of the redox step would seem to account for the toxic effect of oxygen under high pressure.

References

For a bibliography see:

1. SICKEL, W.: Retinal metabolism in dark and light. In: Handbook of Sensory Physiology, Vol. VII/2, ed. M.G.F. FUORTES, pp. 667-727 Springer, Berlin-Heidelberg-New York (1972).
2. SICKEL, W.: Electrical and metabolic manifestations of receptor and higher-order neuron activity in vertebrate retina. Adv. Exp. Med. Biol. 24, 101-118 (1972).

Special reference is made to the contributions:

3. BRIDGES, C.D.B.: Interrelations of visual pigments and "vitamin A" in fish and amphibia. This volume, pp. 115-121.
4. FUTTERMAN, S., M.H. ROLLINS: A catalytic role of dihydroriboflavin in the geometrical isomerization of all-trans-retinal. This volume, pp. 123-130.
5. WALD, G.: Visual pigments and photoreceptor physiology. This volume, pp. 1-13.
6. YOSHIKAMI, S., W.A. HAGINS: Control of the dark current in vertebrate rods and cones. This volume, pp. 245-255.

Discussion

D. Bownds: Do you think that a correlation between receptor activity and O_2 consumption could be made more effectively in an aspartate treated retina?

W. Sickel: The experiments employing aspartate corroborated the conclusions drawn. However, a concentration of aspartate adequate to isolate receptor responses ($3 \cdot 10^{-4}$ mol/l) would increase the overall O_2-uptake and make the light depression of oxygen uptake correspondingly smaller. Higher concentration of aspartate proved poorly reversible, and were considered suspect for this reason.

A.I. Cohen: It cannot be the case that all non-transient responses are from photoreceptors since horizontal and bipolar neurons as well as photoreceptors show sustained responses. A better way to isolate photoreceptor activity is to use retinas from rodents treated postnatally for 10 days with mono-sodium glutamate. These retinas at three months of age consist largely of photoreceptors and Müller cells and only a small surviving population of inner retinal neurons.

W. Sickel: Thank you, you mentioned the rarefied retina, which I now feel very much encouraged to try. I had heard of it from Albert Potts. These animals are blind, and apart from the morphological criteria - which have improved - one had little to be sure of the normalcy of the receptor function at that time. - The distinction between photoreceptor and higher-order neuron activity I proposed was to be immune from possible consequences of a chemical or surgical interference. It is based, therefore, primarily on the electroretinogram, which from more recent experimentation seems to faithfully represent integral activity of the retina - which is what is needed for a correlation with metabolic activities; far better than it is reputed to be: It correlates well with activity transmitted to the tectum (Pickering, Varjú) or cortex (Crescitelli), and both from microprobings and double-stimulus explorations it allows to single out receptor contribution. Perhaps the metabolic correlation could add further credit to the ERG. Bipolars have a PD-characteristic, but the P-part is small and certainly not good for sustained activity of a duration discussed here. The horizontal cell's contribution should be taken care of by the aspartate experiments.

H. Langer: By what means could you be sure of measuring NAD-redox stage by ΔE at 340 nm in the retina? Could change of absorption of rhodopsin in bleaching be excluded?

W. Sickel: I went to great pains initially to distinguish absorption by photopigments and respiratory pigments, respectively. The most straightforward argument seems that respiratory responses can be elicited with light stimuli so small (near threshold) that no measurable change in photopigments is to be expected, and that they can be obtained the same on "dark flashes", when certainly something different should happen to the photopigments.

S.L. Bonting: With regard to the proof of a non-electrogenic cation pump: 1) Is your time resolution of change of oxygen uptake and electrical effects close enough? 2) Is the distinction between receptor O_2-uptake and other retinal O_2-uptake clear enough for such a conclusion?

W. Sickel: Other than the reasoning from the rapidity of drug actions, mine was essentially time-independent: two kinds of retinal activity - O_2-uptake and off-responses - exhibited a high temperature dependency, and one other - the on-response - did not. Therefore, whatever processes were involved to produce the on-reaction - and this includes receptor function -, none would seem to draw directly on metabolic energy. What did were the processes to re-cock the system for the next action, the responding to the off.

G. Wald: The frog retina has a very large capacity for glycolysis, and large stores of glycogen to supply it. When respiration fails can't this fulfill all energy demands? May not the increased oxygen consumption after light is shut off be to remove lactic acid performed by (aerobic) fermentation in the light?

W. Sickel: Lactic acid formation in the retina is high (although under operating conditions not quite as high as would appear from classical Warburg measurements). It is essential, but cannot substitute for oxidation, as can be seen from ERG responses, which signal an anoxic core immediately. Total acid formation of the retina is reduced for the duration of a prolonged exposure to light, as is $^{14}CO_2$ production. Therefore, it does not seem to be lactic acid that has accumulated, because, at the quantity to b accounted for, it should either have leaked out (and become measurable) or affected the light responses (in the sense of a progressively reduced pH).

Rod Dark-Adaptation and Visual Pigment Photoproducts

K. O. Donner
Department of Physiological Zoology, University of Helsinki, Helsinki/Finland

Measurement of the dark-adaptation of single retinal units (ganglion cells) in the frog has shown that the rod branch of the dark-adaptation curve after bleaching small amounts of rhodopsin is approximately exponential when thresholds are given in log units (1). This applies only when the light used to test the sensitivity of the units strikes those rods that have been exposed to the bleaching light, suggesting that the adaptation process takes place in the receptors (2). Measurements of the decay of the photoproducts produced by the bleaching of rhodopsin show a correlation between the time course of rod dark-adaptation (when thresholds are plotted on a log scale) and the decay of a 380-nm product, assumed to be metarhodopsin II (MII). This applies, though less well, also to metarhodopsin III (λ_{max} 470-475 nm) (3). These observations suggest that the 380-nm product (MII?) or photoproducts in general are regulating rod sensitivity in dark-adaptation in such a way that the log threshold of the rods is directly proportional to the amount of these substances present in the rods (1,2,4). It should be stressed that these experiments were carried out so as to exclude other factors, such as changes in the organization of the receptive field and too rapid application of the test flash used to measure sensitivity. Both of these factors can be shown to affect the final result obtained. Green light was used as a test stimulus in order to favour stimulation of the rhodopsin rods (5).

Two main objections can be raised against the validity of this hypothesis. First, in view of the fact that the actual experiments were carried out at only two temperatures the observed correlation between metarhodopsin decay and dark-adaptation might be fortuitous and not based on any causal relationship between both processes. Second, the kinetics of the decay of the photoproducts of frog rhodopsin *in situ* were not at that time known well enough to justify attributing the adaptation effect specifically to MII.

Regarding the first point, additional experimental evidence has been obtained on the isolated frog retina (the previous experiments were carried out on the excised and opened eye). Furthermore, similar experiments to those on the frog have been carried out on the opened eye of the crucian carp, which has an all-porphyropsin retina. A typical result is shown for the crucian carp in Fig. 1, which gives the dark-adaptation curves for two units after bleaching about 2.5 per cent of the porphyropsin. The upper curve refers to an experiment performed at +17 °C and the lower at +7 °C. The triangles indicate values of the cone threshold, and the dots indicate values of the rod threshold. The open circles show the decay of the 400-nm photoproduct formed by bleaching the porphyropsin, scaled from measurements on an isolated retina according to the right hand ordinate and with a time scale beginning from the arrows in the figure. Though these latter data refer to a total bleach of the porphyropsin it is obvious that both processes - the rod adaptation and the decay of the photoproduct proceed at the same rate. Thus for the crucian carp the same kind of result is obtained as in the case of the frog. In the crucian carp retina this result is more straightforward, because there is only a single 400-nm photoproduct (cf. Reuter, 6) that is more long-lasting, and there is no trace of a MIII-like product.

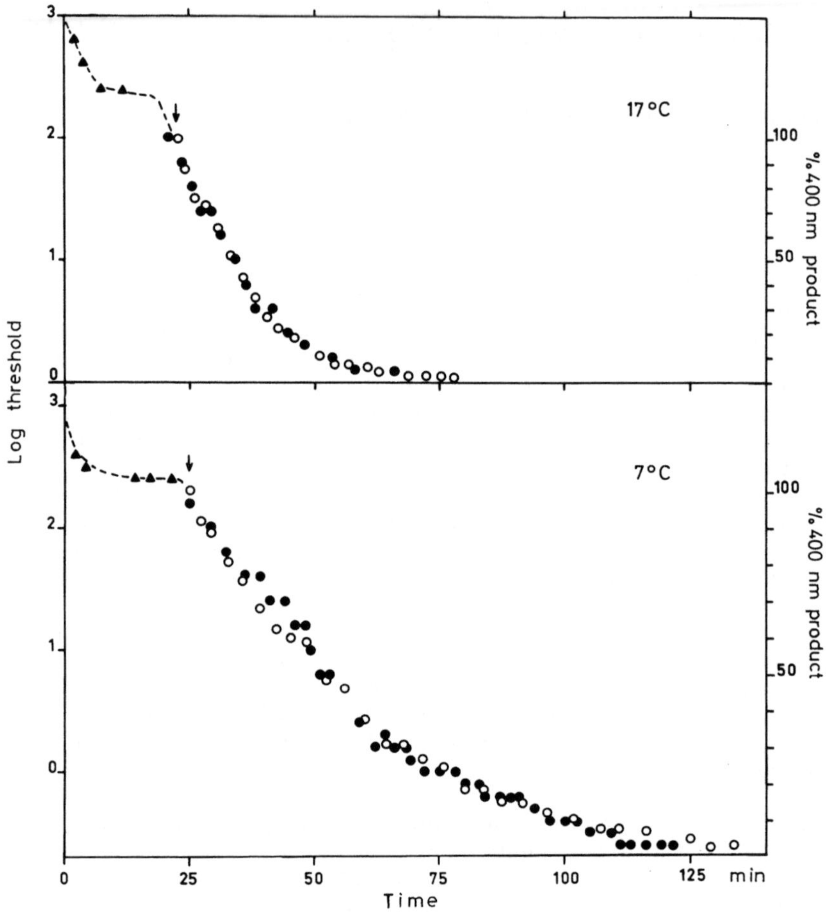

Fig. 1. Dark-adaptation and photoproduct decay in the crucian carp retina. Triangles (cones) and dots (rods) give thresholds of single units after bleaching 2.5% of the porphyropsin. Test light 513 nm, 0.35 mm field centered on the receptive field. The open circles give measurements of the decay of the 400 nm product (ordinate to the right) as measured on the isolated retina. The measurements have been scaled to start from the point indicated by the arrow. Experiments at +17 °C (upper data) and +7 °C (lower). Data from Donner, Hongell, and Reuter (in preparation)

The data so far obtained have been used to derive rate constants for both the rod phase of the adaptation curve and for the decay of the 380- or 400-nm photoproducts. These data are given in Fig. 2. For the frog the dots refer to experiments on the isolated retina (Bäckström, Donner, and Virtanen, unpublished) and for the crucian carp on the excised and opened eye. The dark-adaptation curve was recorded as described. The open symbols, on the other hand, give rate constants for the decay of the 380- and 400-nm photoproducts respectively from measurements on the isolated retina. The values for the frog are from Donner and Reuter (1). All the data for the crucian carp are from Donner, Hongell, and Reuter (unpublished).

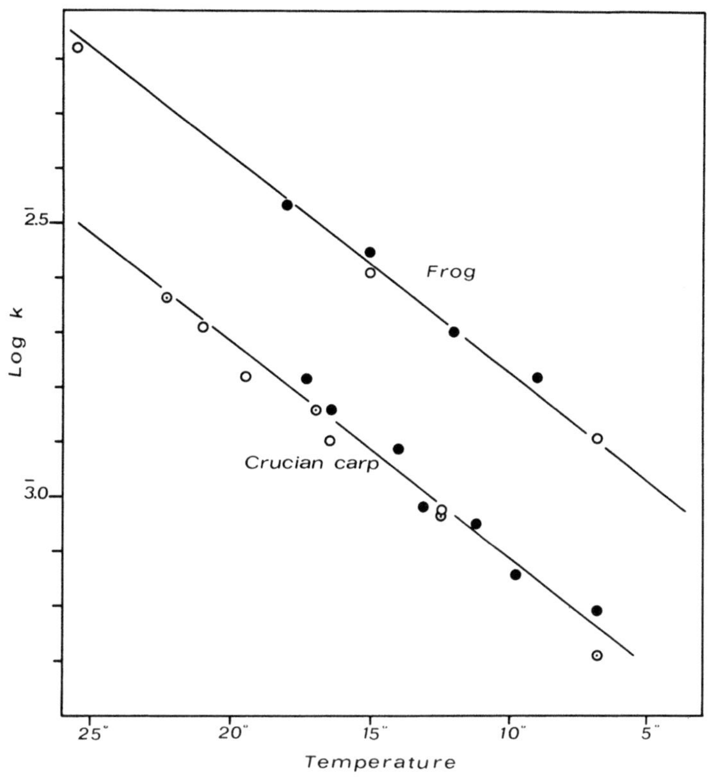

Fig. 2. The rate constants (k) given as a function of temperature for rod dark-adaptation (filled symbols) and 380 - 400 nm photoproduct decay (open symbols). Data for the frog from Bäckström, Donner, and Virtanen (in preparation) and from Donner and Reuter (1), data for the crucian carp from Donner, Hongell, and Reuter (in preparation)

There is a fair agreement between the rates of photoproduct decay and rod adaptation at different temperatures. It therefore seems justified to conclude that some causal link exists between rod sensitivity and the presence of MII - or metaporphyropsin II - like photoproducts. Quantitatively, a linear relation between the log threshold and the amount of "MII" pigment is obtained (1,2). This relation has, however, been experimentally verified only for photoproduct amounts less than about 5% of the total rod pigment.

As to the second question - the identification of the photoproduct responsible for this effect - the original suggestion was that in the frog this photoproduct was MII. This was based on the identification of the absorption increase maximal at 380 nm as that appeared immediately after bleaching and decayed slowly (1,7). In Baumann's (3) recent analysis of the kinetics of rhodopsin photoproducts in the frog this near-ultraviolet absorption is assumed to be caused by MII initially (corresponding to the stabilized MII of Donner and Reuter, 1) and later by retinal plus opsin. The kinetic model used by Baumann in a very satisfactory way describes the decay of the photoproducts, but gives no clue that would make a direct identification of the 380 nm substance possible, since both MII and retinal show only minor differences in their absorption spectra. Further, the model used by Baumann assumes that retinal is still oriented in the plane of the disc membranes of the rods in the same way as when attached to the opsin. It is obvious that the measurements and rate constants referring to the frog retina as given in Fig. 2 represent both MII and retinal (according to Baumann's model). According to our previous model they represent stabilized MII and MII (1). It is thus fairly clear that two substances are involved, and possibly both of them have a desensitizing effect on the rods.

In the crucian carp retina the situation is more simple, because only a single, more long-lasting photoproduct seems to occur after bleaching porphyropsin. This photoproduct absorbs maximally at about 400 nm and thus corresponds to the absorption of metarhodopsin II but also to that of free retinal$_2$. The decay can be described by a first-order reaction. However, a positive identification of this substance with metaporphyropsin II or free retinal$_2$ cannot be made (cf. Reuter, 6).

In the frog it was found (2) that when the amount of photoproducts was 1% of the total rhodopsin the threshold was elevated by about 1 log unit. In the crucian carp the corresponding figure is around 2.5 log units. This significant difference may be caused by the fact that in the crucian carp all bleached porphyropsin molecules occur as the 400-nm product, and presumably all cause rod desensitization. However, in the frog roughly half of the molecules decay over the parallel pathway of MIII (according to Baumann's model) and perhaps do not, in that state, affect the adaptation process. This would mean that the figure for the frog should be given in terms of the 380-nm products only. Then 1% of the total as 380-nm products would raise the threshold roughly 2 log units, bringing the value close to that obtained for the crucian carp.

References

1. DONNER, K.O., T. REUTER: Dark-adaptation processes in the rhodopsin rods of the frog's retina. Vision Res. 7, 17-41 (1967).
2. DONNER, K.O., T. REUTER: Visual adaptation of the rhodopsin rods in the frog's retina. J.Physiol. 199, 59-87 (1968).
3. BAUMANN, C.: Kinetics of slow thermal reactions during the bleaching of rhodopsin in the perfused frog retina. J.Physiol. 222, 643-663 (1972).
4. DONNER, K.O., T. REUTER: The effect of metarhodopsin on the sensitivity of the rhodopsin rods in the frog during dark-adaptation. Acta physiol. scand. 68, Suppl. 277, 40 (1966).
5. DONNER, K.O., T. REUTER: The dark-adaptation of single units in the frog's retina and its relation to the regeneration of rhodopsin. Vision Res. 5, 615-632 (1965).
6. REUTER, T.: Long-lived photoproducts in the retinae of the frog and the crucian carp. This volume pp. 211-217.
7. DONNER, K.O., T. REUTER: The photoproducts of rhodopsin in the isolated retina of the frog. Vision Res. 9, 815-847 (1969).
8. BAUMANN, C.: Regeneration of porphyropsin in vivo. Nature (London) 233, 484-485 (1971).

Discussion

T.P. Williams: Have you measured the activation energy for the disappearance of the meta's in the fish and the frog?

K.O. Donner: For the frog our data actually refer to the decay of MIII that is, however, approximately paralleled to the MII - retinal decay. For the crucian carp, figures for the decay refer to the 400-nm product. In both cases the activation energy appears to be roughly the same, a calculation gives a value around 17 kcal/ Mole.

G. Wald: What does the 400-nm pigment go to as it decays in the dark in the crucian carp? There seem to me to be only two choices: in the isolated retina it will end up mainly as $retinol_2$, in the living animal or opened eye mainly as porphyropsin. Which is it doing in these experiments?

K.O. Donner: Our measurements of the decay of the 400-nm pigment were carried out on the isolated retina where it goes to $retinol_2$. As to the situation in the opened eye or in the living animal we have no direct evidence on this point. However, Baumann's (8) measurements of the regeneration of porphyropsin in the crucian carp after bleaching about 70% of the pigment show an extremely slow time-course of this process, with an initial delay of about 1 hr at $+15^{\circ}C$. Then it does not seem likely that the 400-nm pigment would go directly back to porphyropsin in the opened eye.

S.L. Bonting: In connection with Professor Wald's question on role of the MII decay in dark adaptation, would you comment on the following tentative explanation: Conversion of pigment to MII causes a change in the disc membrane, e.g. permeability decrease with Ca^{++} release, leading to the Na^{+} permeability decrease in the outer membrane. As long as the MII is present in the disc membrane, this disc is out of action. Removal of the MII (without necessarily returning to photopigment immediately) would reactivate the disc.

K.O. Donner: I agree that some effect of the kind suggested by you seems to be indicated in the presence of the photoproducts we have studied, perhaps not resulting in a complete inactivation of single discs but rather in a reduction of their response to light.

Long-Lived Photoproducts in the Retinae of the Frog and the Crucian Carp

Tom Reuter
Department of Physiological Zoology, University of Helsinki, Helsinki/Finland

This paper is a summary describing the results of recording absorption spectra and absorption changes after irradiation of isolated frog and crucian carp retinae mounted in a transparent chamber inside the sample compartment of a spectrophotometer. Detailed descriptions of the experiments will be given in three separate papers in preparation.

Photoproducts in the Retina of the Frog (Rana temporaria and R. pipiens)

The photoproducts pre-lumirhodopsin, lumirhodopsin, and metarhodopsin I (MI) are very short-lived in the frog's retina. Even at 4 °C the photoactivated rhodopsin is converted into metarhodopsin II (λ_{max}=380 nm; MII) within a few seconds. MII is then relatively slowly (half-time several minutes at 4 °C) converted into a mixture of another 380 nm intermediate and a 470 nm pigment. Mainly because a low pH favours the 380 nm component in this mixture we assumed a few years ago (1,2) that these pigments exist in a pH-dependent equilibrium very similar to the MI-II equilibrium observed after irradiation of cattle rhodopsin in solution (3). However, Baumann (4) has recently provided evidence confirming the views of several earlier authors (3,5,6,7), that these two pigments represent two separate pathways for the decay of MII. Following Weale (8) Baumann names the 470 nm pigment MIII (it is called pararhodopsin by Wald, 5) and he identifies the 380 nm component in the mixture with free retinal. Baumann (4) finds that his data are best fitted by the following sequence:

MII (380 nm) $\xrightarrow{k_1}$ MIII (470 nm)

MIII (470 nm) $\xrightarrow{k_3}$ retinal (380 nm)

MII (380 nm) $\xrightarrow{k_2}$ retinal (380 nm) $\xrightarrow{k_4}$ vitamin A (330 nm)

We have obtained independent evidence against the hypothesis that a pH-dependent equilibrium of the MI - MII - type exists between the pigments called MIII and retinal in this scheme. At a time when considerable amounts of both these products had accumulated, we produced rapid pH-changes in the solution surrounding the retina, and found that these pH-changes did not result in sudden and dramatic changes in the relative amounts of the two pigments (Gyllenberg, Reuter, and Sippel, in prep.).

It was still necessary to explain why a retina, which during the whole bleaching process bathed in an acid medium, produced much less MIII than one in alkaline medium. In order to clarify this we have followed the density changes at 480 and 390 nm during the formation and decay of MIII in a series of experiments at 4 °C and pH 6.3, 7.0, and 7.9. In agreement with Baumann (4) we found that the above scheme fits the data much better than a model, in which all MII decays through MIII (Gyllenberg et al., in prep.). The rate constants k_1 and k_3 were empirically determined from the first part of the absorbance change at 390 nm and the later part of the absorbance change at 480 nm respectively, while appropriate k_2- and k_4-values were selected with a computer simulation technique. The accumulation of large amounts of MIII in retinae in alkaline media is explained by the fact that k_1 is larger than k_2 at pH 7.9 while the reverse is true at pH 6.3 (Gyllenberg et al., in prep.). The fact that the k_3-value decreases with increasing pH in the pH 6-8 region (1) also helps to account for the accumulation of MIII at pH 7.9.

Very little is known about the chemical nature of MIII. Matthews et al. (3) and Williams (9) have found that irradiation of MII rapidly produces large amounts of MIII (or a very similar compound), and it is also claimed that irradiation of MIII drives it back to a MII-like substance (3,6,7,9). These findings suggest that the chromophore of MIII could be a cis-isomer of retinal (3,9). On the other hand, we (10,2) have claimed that an isomerization of this intermediate in the first place produces rhodopsin and isorhodopsin, which then of course is converted into MII if the pigment molecules absorb several photons. This latter observation, which now is confirmed (Reuter, in prep.), rather suggests that the chromophore of MIII is all-trans-retinal.

The confirmatory evidence was obtained as follows during work in G. Wald's laboratory at Harvard University. Both retinae from a dark-adapted frog (Rana pipiens) were isolated, spread out on a glass surface and exposed to a strong 40 sec orange irradiation, which transformed 90% of the rhodopsin into MII leaving about 3-4% rhodopsin and 6-7% isorhodopsin. The two retinae were then left in darkness for 17 min, the time needed for accumulating MIII to reach its peak (pH 8, the whole experiment was performed at 4°C). Then one of the retinae was exposed to a 15 sec 492 nm irradiation, a wavelength absorbed by MIII but not by MII or free retinal, and immediately afterwards washed in an acetate buffer containing hydroxylamine, which prevented further synthesis of visual pigments. The other retina was not exposed to this second irradiation, but was also washed with hydroxylamine 17 min after the first bleach. Both retinae were then further washed and the pigments extracted with digitonin using the methods described by Reuter, White, and Wald (11). Absorption spectra of the extracts were recorded with a Cary Model 11 spectrophotometer.

It was found that the retina, which was exposed to the 15 sec 492 nm irradiation, contained on an average twice as much rhodopsin + isorhodopsin as the other retina, and if the 492 nm irradiation continued for 30 sec the amount of rhodopsin + isorhodopsin was still higher in the irradiated retina. This is shown in Fig. 1, which compares the amount of rhodopsin and isorhodopsin (the difference between curves 1 and 3) extracted from a retina exposed 30 sec to 492 nm (A) with the amount extracted from the other retina exposed only to the first orange bleach (B) (cf. also ref. 2, Table I).

It is thus clear that irradiation of MIII increases the amount of rhodopsin and isorhodopsin, and the mechanism is probably an isomerization of the all-trans-chromophore of MIII to 11-cis and 9-cis. Using these and similar experiments it was possible to calculate that MIII is converted into rhodopsin and isorhodopsin with the relative quantum efficiencies 0.4 and 0.3 respectively, when the quantum efficiency for isomerizing rhodopsin was taken as 1.0 (Reuter, in prep.). This can be compared with the relative quantum efficiencies 0.5 and 0.1 obtained by Hubbard and Kropf (12) for the reisomerization in solution at -20°C of cattle MI to rhodopsin and isorhodopsin respectively. The intensity of the 492 nm light was obtained by directly measuring its bleaching effect on dark-adapted retinae.

Photoproducts in the Retina of the Crucian Carp

The crucian carp (Carassius carassius) is a freshwater fish whose rods contain a pure vitamin A_2-based visual pigment, porphyropsin, with λ_{max} 522 nm (13,14).

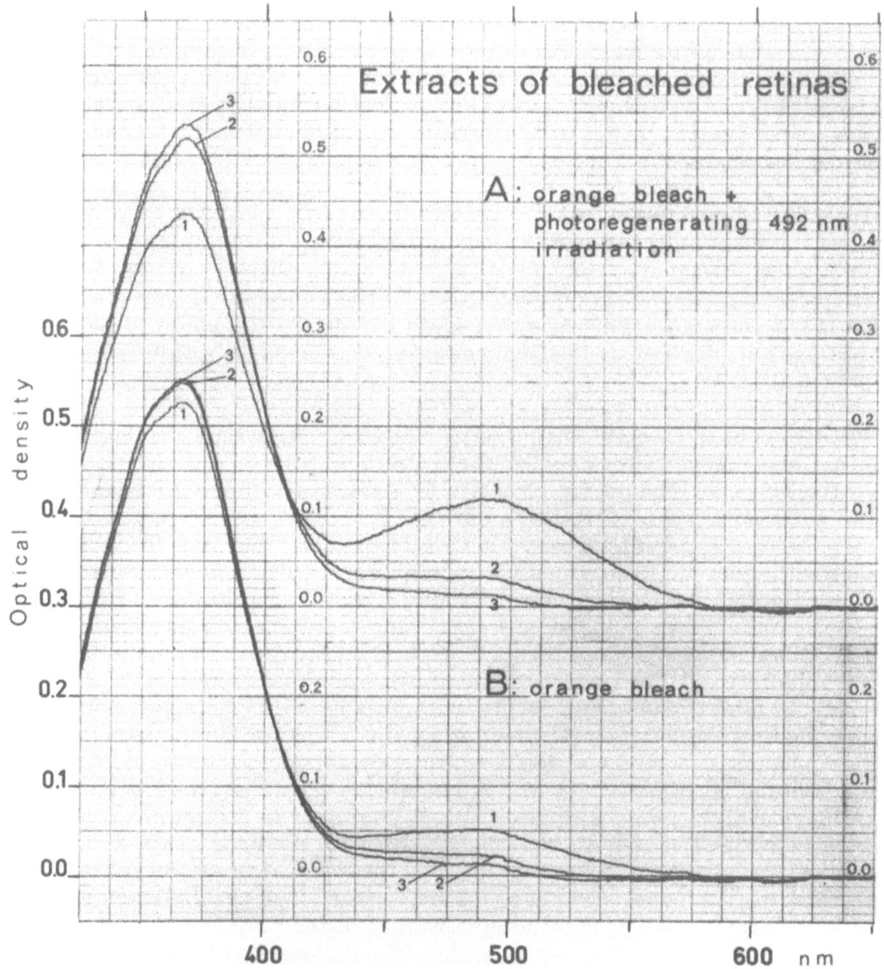

Fig. 1. Absorbance spectra of extracts of two retinae from the same frog. Before the preparation of the extracts these retinae were exposed to the following bleaching program: Retina A was exposed first 40 sec to a strong orange light (Schott cutoff filter OG-5) and 17 min later to a 30 sec 492 nm irradiation (interference filter). Retina B was exposed only to the first orange light. The spectra shown represent the pigments remaining after these exposures. Both extracts contained 0.05 M hydroxylamine and were bleached with the same exposures to orange and yellow lights. Curve 1 is the spectrum of the unbleached extract, curve 2 after 10 min irradiation with an orange light (Schott OG-2) removing all rhodopsin and some isorhodopsin, curve 3 after 10 min yellow irradiation (Schott OG-4) removing all remaining isorhodopsin. The difference between curves 1 and 3 gives the total amount of rhodopsin+isorhodopsin present in the extracts, the exact difference spectra can be used to calculate the relative amounts of these pigments. The high retinal oxime peaks at 367 nm shown already by curve 1 indicate the original amounts of rhodopsin in the retinae, because the retinal oxime formed during the washing of the exposed retinae is quantitatively extracted

The sequence of long-lived photoproducts of this pigment was found to be very simple. One minute after the bleach the only product left is a yellow pigment (λ_{max} 390-400 nm), which during the next few hours undergoes a slow and more or less exponential decay to vitamin A_2 (3-dehydroretinol) and opsin (see Fig. 2A; Donner, Hongell, and Reuter, in prep.). The rate of this decay was measured in the temperature region 7-22°C (see ref. 15).

No "metaporphyropsin III" was observed in the retina of this fish. This difference between the bleaching patterns of crucian carp porphyropsin and frog rhodopsin does not, however, originate from the different chromophores, but rather from some characteristic of the opsin or peculiarity of the rod outer segments of this fish, because large amounts of metaporphyropsin III have been observed in the most dorsal zone of the bullfrog retina (Gyllenberg et al., in prep.), which may contain 80-90% porphyropsin (11).

We were unable to identify unequivocally the yellow 400 nm product observed in the crucian carp retina with either metaporphyropsin II or "free" 3-dehydroretinal. It is well known that irradiation of cattle MII results within milliseconds in the formation of rhodopsin and a MIII-like pigment (3,9). One would expect metaporphyropsin II to behave in a similar way, but we have found that irradiation of the 400 nm product with near UV light does not cause an immediate photoconversion to other pigments, but rather a slow conversion of a small fraction of the 400 nm product to porphyropsin and isoporphyropsin, while most of it decays to vitamin A_2 in the normal way (Donner et al., in prep.). This finding, together with the fact that no metaporphyropsin III-like product has been observed during the decay of the 400 nm product, would tend to favour an identification with "free" 3-dehydroretinal. However, several other observations (Donner et al., in prep.) do not support this idea: The 400 nm product is formed within one minute after illumination at 0°C, a temperature at which cleavage of MII into opsin and free retinal takes several hours not only in the retinae of frogs and rats (1,2,16) but also in digitonin extracts of crucian carp retinae (see Fig. 2B). Like MII and MIII in the rods of the frog (1,2,17,18,19, 20) the 400 nm product is still oriented in the same way as the parent visual pigment, while in aged frog rods, however, we know that "free" retinal tends to assume an axial orientation, i.e. at right angles to that of the original rhodopsin (18). Moreover, if the 400 nm product were "free" 3-dehydroretinal we would expect an acceleration of vitamin A_2 formation when $NADH_2$ and $NADPH_2$ are added to the bathing medium, yet we have found that neither of these compounds has any marked effect on the rate of decay of the 400 nm product.

The Bleaching Sequence in situ and in vitro

It has often been said that the sequence of bleaching products is essentially the same in digitonin solution as *in situ* in the rod outer segments, apart from the enzymatic reduction of retinal to vitamin A in a fresh retina (3,16,21). However, Donner and Reuter (2) have pointed out that there is an interesting difference between the *in situ* and *in vitro* sequences at the stage believed to be associated with visual excitation, i.e. conversion of MI to MII. Even at low temperatures and alkaline pH the only intermediate present in a frog retina a few seconds after a short exposure to strong light (i.e. before the accumulation of MIII) is MII, while an exposure of a digitonin extract of frog retinae directly results in a mixture of 380 and 470 nm pigments (22,23; Gyllenberg et al., in prep.). At present a MI - MII equilibrium of the kind described by Matthews et al. (3) has not been observed in a retina.

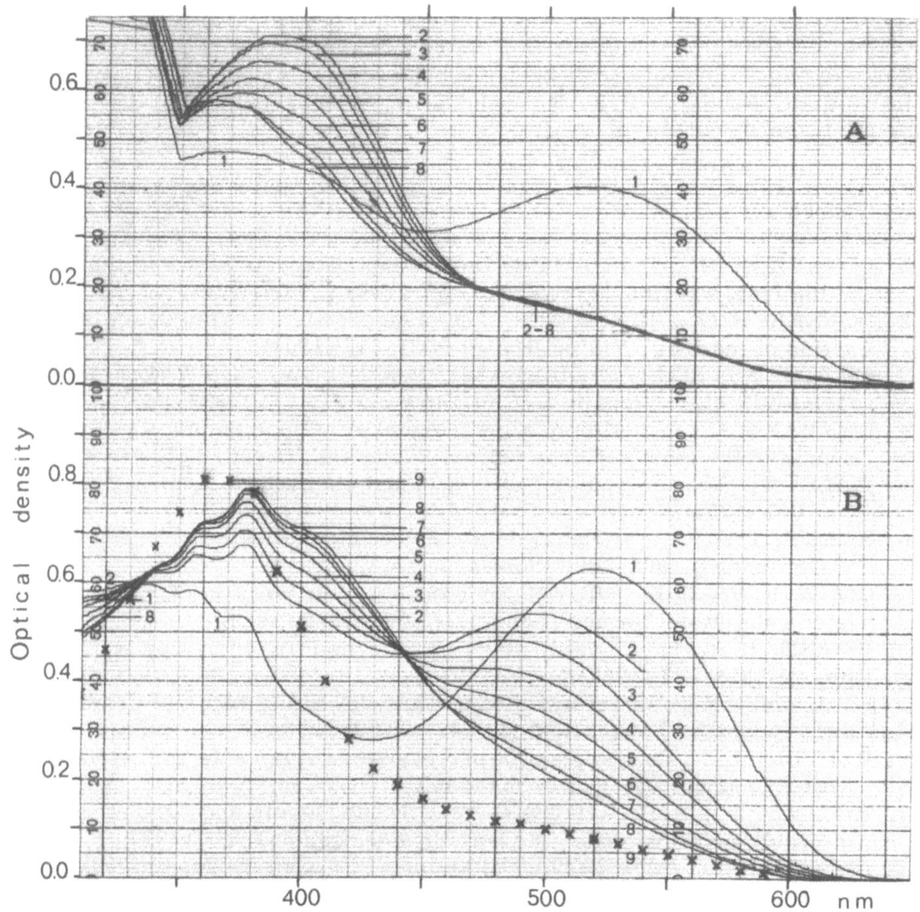

Fig. 2.(A) Absorbance spectra of an isolated crucian carp retina at 0.5°C and pH 7.6. Curve 1 is the spectrum of the unbleached retina. Curve 2 is started from 540 nm 20 sec after the end of a 20 sec yellow irradiation (Schott cutoff filter GG-14). Curves 3-8 are started from 650 nm 8 (3), 32 (4), 65 (5), 105 (6), 163 (7), and 230 (8) min after the irradiation. (B) Absorbance spectra of a retinal extract (3% digitonin) of crucian carp at 0.5°C and pH 7.6. Curve 1 is the spectrum of the unbleached extract. Curve 2 is started from 540 nm 10 sec after the end of a 20 sec yellow irradiation (Schott OG-1). Curves 3-8 are started from 660 nm 7 (3), 25 (4), 56 (5), 104 (6), 181 (7), and 280 (8) min after the irradiation. At 308 min a very small granule of sodium borohydride was stirred into the solution. Curve 9, here plotted with small crosses, was started from 660 nm 310 min after the bleach. - The curves in both 2 A and B were recorded with a Beckman Model DB spectrophotometer (speed 40 nm/min)

The difference between the sequences _in situ_ and _in vitro_ is unusually striking in the case of crucian carp porphyropsin. _In situ_ illumination converts it rapidly into a 400 nm product, while in digitonin solution a very long-lived 480 nm product is observed instead (Fig. 2A and B and Donner _et al._ , in prep.).

Acknowledgement. This work was supported in part by grants from the National Research Council for Sciences in Finland. During part of the investigation the author was on a fellowship from the European Molecular Biology Organization. I thank G. Wald, R. Hubbard, P.K. Brown, and W. Wright for helpful discussions concerning the photoconversion of MIII.

References

1. DONNER, K.O., T. REUTER: Dark-adaptation processes in the rhodopsin rods of the frog's retina. Vision Res. 7, 17-41 (1967).
2. DONNER, K.O., T. REUTER: The photoproducts of rhodopsin in the isolated retina of the frog. Vision Res. 9, 815-847 (1969).
3. MATTHEWS, R.G., R. HUBBARD, P.K. BROWN, G. WALD: Tautomeric forms of metarhodopsin. J.Gen.Physiol. 47, 215-240 (1963).
4. BAUMANN, C.: Kinetics of slow thermal reactions during the bleaching of rhodopsin in the perfused frog retina. J.Physiol. (Lond.) 222, 643-663 (1972).
5. WALD, G.: The molecular basis of visual excitation. Nature (Lond.) 219, 800-807 (1968).
6. EBREY, T.G.: The thermal decay of the intermediates of rhodopsin in situ. Vision Res. 8, 965-982 (1968).
7. CONE, R.A., P.K. BROWN: Spontaneous regeneration of rhodopsin in the isolated rat retina. Nature (Lond.) 221, 818-820 (1969).
8. WEALE, R.A.: On an early stage of rhodopsin regeneration in man. Vision Res. 7, 819-827 (1967).
9. WILLIAMS, T.P.: Photolysis of metarhodopsin II: Rates of production of P 470 and rhodopsin. Vision Res. 8, 1457-1466 (1968).
10. REUTER, T.: The synthesis of photosensitive pigments in the rods of the frog's retina. Vision Res. 6, 15-38 (1966).
11. REUTER, T., R.H. WHITE, G. WALD: Rhodopsin and porphyropsin fields in the adult bullfrog retina. J.Gen.Physiol. 58, 351-371 (1971).
12. HUBBARD, R., A. KROPF: The action of light on rhodopsin. Proc.Nat.Acad. Sci.(Wash.) 44, 130-139 (1958).
13. DARTNALL, H.J.A.: The photobiology of visual processes. In: The Eye, ed. H. DAVSON, Vol. 2, pp. 321-533. Academic Press, New York (1962).
14. BRIDGES, C.D.B.: Spectroscopic properties of porphyropsins. Vision Res. 7, 349-369 (1967).
15. DONNER, K.O.: Rod dark-adaptation and visual pigment photoproducts. This volume, pp. 205-209.
16. CONE, R.A., W.H. COBBS III: Rhodopsin cycle in the living eye of the rat. Nature (Lond.) 221, 820-822 (1969).
17. DENTON, E.J.: The contributions of the orientated photosensitive and other molecules to the absorption of whole retina. Proc.Roy.Soc.Lond., Ser. B 150, 78-94 (1959).
18. HÁROSI, F.I.: Frog rhodopsin in situ: Orientational and spectral changes in the chromophores of isolated retinal rod cells. Ph.D. thesis, Dept.Biomed. Eng., The Johns Hopkins University (1971).
19. KEMP, C.M.: Dichroism in rods during bleaching. This volume, pp. 309-314.
20. LIEBMAN, P.A.: Microspectrophotometry of photoreceptors. In: Handbook of Sensory Physiology, Vol. VII/1, ed. H.J.A. DARTNALL, pp. 482-528. Springer, Berlin (1972).
21. BRIDGES, C.D.B.: Studies on the flash-photolysis of visual pigments - IV. Dark reactions following the flash-irradiation of frog rhodopsin in suspensions of isolated photoreceptors. Vision Res. 2, 215-232 (1962).

22. LYTHGOE, R.J., J.P. QUILLIAM: The relation of transient orange to visual purple and indicator yellow. J. Physiol. (Lond.) 94, 399-410 (1938).
23. BRIDGES, C.D.B.: Studies on the flash-photolysis of visual pigments - III. Interpretation of the slow thermal reactions following flash-irradiation of frog rhodopsin solutions. Vision Res. 2, 201-214 (1962).
24. DE PONT, J.J.H.H.M., F.J.M. DAEMEN, S.L. BONTING: Biochemical aspects of the visual process. VIII. Enzymatic conversion of retinylidene imines by retinoldehydrogenase from rod outer segments. Arch. Biochem. Biophys. 140, 275-285 (1970).

Discussion

E.W. Abrahamson: Apropos of the question of the identity of the 400 nm product of the photolysis of carp porphyropsin: I would like to suggest that the data of Dr. Donner on the parallelism of the curves of decay rate constants vs. temperature (and hopefully vs. $\frac{1}{T}$) suggest a common activation energy for the decay of frog MII and the carp 400 nm intermediate. If this is indeed the case then it constitutes supportive evidence that the 400 nm intermediate is not retinal$_2$ but rather a true analog of frog MII.

T. Reuter: I agree; see Dr. Donner's reply to Dr. William's question.

S.L. Bonting: Regarding your arguments pro and con identifying the 400 nm product with MII: There is no effect of added NADPH/NADH; one obtaines reasonably fast reduction without adding these cofactors. Hence your retinol DH system may be saturated with endogenous cofactors, thus explaining lack of effect of added cofactors. However, in the light of de Pont et al. (24), this reduction does not argue against the 400 nm product being MII.

T. Reuter: I agree.

Adaptation Properties of Intracellularly Recorded Gekko Photoreceptor Potentials

Jochen Kleinschmidt
The Biological Laboratories, Harvard University, Cambridge, MA 02138/USA

Over the past century, a comprehensive description of the phenomena of visual adaptation in vertebrates has been achieved. More recently, much interest has focused on the identification of the cellular locus of visual adaptation and on the basic mechanisms of this phenomenon. Striking advances in the first problem have been made within the past few years. The adaptation properties of retinal extracellular field potentials and of single units throughout the retina have been studied by electrophysiology, and the evidence strongly suggested that the photoreceptors themselves are the primary site of the basic aspects of visual adaptation (1-5). This expectation has been confirmed now with studies on the isolated extracellular receptor potential (6,7) as well as through intracellular recording from single photoreceptors (8,9). The latter technique has been developed so that now it is possible to obtain stable recordings from single receptors for periods of time of 30 to 60 minutes and to perform long-duration adaptation experiments.

Two problems of current interest which now can be approached are as follows: 1) To what extent do single vertebrate photoreceptors operate as autonomous units in adaptation, and 2) how do changes in the state of adaptation manifest themselves in the electrical activity of the receptor, particularly in terms of changes of steady membrane potential?

In this paper I will describe experimental results bearing on the above questions. Intracellular recordings were obtained from single receptors in the *Gekko gekko* eyecup preparation, using superfine glass micropipette electrodes with tip diameters of less than 0.1 microns. The electrodes were filled with 2M potassium acetate. Only electrodes which had resistances between 200 and 250 megohms gave stable and noise-free recordings. The *Gekko gekko*, a species of nocturnal lizard endemic to Southeast Asia, has an all-rod retina with rather large photoreceptors which are routinely penetrated in these experiments by electrodes of the described properties. White light and diffuse full-field illumination were used throughout the experiments.

Form of the Receptor Response

Penetration of a *Gekko* photoreceptor with a micropipette electrode is signaled by a sudden potential drop of 30-40 mV and the appearance of a hyperpolarizing potential change in response to a flash of light. The light-evoked potential change is associated with an increase in the resistance of the cell (10,11). Fig. 1A shows the response of a *Gekko* receptor to short flashes of increasing intensity. The response is purely monophasic up to intensities about 2.5 log units above threshold. At brighter intensities a fast initial transient appears which becomes very prominent at and above saturating intensities. Flashes of supersaturating intensity evoke responses with no larger amplitudes; however, the absolute latency and the latency-to-peak of the response further decrease, and the decay back to the dark resting potential is much more delayed.

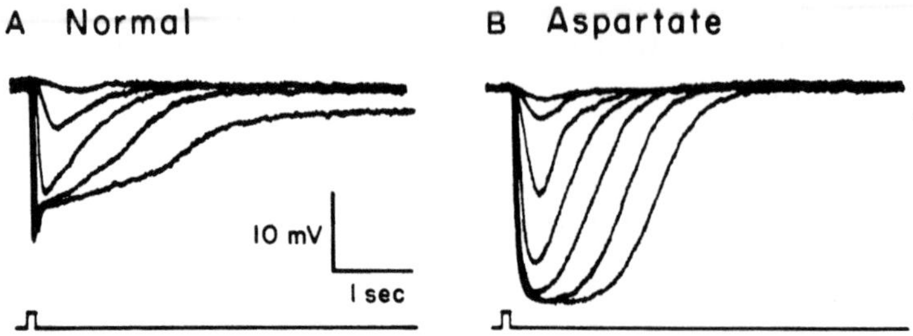

Fig. 1. Superimposed tracings of responses of the dark-adapted receptor to flashes of 100 msec duration and of different intensities. A. Normal receptor. Light intensity increasing in steps of one log unit. B. Receptor from the aspartate-treated eyecup. Light intensity increasing in steps of 0.5 log unit. The onset and cessation of the flash is indicated by the positive pulses on the bottom traces

---

The intensity-response relation of a Gekko photoreceptor in a semilogarithmic plot is displayed in Fig. 2. The data points are well fitted by the empirical hyperbolic tangent function $V/V_{max} = I/(I+K)$, first formulated by Naka and Rushton (1) to describe the intensity-response relation of tench S-Potentials. This function appears to be the basic sensory transduction function for brief stimuli in many types of sensory receptors (12).

As shown in Fig. 1A, the Gekko receptor response to bright light intensities has a rather complex wave form. There is reason to suspect that this complexity does not entirely stem from receptor events. For example, Baylor, Fuortes, and O'Bryan (13) have shown that in the turtle retina horizontal cells feed back onto receptors. They suggest that the sequence of electrical events in the generation of the turtle receptor potential is as follows: Light absorbed by the receptor initiates a hyperpolarizing potential change, which with a short delay generates a hyperpolarizing horizontal cell response. During the rising phase of the S-Potential, the horizontal cell negatively feeds back onto the receptors causing a rapid partial turning off of the receptor potential. Thus, the initial transient seen in the response of turtle cones to large spots of light (which mode of stimulation causes a strong excitation of the horizontal cells) is an expression of this horizontal cell feedback. The initial transient observed at bright light intensities in the Gekko receptor response appears similar.

The strategy adopted here to study the initial transient and its possible involvement in receptor adaptation is based on the findings of Cervetto and MacNichol (14) and Dowling and Ripps (15). These investigators showed that sodium L-aspartate strongly depolarizes horizontal cells and renders them unresponsive to light stimuli. Effectively, aspartate uncouples horizontal cells from receptors. In the aspartate-treated retina, no S-Potentials will be generated, and there will be no light-induced horizontal cell feedback onto receptors. If the initial transient in the Gekko receptor response is due to this horizontal cell feedback, then in the aspartate-treated retina no such transient should be observed. This expectation is born out by experiments in which the eyecup was treated with 50 mM aspartate in a Ringer for 5 minutes, and then recorded from single receptors.

Fig. 1B shows the potentials obtained under these conditions. There is a dramatic change in the waveform of the receptor response. In the presence of aspartate, the response is purely monophasic; there is no trace of an initial transient, not even at saturating intensities. Along with the squaring of the response there is an increase in the maximum amplitude. The maximum amplitude of the normal receptor potential is usually never larger than 20 mV; with aspartate it often exceeds 30 mV.

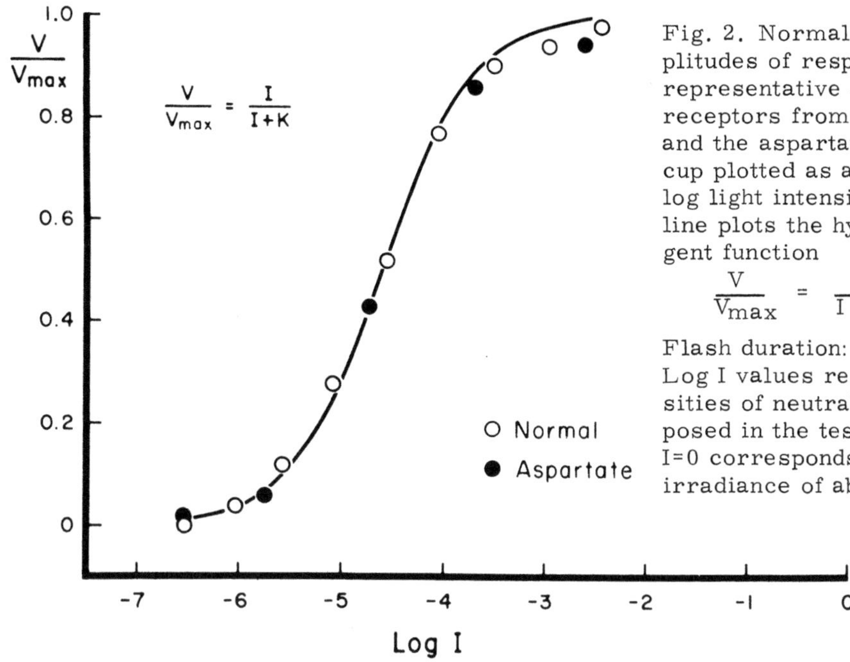

Fig. 2. Normalized peak amplitudes of responses of two representative dark-adapted receptors from the normal and the aspartate-treated eyecup plotted as a function of log light intensity. The solid line plots the hyperbolic tangent function

$$\frac{V}{V_{max}} = \frac{I}{I+K}.$$

Flash duration: 100 msec. Log I values refer to the densities of neutral filters interposed in the test beam; log I=0 corresponds to a retinal irradiance of about 1 mW/cm^2

It appears however, that the membrane potential reached by the receptor in response to saturating flashes is the same under both conditions, and that the larger maximum amplitude under aspartate is the outcome of an expanded deviation range due to a lower dark resting potential. In all experiments to date, the dark resting potential of the receptor under aspartate was found to be 15-25 mV inside negative with respect to outside in contrast to values of 30-40 mV inside negative for the normal receptor. There appears to be no difference in the absolute latency under the two conditions; however, the maximum slope of the initial phase of the response is less steep in the aspartate-treated retina.

Although aspartate appears to have some direct action on the receptor, the intensity-response relation appears unaffected. Fig. 2 shows the normalized intensity-response relation for a representative receptor recorded from the aspartate-treated eyecup. The hyperbolic tangent function equally well describes the behaviour of receptors uncoupled from horizontal cells by the action of aspartate. The simple monophasic response of the aspartate-treated receptor, unmodified by feedback from a second-order neuron, may more directly reflect the events of sensory transduction.

Light Adaptation of Gekko Photoreceptors

Gekko photoreceptors respond to steady background lights with maintained hyperpolarizing potentials. However, the maintained level of membrane potential is always less hyperpolarized than the initial potential induced by the adapting light. In steady light, the membrane potential slowly creeps upward from the initial peak to the maintained level. The duration of this relaxation phase increases with background intensity and is in the order of seconds. Test flashes of fixed intensity superimposed upon the steady background during this transition phase give incremental responses whose amplitude grows in direct proportion to the upward movement

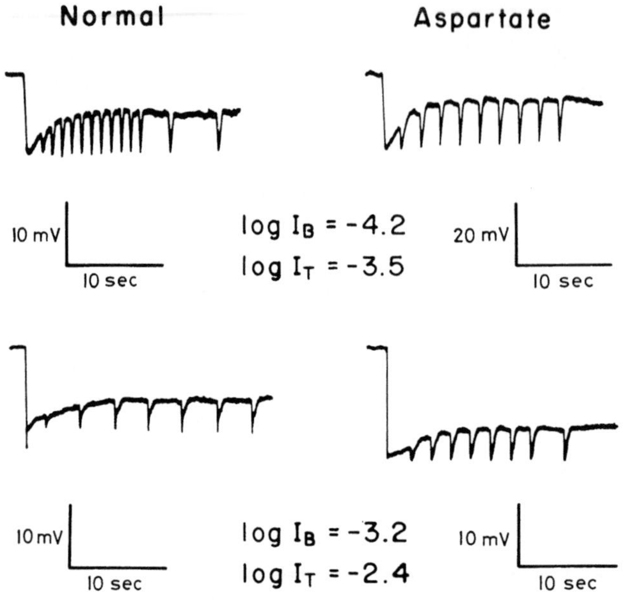

Fig. 3. Dynamics of light adaptation of Gekko receptors in the normal and the aspartate-treated eyecup at two different intensities of background light (I_B) and of test flashes (I_T). The steady background light was turned on immediately before the strong hyperpolarization. Test flashes of 100 msec duration and of fixed intensity were superimposed subsequently at a frequency of one every sec or one every two sec. Note the abolition of the fast initial transient at the brighter adapting intensity in the aspartate-treated preparation. The time course of relaxation of membrane potential and growth of increment responses appear unaffected by aspartate

of membrane potential. This is shown in Fig. 3 at two different intensities of background light and superimposed test flashes.

By the time the maintained level of membrane potential is reached, thresholds have assumed constant values. If one now superimposes a series of test flashes of increasing intensity upon steady background lights of different intensities, one obtains a set of increment intensity-response curves. These curves are shifted to the right along the abscissa in the V-logI plot in proportion to the intensity of the adapting light. Their saturation levels lie at much higher light intensities than the saturation level of the dark-adapted intensity-response curve. Light adaptation extends the dynamic range of the receptor to at least six log units, the maximum range tested in these experiments.

Fig. 4 shows increment thresholds as a function of background intensity measured during the steady phase of light adaptation. Weber's law is adhered to over a range of at least 3 log units. Fig. 3 and Fig. 4 also show the results of adaptation studies on aspartate-treated receptors. The dynamics of light adaptation, the increment-threshold function, and the increment intensity-response curves appear identical in normal receptors and aspartate-treated receptors. That the adaptation behavior of receptors is unaltered by an agent which uncouples them from horizontal cells signifies that receptors adapt as autonomous units and that the basic mechanism of intensity adaptation must reside in them.

Discussion

The realization that the basic aspects of visual adaptation in the vertebrate retina originate in the receptors is rather recent. The experiments described in this paper confirm this fact and further show that receptors, apparently uncoupled from proximal neurons, exhibit normal adaptation properties. Furthermore, they demonstrate the existence in the receptors of a mechanism which increases their responsiveness with time in the light after an initial strong desensitization. This

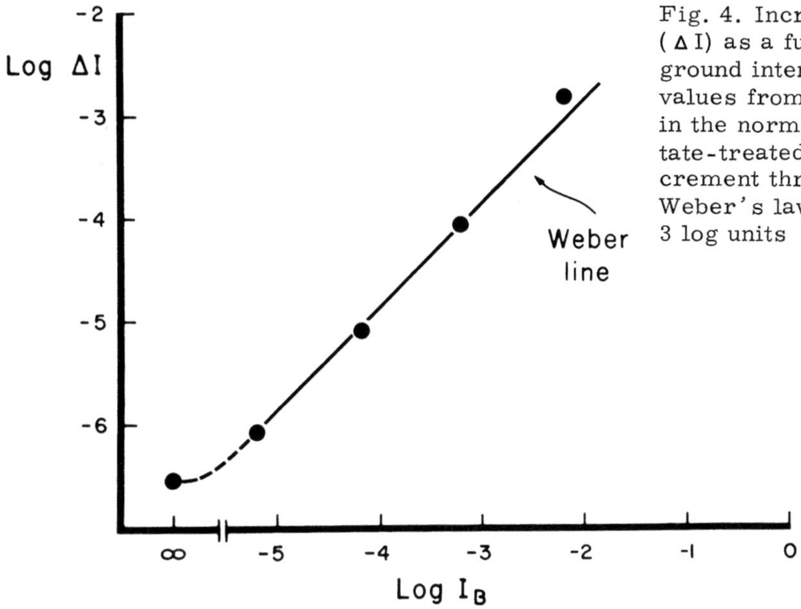

Fig. 4. Increment thresholds (ΔI) as a function of background intensity (I_B). Average values from several receptors in the normal and the aspartate-treated eyecup. The increment threshold follows Weber's law over at least 3 log units

mechanism prevents saturation of the receptor and expands its dynamic range. Electrically it expresses itself as a recovery of membrane potential in the light.

The only model of adaptation in vertebrate photoreceptors that has been put forward so far is the one by Boynton and Whitten (6) for monkey cones. These authors propose that the hyperbolic tangent intensity-response function holds for the steady membrane potential associated with steady background lights as well as for the peak amplitude of responses to short flashes. Background and test lights are assumed to add linearly in causing a total response of a given magnitude. In this scheme, a steady background light hyperpolarizes the receptor to a certain level - graded with background intensity in conformity with the intensity-response function - at which the membrane potential then remains clamped for the duration of the illumination. Responses to superimposed flashes rise from progressively lower-sloper portions of the intensity-response curve. This results in a response compression: more light has to be put in to give a constant criterion incremental response as the background intensity increases and the value of steady membrane potential moves up the intensity-response curve. Ultimately, the receptor would saturate, at the same intensity of the sum of background and test light that saturates it if given in a brief flash to the dark-adapted cell. Boynton and Whitten (6) propose that in monkey cones, however, saturation is prevented by the circumstance that the intensities in the upper range of the intensity-response curve are sufficiently bright to bleach significant amounts of pigment. The reduction of quantum catching power of the receptor through bleaching of pigment by bright adapting lights is a second adaptive mechanism that is postulated to take over at an intensity at which the electrical mechanism would otherwise have saturated.

Clearly, the results from Gekko photoreceptors are at variance with this simple scheme. The intensity-response curve for the maintained part of the response to steady background lights is much flatter than the one for the initial peak. The transition between these two functions in the first seconds of light adaptation signifies a true receptor adaptation which electrically manifests itself in the relaxation of membrane potential. Saturation of the receptor is prevented by this mechanism, not by photopigment bleaching which probably is negligible at light intensities that initially saturate these receptors.

Very similar results have recently been obtained by Dowling and Ripps (7) in studies on the adaptation behavior of the aspartate-isolated extracellular receptor potential of the skate. They too have found that with backgrounds too weak to bleach significant amounts of visual pigment, skate receptors show an adaptive mechanism that increases their responsiveness in the light by opposing the voltage loss initially imposed by response compression. The nature and mechanism of this receptor adaptation is an intriguing problem, one that may well engage both the biochemist and the physiologist.

References

1. NAKA, K.I., W.A.H. RUSHTON: S-potential and dark adaptation in fish. J.Physiol.(London) 194, 259-269 (1968).
2. DOWLING, J.E., H. RIPPS: Visual adaptation in the retina of the skate. J.Gen.Physiol. 56, 491-520 (1970).
3. DOWLING, J.E., H. RIPPS: S-potentials in the skate retina. Intracellular recordings during light and dark adaptation. J.Gen.Physiol. 58, 163-189 (1971).
4. FRANK, R.N.: Properties of "neural" adaptation in components of the frog electroretinogram. Vision Res. 11, 1113-1123 (1971).
5. WERBLIN, F.S.: Adaptation in a vertebrate retina: intracellular recording in Necturus. J.Neurophysiol. 34, 228-241 (1971).
6. BOYNTON, R.M., D.N. WHITTEN: Visual adaptation in monkey cones: recordings of late receptor potentials. Science (Washington) 170, 1423-1426 (1970).
7. DOWLING, J.E., H. RIPPS: Adaptation in skate photoreceptors. In press (1972)
8. GRABOWSKI, S.R., L.H. PINTO, W.L. PAK: Adaptation in retinal rods of axolotl: Intracellular recordings. Science (Wash.) 176, 1240-1243 (1972).
9. KLEINSCHMIDT, J.: Paper presented at the ARVO spring meeting, Sarasota, Fla. (1972).
10. TOYODA, J., H. NOSAKI, T. TOMITA: Light-induced resistance changes in single photoreceptors of Necturus and Gekko. Vision Res. 9, 453-463 (1969).
11. KLEINSCHMIDT, J.: Unpublished observations. (1971).
12. LIPETZ, L.E.: The relation of physiological and psychological aspects of sensory intensity. Handbook of Sensory Physiology, Vol. I, ed. W.R. LOEWENSTEIN, pp. 191-225. Springer, Berlin-Heidelberg-New York (1971).
13. BAYLOR, D.A., M.G.F. FUORTES, P.M. O'BRYAN: Receptive fields of cones in the retina of the turtle. J.Physiol. (London) 214, 265-294 (1971).
14. CERVETTO, L., E.F. MACNICHOL, jr.: Pharmacology of horizontal cells in the isolated perfused skate retina. Biol.Bull. 141, 381 (1971).
15. DOWLING, J.E., H. RIPPS: Aspartate isolation of receptor potentials in the skate retina. Biol.Bull. 141, 384 (1971).

Discussion

H. Stieve: Do the responses of the dark-adapted receptor saturate at the same absolute voltage as the responses of the light-adapted receptor?

J. Kleinschmidt: At the present time it appears as if the responses of the light-adapted receptor don't quite saturate at the same voltage. They appear to saturate at levels of membrane potential that are the less hyperpolarized the brighter the background intensity and the more hyperpolarized the steady membrane potential is.

Receptor Adaptation and Receptor Interactions: Some Results of Intracellular Recordings (Preliminary Note)

W. L. Pak, S. R. Grabowski, and L. H. Pinto
Purdue University, Lafayette, IN/USA

Regeneration of the light sensitive visual pigment contained in the receptor outer segments apparently can account for only a portion of the sensitivity changes observed during dark adaptation (1,2). The remaining sensitivity changes (the "neural" adaptation) take place even under conditions in which the bleached visual pigment does not undergo regeneration (3). In the past few years it has become increasingly evident that at least a portion of even "neural" adaptation occurs at the receptor level (4-9). In this work our aim was to demonstrate directly the existence of a receptor adaptation mechanism that is independent of visual pigment regeneration. We accomplished this by recording intracellularly from individual rods in the isolated retina of the axolotl (Ambystoma mexicanum) before and after bleaching varying amounts of the rod pigment (10). In our preparation no regeneration of the rod pigment could be detected in parallel spectrophotometric measurements (10).

Immediately upon obtaining a receptor potential, monochromatic stimuli at 430, 525, and 580 nm were presented sequentially at several intensities to establish which of the three known types of axolotl photoreceptors had been penetrated: red rods ($\lambda_{max} \sim 515$ nm), green rods ($\lambda_{max} \sim 430$ nm), or cones ($\lambda_{max} \sim 575$ nm) (10-13). The results reported here were obtained from the red rods, which are by far the most preponderant of receptor types in the axolotl retina and which gave the most stable responses, occasionally lasting for 2 or more hours. First, the threshold (0.5 mV criterion response) and dynamic range of the dark adapted cell was obtained using 200 msec stimuli of 525 nm wavelength, and then after bleaching approximately 10, 20, 50, or 80% of the rod pigment, the change in rod sensitivity was followed as a function of time.

Both pre- and post-bleach response amplitude (V) can be described as a function of stimulus intensity (I) by a modified version of the equation first introduced by Naka and Rushton (14):

$$V = \frac{I^n}{I^n + \sigma^n} \cdot (V_{max})$$

where V_{max} is the saturated response amplitude, and σ is the half saturation constant. A value of n equal to 0.9 ± 0.05 (standard error of the mean) appears to best fit our data. Bleaching dramatically reduces the saturated response amplitude V_{max} and increases the value of σ. Subsequently in the dark, both quantities partially recover, i.e. dark adaptation takes place so that both quantities finally stabilize to new levels. From our results, the final stabilized values of V_{max} and σ after the bleach can be related to the dark adapted values of V_{max} and σ and the fraction of rod pigment remaining p by the following empirical equations.

$$V_{max,\,b} = p \cdot V_{max,\,d}$$

$$\sigma_b = \sigma_d \, 10^{2.0\,(1-p)} \quad \text{for} \quad (1-p) \le 0.5$$

where $V_{max, d}$ and $V_{max, b}$ are the saturated response amplitude obtained from the dark-adapted cell and the stabilized value obtained after a bleach, respectively. Similarly, σ_d and σ_b refer to the corresponding values of the half saturation constant before and after the bleach. Partial recovery of the threshold after a bleach takes place over a period ranging from about 7 minutes (~ 10% bleach) to over 20 minutes (~ 50% bleach) depending on the amount of pigment bleached (temperature $24 \pm 2^{\circ}C$). As more and more pigment is bleached away, it takes longer and longer for the threshold to settle down to a steady state value.

These results demonstrate that a substantial portion of the dark adaptation process occurs at the receptor level. The previous speaker (15) has already shown that light adaptation in gecko photoreceptors is a property of the receptors themselves and does not depend on feedback interactions through the horizontal cells described by Baylor et al. (16) for the cones of the turtle retina. Similarly, dark adaptation, too, appears to originate mainly from the receptors themselves. Our results described above and in reference (10) are remarkably similar to those reported by Hood et al. (17) on the adaptation properties of the aspartate isolated P_{III} component of the isolated frog retina. Since aspartate has been shown to block the horizontal cells (18), the results of Hood et al. are not likely to depend on feedback inputs from the horizontal cells. Consequently, the dark adaptation we observe within an individual rod also is very likely to be controlled mainly by that cell.

A question then arises as to what role, if any, the receptor-receptor interactions of Baylor et al. (16) might play in the control of receptor sensitivity. A partial answer to this question has been obtained in our experiments comparing the time course of the light-induced change in membrane resistance with the time course of the potential change in the same cell.

Most of the measurements were made upon single photoreceptor cells of the gecko (Gekko gekko), but some experiments were also done on the axolotl and the tiger salamander (Ambystoma tigrinum). A feedback regulator forced sinusoidally modulated current (1.7 nA RMS, f=16-100 Hz) through the cell membrane via the electrode. Changes in membrane resistance caused changes in the component of the transmembrane voltage in phase with the current and were detected by a synchronous detector. Both the resistance change and the potential change were averaged over several stimulus cycles.

If the potential recorded intracellularly from the gecko photoreceptors consists only of the receptor potential generated by the decrease in the conductance for the inward flow of the sodium current (19, 20, 21), one would expect the increase in membrane resistance and decrease in membrane voltage to be proportional to each other throughout the stimulus cycle. The proportionality need not hold if receptor interactions contribute additional ionic processes to the potential recorded from the receptors.

In fact, the time course of the resistance change and the time course of the potential change do not, as a rule, parallel each other. In general, the membrane resistance change tends to lag behind the potential change, particularly at the offset, if stimuli of greater than 300 μm in diameter are used. The following results suggest that the lag is probably due to inputs from receptors in the periphery (~100-500 μm from the impaled cell), presumably through the horizontal cells:
1) The lag disappears if restricted stimuli (75 μm ϕ) are used on excised retinas of the gecko. This result can be readily explained, since with restricted stimuli not as many of the receptors in the periphery are illuminated. Moreover, excision

of the gecko retina is expected to break many of the far surround interactions through injury.
2) The resistance change due to input from the periphery can be isolated by desensitizing the cell from which the recording is made using a steady 10 μm ϕ spot and by stimulating the receptors in the periphery using a 950 μm ϕ stimulus of 390 msec duration. Effects of the peripheral input, thus isolated, result in a decrease in the membrane resistance, i.e. the peripheral inputs are antagonistic to the effects of the center illumination.

The fact that the peripheral inputs are antagonistic to the effects of the center illumination suggest that they may act to reduce the amplitude of the response of the impaled receptor. This expectation is borne out in the following experiment. The response to a stimulus of 10 μm ϕ (200 msec) directed toward the impaled cell is greatly diminished by presentation of a concentric annulus (165 μm inner ϕ). Under these conditions the amplitude of the receptor potential elicited by the 10 μm ϕ spot plus the scattered light from the annulus is less than that due to the spot alone. In other words, illumination of the periphery by the annulus suppresses the amplitude of the receptor potential.

Thus, in addition to the dark adaptation mechanism discussed previously, there appears to be a form of "field adaptation" mechanism operating at the receptor level. This form of adaptation manifests itself only when a population of receptor cells are illuminated, and appears to exert its effect through interactions among receptors over a considerable distance.

Acknowledgement. This work was supported by PHS research grant EY 00033 to W.L. Pak and PHS predoctoral fellowship GM 45860 to S.R. Grabowski.

References

1. DOWLING, J.E.: Neural and photochemical mechanisms of visual adaptation in the rat. J.Gen.Physiol. 46, 1287-1301 (1963).
2. RUSHTON, W.A.H., D. SPITZER POWELL: The rhodopsin content of the visual threshold of human rods. Vision Res. 12, 1073-1081 (1972).
3. WEINSTEIN, G.W., R. HOBSON, J.E. DOWLING: Light and dark adaptation in the isolated rat retina. Nature (Lond.) 215, 134-138 (1967).
4. DONNER, K.O., T. REUTER: Dark adaptation processes in the rhodopsin rods of the frog's retina. Vision Res. 7, 17-41 (1967).
5. NAKA, K.I., W.A.H. RUSHTON: S-potential and dark adaptation in fish. J.Physiol. (Lond.) 194, 159-169 (1968).
6. DOWLING, J.E., H. RIPPS: S-potentials in the skate retina. Intracellular recordings during light and dark adaptation. J.Gen.Physiol. 58, 163-189 (1971).
7. ERNST,W., C.M. KEMP: Dark adpatation in the isolated rat retina. Vision Res. 11, 1197-1198 (1971).
8. FRANK, R.N.: Properties of "neural" adaptation in components of the frog electroretinogram. Vision Res. 11, 1113-1123 (1971).
9. RUSHTON, W.A.H., D. SPITZER POWELL: The early phase of dark adaptation. Vision Res. 12, 1083-1093 (1972).
10. GRABOWSKI, S.R., L.H. PINTO, W.L. PAK: Adaptation in retinal rods of axolotl: Intracellular recordings. Science 176, 1240-1243 (1972).
11. JONOVÁ, M.: Vergleichende Untersuchungen über die Sehorgane der Larve von Siredon pisciformis und ihre erwachsene Form Amblystoma mexicanum. Zool. Jahrb., Abt.Anat.Ontog.Tiere 57, 296-350 (1933).

12. LIEBMAN, P.A., J.A. PARKES: personal communication.
13. CUSTER, N.V.: Anatomical pathways of information flow in the axolotl retina. Ph.D. thesis: Purdue University, Lafayette, Indiana (1972).
14. NAKA, K.I., W.A.H. RUSHTON: S-potentials from color units in the retina of fish (Cyprinidae). J. Physiol. (Lond.) 185, 536-555 (1966).
15. KLEINSCHMIDT, J.: Adaptation properties of intracellularly recorded Gekko photoreceptor potentials. This volume, pp. 219-224.
16. BAYLOR, D.A., M.G. FUORTES, P.M. O'BRYAN: Receptive fields of cones in the retina of the turtle. J. Physiol. (Lond.) 214, 265-294 (1971).
17. HOOD, D.C., B.G. GROVER, P.A. HOCK: Recovery of rod and cone activity after light adaptation of the NaAsp treated isolated frog retina. Assoc. for Res. Vision Ophthalm., Spring Meeting, Sarasota, Florida (1972).
18. CERVETTO, L., E.F. MAC NICHOL JR.: A study of intracellular responses from the perfused turtle retina: Supression of synaptic transmission between receptors and horizontal cells by acidic amino acids. Assoc. for Res. Vision Ophthalm., Spring Meeting, Sarasota, Florida (1972).
19. PENN, R.D., W.A. HAGINS: Signal transmission along retinal rods and the origin of the electroretinographic a-wave. Nature (Lond.) 223, 201-205 (1969).
20. SILLMAN, A.J., H. ITO, T. TOMITA: Studies on the mass receptor potential of the isolated frog retina. II. On the basis of the ionic mechanisms. Vision Res. 9, 1443-1457 (1969).
21. TOYODA, J., H. NOSAKI, T. TOMITA: Light induced resistance changes in single photoreceptors of Necturus and Gekko. Vision Res. 9, 453-462 (1969).

Excitation and Adaptation in the Cephalopod Retina: An Equivalent Circuit Model

George Duncan and Peter C. Croghan
School of Biological Sciences, University of East Anglia, Norwich, NOR 88C/Great Britain

There are three basic questions in photoreceptor physiology that have not yet been fully answered: the problems of latency, excitation and adaptation. Any model of a photoreceptor cell, if it is to be useful, must help to resolve these and it must also predict a non-linear, ultimately saturating response to light.

The most promising model hitherto is probably that of Fuortes and Hodgkin (1) who were able to fit the time course of the photoreceptor potentials in _Limulus_ using a ten stage linear network. However, an additional feedback loop had to be incorporated to produce non-linear responses at high light intensities, and a further drawback to their model is that there are no structures in the photoreceptors that the networks can be ascribed to. Nor does the model say what the actual membrane conductances or even resting potentials should be. We therefore set out to construct a model based solely on known, or at least estimable membranes parameters.

The physical basis for our model is the photoreceptor unit described by Young (2) and Zonana (3) modified by our own observations on _Sepiola_ using light microscopy (Fig. 1). It is at the outer segment membranes which contain the rhodopsin (4) that the primary light-stimulated sodium permeability increase is believed to occur (5) and this we call the Active Membrane. The inner segment is regarded as insensitive to light and this we term the Inactive Membrane. We regard the membrane parameters of the active and inactive membranes as identical except that the light flash results in an increase in the sodium conductance of the active membrane. The increased sodium conductance is considered to decay exponentially with time. The potassium conductance is regarded as constant. The equivalent circuit of this model is represented in Fig. 2 and its behaviour is described by the following equations:

$$C \frac{dV_A}{dt} = - I - G^o_K (V_A - E_K) - G_{Na} (V_A - E_{Na}) \qquad 1$$

$$C \frac{dV_I}{dt} = \frac{A_A}{A_I} I - G^o_K (V_I - E_K) - G^o_{Na} (V_I - E_{Na}) \qquad 2$$

$$G_{Na} = G^o_{Na} (1 + A \exp(-kt)) \qquad 3$$

$$I = (V_A - V_I) / (R_E + R_I) \qquad 4$$

$$V_E = - I \cdot R_E \; ; \qquad 5$$

where C is the membrane capacitance, V_A and V_I are the potentials across the active and inactive membranes respectively, E_K and E_{Na} are the Nernst potentials of the potassium and sodium respectively, G^o_K is the potassium conductance, G^o_{Na} and G_{Na} are the sodium conductances of the inactive and the active membrane, I is the current flowing normal to the retina (initially zero in cephalopods, 5), k is decay rate constant of the increased sodium conductance, A_A/A_I is the ratio of the area of active membranes, R_E and R_I are the extracellular and R_I the intracellular

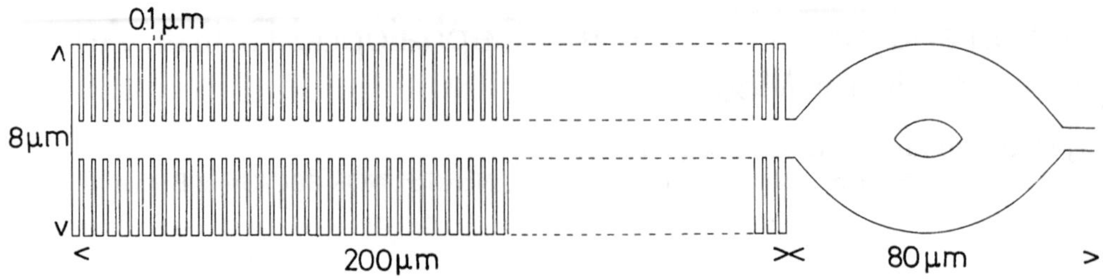

Fig. 1. Diagramatic representation of Sepiola photoreceptor cell

resistances, and V_E is the potential difference between the two sides of the retina. A is the fractional increase in sodium conductance. The solutions to these equations have been computed digitally.

The basis for our model lies in equation 3 which states that, after the latency period, the sodium conductance increases instantaneously to a new value and the conductance then returns exponentially back to the resting, dark state. The amount of conductance increase is determined by the parameter A which must ultimately be related to the stimulus intensity.

We have used the retinas from the cephalopod Sepiola atlantica to test our model, and the extracellular potential (V_E) is very readily obtained from the isolated retina (Duncan and Croghan, in preparation). We chose to stimulate with very short flashes (10 µsec from a Dawes Strobotorch) so that the light pulses ended well before the electrical signals were initiated and in this way there was no confusion between on and off responses.

Fig. 3 (bottom trace) shows the response to a single light flash. It consists of a 25 msec latency period, followed by a steeply rising initial phase. The recovery phase is more prolonged. The equivalent circuit model can generate such a response from the parameters given (Fig. 4). The estimates of membrane area were computed from Zonana's (3) electron micrographs; membrane conductances and

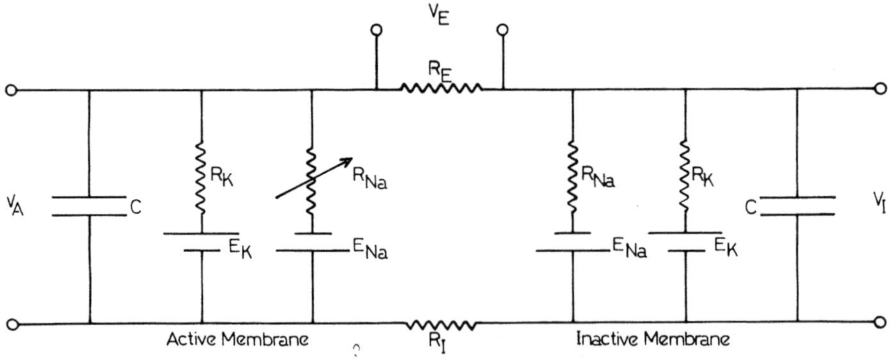

Fig. 2. Equivalent circuit model of cephalopod photoreceptor cell. E_K and E_{Na} are the Nernst potentials for Na and K and R_K and R_{Na} are the resistances of sodium and potassium channels. V_A and V_I are the potentials across the active and inactive membranes

Fig. 3. Examples of experimental response of the potential difference across Sepiola retina (V_E) to 10 μsec light flashes. Bottom trace, single flash; middle trace, two flashes 20 msec apart; upper trace, two flashes 40 msec apart

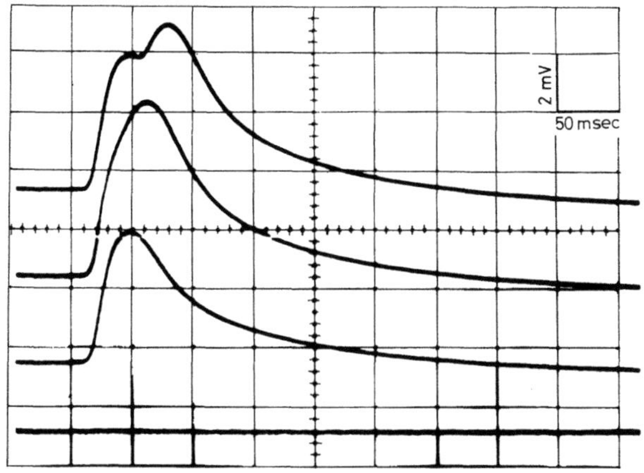

Nernst potentials were calculated from Sepia data (6; Duncan and Croghan, in preparation). The extracellular resistance (R_E) was computed from the transverse resistance value of $10^{-2}\Omega\,m^2$ (5; Duncan and Weeks, unpublished). Values for A, k, and C were adjusted to give the best fit to the experimental data.

The model was further tested by using data relating the height of a second flash to the time between the peaks (Fig. 3, top two traces). The ratios of the height of second flash to height of first are plotted on Fig. 5 and the computed response to the second flash was obtained from the relationship

$$G_{Na} = G^{o}_{Na}\,(1 + A\exp(-kt) + A\exp(-k(t-x))),\ \text{if}\ t > x; \qquad 6$$

where x is the interval between the two light flashes. This assumes that the effects of the two light flashes on sodium conductance were simply additive and the assump-

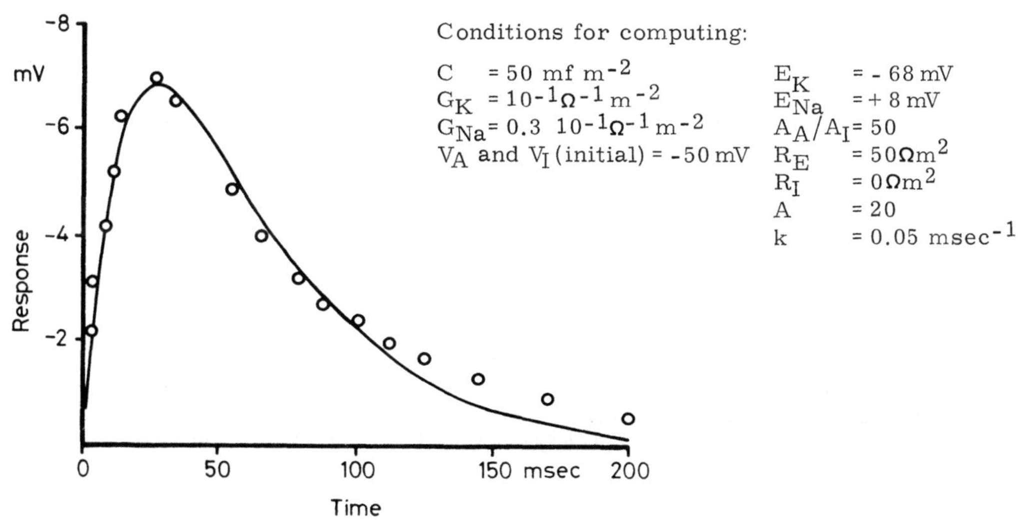

Fig. 4. Photoreceptor response to single light flash. o, experimental response derived from Fig. 1. Solid line, response of model computed using parameters given. The computed response was timed to start at end of latency period

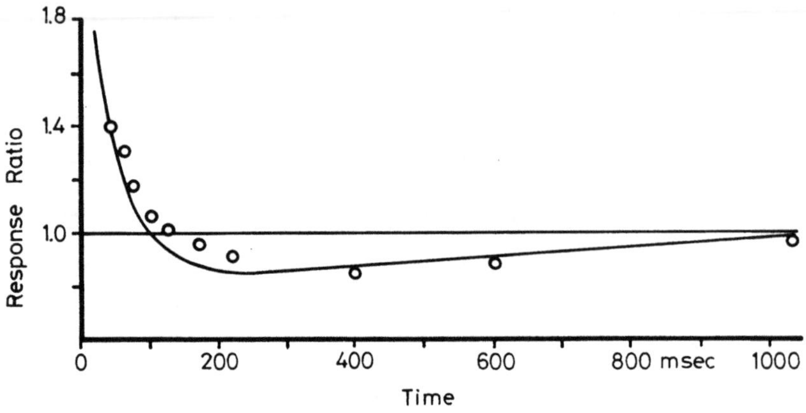

Fig. 5. Response to two light flashes. Ratio represents height of second peak (or peak if two responses not discriminated) to height of control response to single flash. Zero time represents onset of first response and time is measured from this point to time of second peak. o, experimental response (first two points are from Fig. 3); solid line, computed response

tion involved is that no significant bleaching of rhodopsin occured in the first flash. Such adaptation as occurs is seen purely as of electrical rather than chemical in origin and the model provides a possible explanation for the fast adaptation (and recovery) process that is known to occur in both invertebrate (1) and vertebrate retinas (7). In the latter case the fast component was shown to be independent of

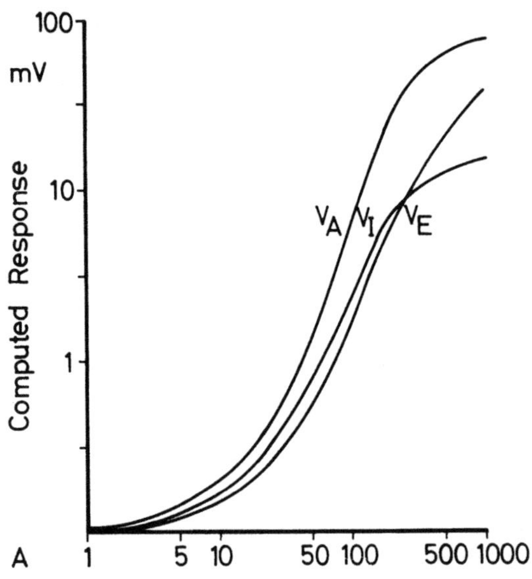

Fig. 6. Computed relationships of response peak to parameter A. V_A, V_I, and V_E are potentials across active membrane, space respectively. Apart from veriable A, inactive membrane, and extracellular parameters same as those in Fig. 4

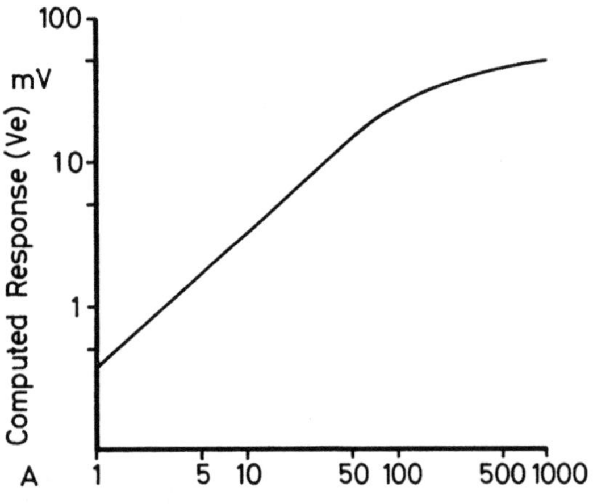

Fig. 7. Computed relationships of response peak to parameter A, both plotted on a logarithmic scale. This indicates that the response obeys the relationship $V_E=A^x$ (power law) for a considerable range of A

the rhodopsin level in the retina and it was suggested that it was "neural" in origin. We believe that the explanation for this adaptation lies in the membrane time constants of the photorecptor cell at least in Sepiola.

The equivalent circuit model further predicts a linear response for low values of A and a non-linear response for higher values. The response ultimately saturates when the Nernst potential for sodium is reached. The model predicts that the extracellular potential (V_E) obeys the Weber-Fechner law over a range of values for A (Fig. 6) and it obeys the exponent or power law over a wider range (Fig. 7). Experiments are at present underway with high intensity flash sources to find out whether the actual responses fit the computed values.

One drawback of our model is that it gives no explanation of the latency phenomenon, although this might not be too serious a defect as it is currently believed that the latency is photochemical rather than electrical in origin (8). This implies that the increase in sodium conductance is the result of the appearance of one of the photoproducts of rhodopsin. The fall in the conductance might then be due to the disappearance of this critical photoproduct. However, the model does elucidate the generation mechanism and the fast adaptation phenomenon. It also predicts a non-linear, ultimately saturating response.

Acknowledgement. G. D. wishes to thank the Wellcome Trust for a travel grant.

References

1. FUORTES, M.G.F., A.L. HODGKIN: Changes in time scale and sensitivity in the ommatidia of Limulus. J. Physiol. 172, 239-264 (1964).
2. YOUNG, J.Z.: The visual system of Octopus. Nature (Lond.) 186, 836-839 (1960).
3. ZONANA, H.V.: Fine structure of the squid retina. Bull. Johns Hopkins Hosp. 109, 185-205 (1961).
4. HARA, T., R. HARA: Regeneration of squid retinochrome. Nature (Lond.) 219, 450-454 (1968).
5. HAGINS, W.A.: Electrical signs of information flow in photoreceptors. Cold Spring Harbor Symp. on Quant. Biol. 30, 403-418 (1965).
6. DUNCAN, G., J.J.H.H.M. DE PONT, S.L. BONTING: Distribution and movement of ions in the dark-adapted retina of the cuttlefish (Sepia officinalis). Pflüger's Arch. 322, 264-277 (1971).
7. WEINSTEIN, G.W., R.R. HOBSON, J.E. DOWLING: Light and dark adaptation in the isolated rat retina. Nature (Lond.) 215, 134-138 (1967).
8. BONTING, S.L.: The mechanism of the visual process. Curr. Topics Bioenerg. 3, 351-415 (1969).

V. Ionic Aspects of Excitation and Regeneration

Experiments on the Ionic Mechanism of the Receptor Potential of *Limulus* and Crayfish Photoreceptor

H. Stieve
Institut für Neurobiologie der Kernforschungsanlage Jülich, Jülich/W. Germany

The ionic mechanism of the receptor potential (ReP) of the invertebrate photoreceptor is not entirely understood. In our experimental study of this problem we worked with the eye of the horseshoecrab, Limulus, or the crayfish, Astacus. There are indications that in these two species the ionic mechanism is not entirely similar.

Method

A slice of the retina of the Limulus lateral eye or of the Astacus eye is kept in streaming saline.

a) The dark potential (DaP), the ReP, and the impedance of the measuring circuit (cell membrane, cell, and electrode in series) can all be measured with one intracellular electrode. The DaP is measured and corrected to allow for potential drift during the experiment (5).

b) The ReP of just the receptor cell layer is measured by external electrodes.

The preparation was stimulated with the same light stimulus every 10 minutes.

Results

A. Ouabain. In a number of experiments we could show that the ReP is not caused by a change in the activity or electrogenicety of an electrogenic ion pump (6). The observed change in conductivity of the photoreceptor cell membrane seems to be the cause for the ReP. However, under certain conditions, an electrogenic pump can contribute to the DaP of the photoreceptor cell.

Smith et al. (2) briefly reported an experiment in which they poisoned the visual cell of Limulus with ouabain and observed that its excitability was lost, although the resting potential had decreased only to about 50% of its original value. This result is more easily explained by an electrogenic pump mechanism causing the ReP than by a conductance increase mechanism. The effect seemed to be important enough for a more thorough investigation.

The effect of ouabain on the membrane potential of the photoreceptor cell was measured in 5 experiments on a slice of the Limulus lateral eye. Fig. 1 shows a typical experiment. After a preperiod of 60 min the eye was poisoned for a period of 1 - 2 hours with ouabain (1 mM/l) added to the saline. Due to the blocking of the ionic pump by ouabain the ReP and the DaP gradually decrease. The cell becomes inexcitable long before the DaP approaches zero. The observed effects are not reversible. Washing out the poison does not lead to a recovery of the cell in contrast to our findings with unpenetrated cells (external electrodes). The dark potential decreases further in the after-period (after ouabain application).

The evaluation of all 5 experiments shows that under the influence of ouabain the ReP is already zero when the average DaP is 39 ± 10% of its original value. These

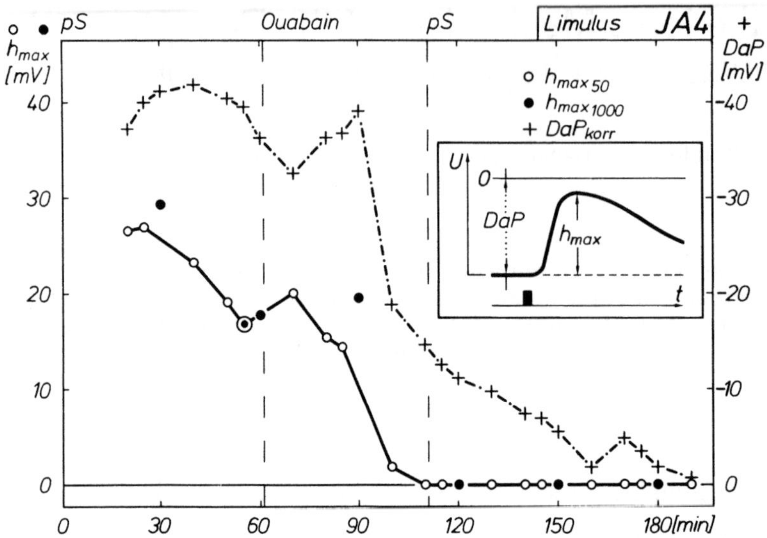

Fig. 1. Plot of the amplitudes (h_{max}) of the receptor potential and the corrected dark potential (DaP_{korr}) under the influence of ouabain. Stimulus duration τ = 50 msec and 1000 msec, respectively. Limulus lateral eye, intracellular recording

---

results confirming the findings of Smith et al. (2) could have two possible explanations.

1. The conductance change of the photoreceptor cell membrane does not occur when the visual cell is sufficiently poisoned by ouabain (or sufficiently depolarized). This possibility will be tested but does not seem to be very probable.

Preliminary measurements of the impedance indicate that the resistance of the cell membrane in the dark increases under the action of ouabain and is still increasing in the after-period in physiological saline. The conductance change caused by light has not yet been measured under the conditions of ouabain poisoning.

2. The gradient of the ion species which determines the ReP (probably mainly Na^+) approaches zero earlier than the gradient which determines the DaP (mainly K^+). The second explanation seems probable, since the intracellular volume is very small compared with the extracellular volume (streaming saline).

The height of the DaP is determined mainly by the gradient for K-ions which is due to a high intracellular and a low extracellular K-concentration. The height of the ReP is determined by other cations (mainly Na^+, see below) which have a high concentration extracellularly and a low one intracellularly. An ionic gradient is much more changed when a certain number of ions is added on the side of the lower concentration than when the same number of ions is subtracted from the higher concentration. If, for instance, Na- and K-ions are both diffusing "downhill" along their concentration gradients at about the same rate, the Na^+ gradient is much faster decreased than the K^+ gradient since the low extracellular concentration of K is almost unchanged in contrast to the low intracellular concentration of Na.

B. Calcium ions: In earlier experiments we showed that Ca is necessary for the excitation of the crustacean eye (4,7). Ca is thought to be presumably released from the cell membrane during excitation, thus causing an increase of membrane conductance for certain ions. We wanted to check how ReP, DaP, and membrane conductance is changed when Ca is bound by a binding agent specific for Ca.

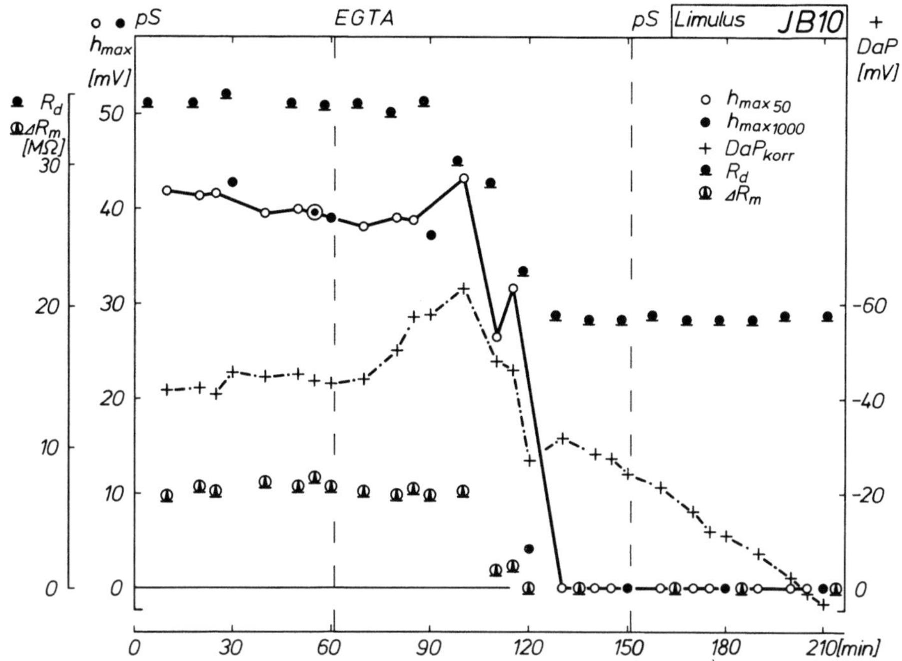

Fig. 2. Plot of the amplitude (h_{max}) of the ReP, the corrected dark potential (DaP_{korr}), the dark resistance (R_d) and the maximal change of that resistance (ΔR_m) under the influence of 1 mM/l EGTA in the presence of 55 mM/l Mg. τ = 50 msec and 1000 msec, respectively. (For definitions cf. inset of Fig. 1.) Limulus lateral eye, intracellular recording

---

Fig. 2 and 3 show typical records of 1 of 5 experiments. During a preperiod of 60 min the retina was in physiological saline. Then the saline was changed for one to which 1 mM/l EGTA [ethylene-glycol bis (β -amino-ethyl-ether)-N,N'-tetra-acetic acid] was added and which contained no Ca but 55 mM/l Mg. During the following 2 hours the receptor cell responded to the same light stimulus with smaller and smaller ReP's. Finally the retina failed to respond. The dark potential also decreased gradually, but approached zero long after the cell was inexcitable.

The measured resistance of cell membrane, cell, and electrode in series in the dark (R_d) decreases under the action of EGTA to about 80% on the average. The maximal change of this resistance induced by light (ΔR_m, measured at a time corresponding to the maximum of the ReP) decreases to zero under the influence of EGTA (Fig. 2).

The changes induced by EGTA could not be reversed by washing it out and restoring Ca to the medium. Again this contrasts with what was seen with extracellular recording. In preliminary experiments EDTA [ethylene-diamine-tetra-acetic acid] was used instead of EGTA in a saline containing neither Ca nor Mg. It appears that the changes observed after this treatment are similar to those after EGTA treatment, except that the sign of the DaP reverses.

The results can be interpreted in the following way.

1. The conductivity of the photoreceptor cell membrane in the dark increases when

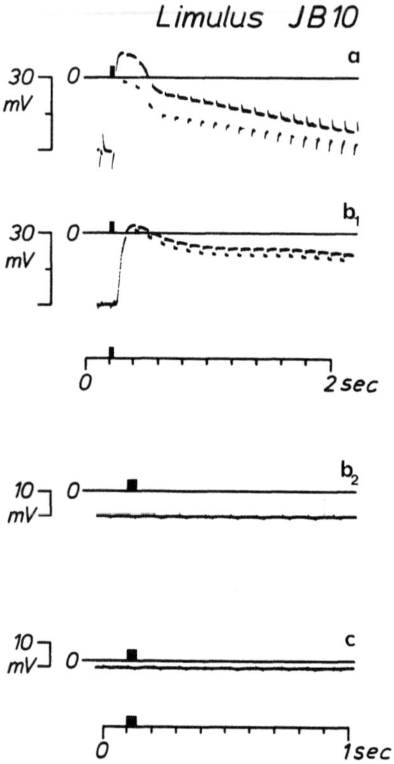

Fig. 3. Receptor potentials under the influence of EGTA. τ = 50 msec. a: reference potential after 60 min in physiological saline; EGTA added at t = 61 min; b_1: at t = 115; b_2: at t = 145; EGTA washed out at t = 151; c = 205 min. Limulus lateral eye, intracellular recording

the extracellular Ca is bound by EGTA or EDTA.

2. There is no conductivity increase caused by light when the external Ca-concentration is very low.

The difference in the action of EGTA and EDTA is probably due to the different degree to which these agents bind Mg^{++}.

The results are consistent with the hypothesis that in the photoreceptor cell the light induced increase of conductivity for certain ions occurs by releasing Ca at the outer surface of the cell membrane and the conductivity decreases when Ca is bound again.

C. Pharmaka: In 3 groups of experiments the effect of tetrodotoxin (TTX), tetraethylammonium (TEA), and veratridin (VERA) on the ReP of the crayfish measured by extracellular electrodes was tested (8). The results can be briefly summarized as follows.

1. TTX blocks the Na-channels in the nerve membrane which are open during excitation. 10^{-6} to 10^{-5} g/l TTX added to the bathing solution of the retina does not influence the size and shape of the ReP to any extent.

2. TEA in the nerve prevents the increase in K permeability during excitation. In the retina 5 mM/l TEA decreases the height of the ReP; the maximum value is more affected than the plateau-value. In contrast to its action on the nerve membrane, it causes no change in the decreasing phase of the ReP.

3. VERA produces a delayed increase of Na-permeability following excitation in the nerve and muscle membrane. 10 mg/l VERA have no measurable effects on the photoreceptor membrane.

These results show that at least in the crayfish photoreceptor cell membrane the nature of the permeability changes is different from that observed for the membranes of the investigated vertebrate and mollusc nerve and muscle.

D. Sodium substitutes: In 5 groups of experiments on the retina of the crayfish, the influence of 5 different Na substitutes on the ReP as measured by external electrodes was investigated (7,8). The Na substitutes were: Li, NH_4, choline, tris-hydroxymethyl-ammonium-ethan-hydrochloride (TRIS), and glucose. The results can be summarized as follows.

The substitutes all influence maximum and steady state values of the ReP to a different degree and all have in some respects different actions. When Na is completely replaced by choline or TRIS the ReP does not disappear but the maximum is de-

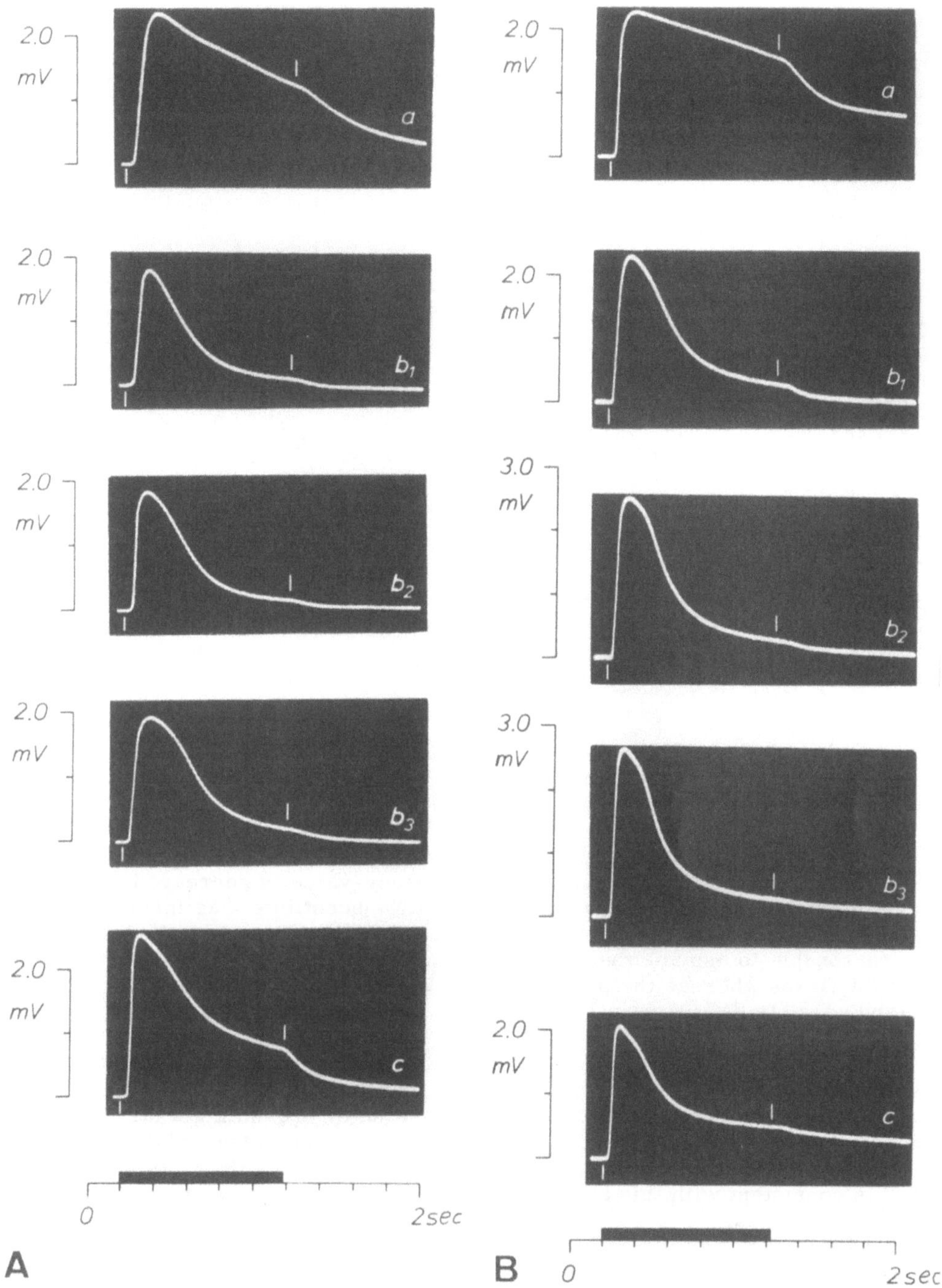

Fig. 4. Receptor potentials under the influence of TRIS (A) and Li (B) substituted for Na-ions. τ about 1000 msec. a: reference potential after 60 min in physiological saline; substituent added at t = 61 min; b_1: at t = 120; b_2: at t = 180; b_3: at t = 240; substituent washed out at t = 241; c: at t = 300 min. Astacus compound eye, extracellular recording, A 106 (left) and A 81 (right)

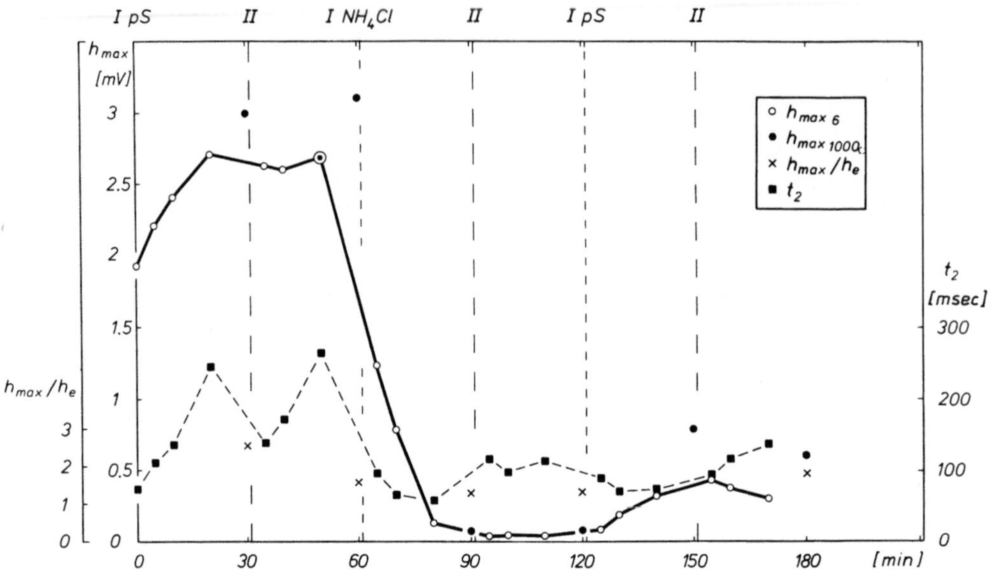

Fig. 5. Plot of the measured quantities of the receptor potential when the extracellular Na is substituted by NH_4. τ = 6 msec and 1000 msec, respectively. Astacus compound eye, extracellular recording, A 99

creased to about 60% and the plateau is even more reduced viz. to 20% (Fig. 4a). This is most probably due to the fact that in the absence of Na the reduced membrane current during the ReP is carried by Ca and Mg.

Fig. 4b shows the action of Li as a substitute for Na. Whereas the maximum of the ReP is slightly increased (to about 120%) the plateau-value is decreased to 20-25%. In the nerve membrane Li is known to be a good Na-substitute - as far as passive ion fluxes are concerned - and a bad substitute for active transport. It therefore seems reasonable to assume that the maximum of the ReP is mainly determined by passive Na-fluxes whereas the plateau is additionally influenced by active transport of Na (and possibly K).

This may also explain why the other Na-substitutes such as choline and TRIS decrease the plateau much more than they do the maximum of the ReP. It is possible that active transport contributes to the plateau-value by preventing a larger decrease in the Na-concentration gradient. The observation that the plateau of the ReP is more strongly reduced by ouabain (6) or cyanide (3) poisoning than the maximum of the ReP is consistent with this assumption.

NH_4-ions show quite a different effect from all the other substitutes in contrast to the findings for the nerve membrane and those of Smith et al. (1) for the ventral eye of Limulus. In our experiments the ReP is decreased strongly and irreversibly (Fig. 5). NH_4 is likely to affect either the membrane structure of the visual cell and thereby the mechanism of the membrane permeability change or even to act directly on the rhodopsin molecule.

Conclusions

The results of the experiments described here can be explained by the following assumptions.

1. The dark potential of the arthropode photoreceptor cell is mainly determined by the potassium gradient and by the high permeability for potassium ions.
2. The receptor potential is caused by a light induced release of Ca from the outer surface of the cell membrane. This in turn causes an increase of membrane conductance.
   a) The maximum of the ReP is caused mainly by an increase in the permeability of the cell membrane to sodium but to some extent Ca- and Mg-currents also contribute to it.
   b) The steady state value of the ReP to a long light stimulus is determined by the sodium concentration gradient, by active transport processes, by the Ca- and Mg-gradients. Furthermore the chloride gradient may contribute to the plateau-value as is suggested by the experiments in which glucose was used as substitute for NaCl (8).
3. The conductance changes which occur in the photoreceptor membrane of the Astacus eye are distinctly different in chemical nature from the conductance changes which occur in the squid nerve membrane. This is seen in the different behaviour of the Astacus photoreceptor membrane in the presence of such pharmacological agents as TTX.

References

1. SMITH, T.G., W.K. STELL, J.E. BROWN: Conductance changes associated with receptor potentials in Limulus photoreceptor. Science (Washington) 162, 454-456 (1968).
2. SMITH, T.G., W.K. STELL, J.E. BROWN, J.A. FREEMAN, G.C. MURRAY: A role for the sodium pump in photoreception in Limulus. Science (Washington) 162, 456-458 (1968).
3. STIEVE, H.: Veränderungen des Belichtungspotentials der Retina des Einsiedlerkrebses (Eupagurus bernhardus L.) bei Vergiftung des Stoffwechsels und bei Adaptation. Z.vergl.Physiol. 47, 17-38 (1963).
4. STIEVE, H.: Das Belichtungspotential der isolierten Retina des Einsiedlerkrebses (Eupagurus bernhardus L.) in Abhängigkeit von den extrazellulären Ionenkonzentrationen. Z.vergl.Physiol. 47, 457-492 (1964).
5. STIEVE, H.: In preparation (1972).
6. STIEVE, H., H. BOLLMANN-FISCHER, B. BRAUN: The significance of metabolic energy and the ion pump for the receptor potential of the crayfish photoreceptor cell. Z.Naturforsch. 26 b, 1311-1321 (1971).
7. STIEVE, H., CHR. WIRTH: Über die Ionen-Abhängigkeit des Rezeptorpotentials der Retina von Astacus leptodactylus. Z.Naturforsch. 26 b, 457-470 (1971).
8. STIEVE, H., H. GAUBE, T. MALINOWSKA: The effect of some sodium substitutes on the receptor potential of the crayfish photoreceptor cell. Zeitschr. f.Naturforsch. In press (1972).
9. FULPIUS, B., F. BAUMANN: Effects of sodium, potassium and calcium ions on slow and spike potentials in single photoreceptor cells. J.Gen.Physiol. 53, 541-561 (1969).
10. MILLECCHIA, R., A. MAURO: The ventral photoreceptor cells of Limulus. II. The basic photoresponse. J.Gen.Physiol. 54, 310-330 (1969).
11. MILLECCHIA, R., A. MAURO: The ventral photoreceptor cells of Limulus. III. A voltage-clamp study. J.Gen.Physiol. 54, 331-351 (1969).

Discussion

T. H. Goldsmith: How do the results on crayfish compare with those of Fulpius and Baumann (9) on retinular cells of drone honeybee (Apis)?

H. Stieve: Most of the results of the experiments of Fulpius and Baumann (9) are in good agreement with our results, which are comparable. Only Li as a substitute for Na shows a somewhat different action. In their experiments Li reduces the height of the transient of the ReP as well as of the plateau. Since the authors did not publish a statistical evaluation of their measurements, we cannot safely conclude that this difference is significant.
Our results are also in good agreement with the results of Millecchia and Mauro (10, 11) on the ventral eye of Limulus. The only important difference seems to be the higher sensitivity of the Limulus eye to Na deficiency as compared to the Astacus.

Control of the Dark Current in Vertebrate Rods and Cones

S. Yoshikami and W. A. Hagins
National Institutes of Health, Bethesda, MD 20014/USA

The excitatory process in a vertebrate rod or cone is known to begin with photochemical cis-trans isomerization of a carotenoid chromophore and to result in transient change in the cell's membrane potential, the latter being the first detectable neural signal. Although the machinery linking these two events has not yet been fully worked out, a number of old and new facts give us clues about it.

First clue: Rods respond to single absorbed photons with high quantum efficiency and low dark noise. From this we can deduce that nearly all of the photopigment molecules in a rod or cone are connected to the sensory output of the cell by an efficient communication system.

Second clue: Vertebrate photoreceptors contain very large numbers of photopigment molecules, at least 10^7 in a rat rod and perhaps 10^9 in that of a frog. Thus the excitatory mechanism is capable of detecting an isomerized chromophore even when it is an "impurity" amounting to less than 0.1 part per million among its unexcited neighbors in an outer segment.

Third clue: In vertebrates as well as many invertebrates, the photopigment molecules seem to be attached to lipid bilayers derived from the plasma membrane. But while the light-absorbing membranes remain confluent with the cell surface in some cones and in parts of some rods, a large part of the rhodopsin of rods is contained in closed, flattened sacs which are apparently separated from the plasma membrane by a space of at least 100-200 nanometers (see 1,2,3,4 for references). Since photons absorbed in the sacs are physiologically effective, sensory signals must traverse some part of the cytoplasm of the outer segments to reach the plasma membrane.

Fourth clue: As in many short neurons, sensory signals are transmitted along vertebrate photoreceptors by passive electrical polarization of the plasma membrane. The unusual feature of the process in a rod or cone is that a large positive current flows inward through the envelope of the outer segments in darkness, re-emerging along the rest of the cell (Fig. 1a). As this "dark current" I_D flows between the rods, it produces an interstitial voltage gradient V_D. Light reduces the dark current by means of some relatively local action in the outer segment (Fig. 1b) (5). At the same time, the cell membrane potential becomes more inside-negative (6,7). The effectiveness of light is so great that a single absorbed photon can transiently reduce a rod's dark current by about 3% (3).

Taken together, these clues pose a set of rather specific tests which any proposed excitatory mechanism must pass if it is to be considered seriously as a part of the sensory process. This paper describes some experiments designed to get at how the dark current of rods and cones might be controlled by light. In all cases, isolated rat retinas were used, and the electrical responses of the rods were recorded with extracellular microelectrodes placed across the receptor layer under direct vision by the same technique described previously (5). A small flow chamber held

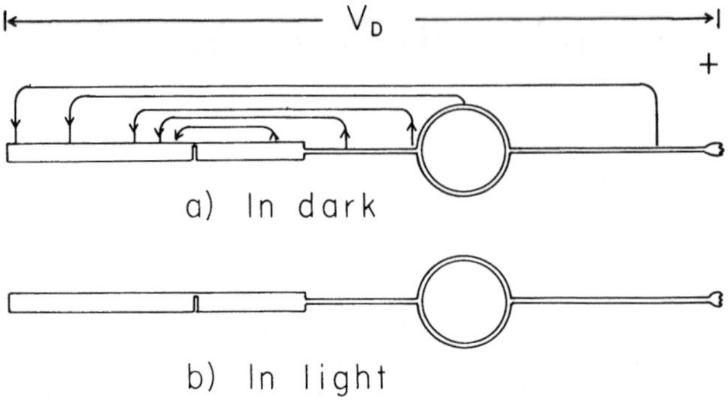

Fig. 1. Pattern of extracellular current around a vertebrate rod a) in darkness and b) in light

the retina in a stream of physiological solution whose composition could be changed in 1-2 sec. All experiments were conducted at 37°C.

The Ionic Nature of the Dark Current

The dark current is believed to be carried through the outer segment envelope as an influx of sodium ions, and light is thought to reduce the Na^+ conductance of the plasma membrane (see model of Fig. 2). Two kinds of evidence support this view. First, the resistance of a microelectrode whose tip is inserted in a rod or cone increases during the light-induced hyperpolarization (8,9). Second, the electrical responsiveness of rods and cones is rapidly and reversibly abolished by removing the Na^+ from the bathing fluid (10,11). Ordinarily these results would be quite convincing, but when examined closely, they leave one uneasy. The resistance changes observed by the microelectrode technique measure the series combination of several resistances including those in the plasma membrane and of the cytoplasm of inner and outer segment. Both structural calculations (5) and estimates from studies of the fast photovoltage of rods (12,13) indicate that the internal longitudinal resistance of outer segments is of the order of several megohms per micrometer of length. Thus changes in longitudinal resistance of a rod or cone, as well as membrane resistance changes, affect the apparent resistance of a microelectrode inserted through the membrane of either an inner or outer segment. Moreover, the published records (8,9) all show that high stimulus intensities produce resistance changes which outlast the observed hyperpolarizations, instead of being coincident with them. Nor can the need for Na^+ in the bathing fluid be taken to mean that con-

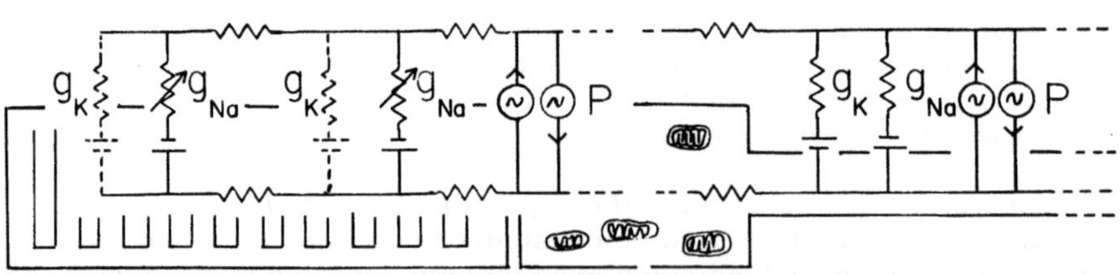

Fig. 2. Equivalent circuit for the probable electrochemical basis of the dark current in a rod. Membrane-based Na:K exchange pumps are indicated at P although the sites of Na^+ extrusion have not yet been localized accurately

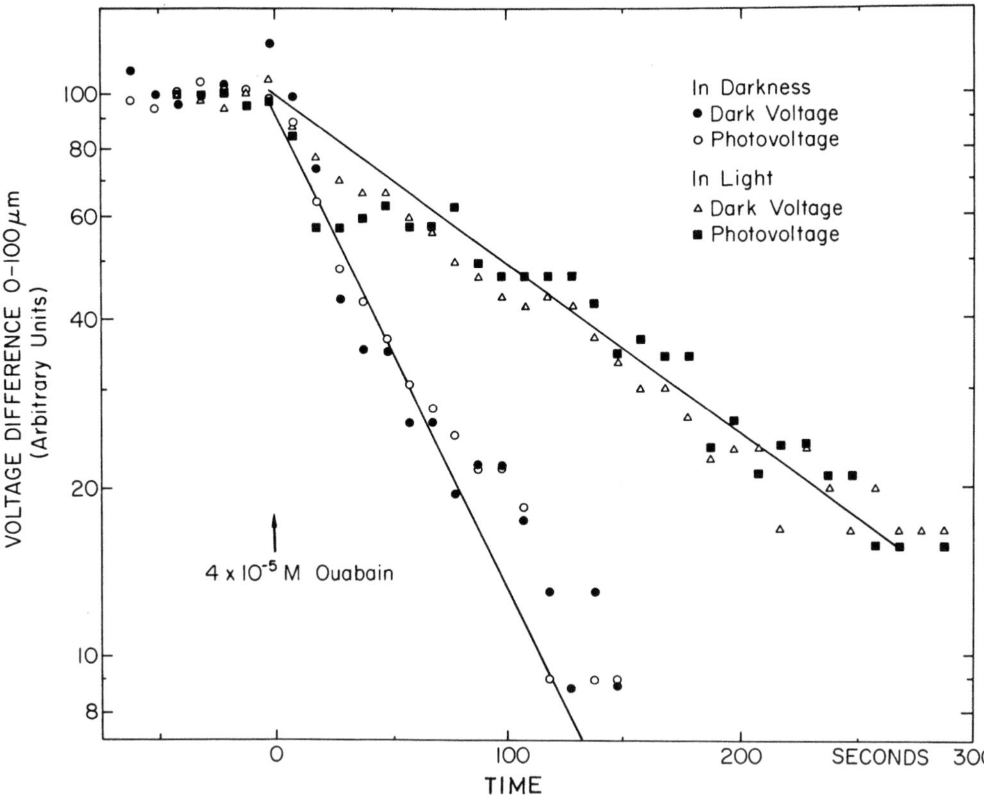

Fig. 3. Effect of light on rate of decline of dark voltage and photovoltage responses of rods when exposed to 40 μM ouabain at t = 0

trol of the inward current in outer segments be by suppression of g_{Na}, for when Na^+ is removed from the external medium, the model of Fig. 2 predicts that I_D will be reduced to zero and thus any effect of light on the conducting paths through the membrane will be concealed. For example, if light were to increase g_K (Fig. 2) the effect on I_D would be qualitatively the same as if its action were to reduce g_{Na}. This is because longitudinal flow of current between two regions on the cell depends upon the ratios g_K:g_{Na} and/or the transmembrane concentration gradients of the two ions being different in the two regions. If one ion, such as Na^+, is removed from the system, the cell becomes isopotential and I_D thus stops, regardless of whether it is g_K or g_{Na} which is primarily changed by light.

Nevertheless, we believe that the conventional view that light decreases g_{Na} is probably correct because of the following experiment. If I_D is to flow (Fig. 2), there must be an influx of Na^+ in the outer segment which must be balanced by an equal outward current along the remainder of the cell. At any instant, this balancing current could be carried as an influx of, say, Cl^- or as a K^+ outflux, but in the steady state, the action of electrolyte transport pumps P must eventually convert the net current along the inner segment and cell body to a Na^+ outflux equal in size to the influx in the outer segment. But if the pumps could be stopped quickly, the Na^+ influx would be expected eventually to discharge the ionic concentration differentials (apart from the Donnan potential due to impermeable cytoplasmic ions). Rough calculations based on the dark current of rat rods (5) predict that electrolyte exchange due to I_D alone should cause V_D to disappear within about 1 min after the pumps are stopped.

This is in fact what happens when the voltage gradient V_D across the receptor layer of the rat retina is measured in the presence of 40 μM ouabain (Fig. 3). In darkness, V_D declines with an exponential course and a time constant of about 1 min. The photocurrent responses to small test flashes decline simultaneously. But if the retina is illuminated with a background light (~1000 photons absorbed sec^{-1} rod^{-1}, 560 nm) 80% of the time, the decline of both V_D and the photocurrent is four times slower. This result suggests that light delays the collapse of the ionic gradient across the plasma membrane of a ouabainized cell and is consistent with the idea that it is g_{Na} which is reduced by light and not g_K which is increased. Qualitative studies of the ionic composition of the photoreceptors of ouabainized rat retinas by electron probe microanalysis (unpublished results) support this conclusion, as do studies of osmosis in rods (14). Nevertheless, we cannot yet establish the absolute values of g_{Na} and g_K of the plasma membranes of either inner or outer segment. The fact that bright stimuli reduce V_D to zero in solutions whose $a_{Ca^{++}}$ are 1.36 mM suggests that g_K in the outer segment envelope must be low relative to g_{Na} in the dark, a conclusion also supported by the experiments of Korenbrot and Cone (14). However, if $a_{Ca^{++}}$ is lowered to $<10^{-7}$ M, bright light stimuli may produce a weak reversal of V_D in some retinas (from, say, -200 μV in dark to +100 μV at the peak of the light response). Thus g_K may be appreciable in the outer segment in such cases.

Similar Action of Light and Calcium

Presumably, agents such as ouabain reduce or abolish the dark current slowly by interfering with ion transport and allowing the transmembrane ionic batteries to be discharged. But if the external $[Ca^{++}]$ is increased, V_D is reduced much more quickly than it is by ouabain. Fig. 4 shows the effect on V_D of ejecting a small amount of a physiological solution 20 mM in Ca^{++} near the point where a pair of pipettes are placed across the rod layer of a rat retina. The effects of several light flashes are also shown. As the high Ca^{++} solution reaches the rod outer segments, I_D is suppressed and V_D almost vanishes for about 1 sec and then resumes as the flowing stream of solution containing 100 μM Ca^{++} washes away the high Ca^{++} solution. Simultaneously, the effect of light on the voltage across the receptor layer is suppressed. Ca^{++} acts with no detectable latency: The delayed fall and rise in V_D can be entirely accounted for by diffusion of ions into and out of the receptor layer. Unlike Ca^{++}, Mg^{++} (20 mM in Ringer's solutions) (as well as K^+, Cl^- and added Na^+) are virtually without immediate effect on I_D.

V_D seems to be suppressed by the same mechanism as that by which light acts. When the decline of V_D and the amplitude of the photocurrent responses to a standard test flash in 40 μM ouabain are measured in Ringer containing 5 mM Ca^{++} it is about four times slower than in our standard solution containing 1.36 mM Ca^{++} (Fig. 5)

There is a second similarity between the effects of light and of externally applied Ca^{++}. If the rods are exposed to a solution whose Ca^{++} activity is buffered to 10^{-5} M with EGTA, V_D becomes very large exceeding the normal value of 100-200 μV by three fold. This large V_D can be sustained indefinitely. If a bright flash of light equivalent to $\sim 10^7$ photons absorbed rod^{-1} is now delivered to the retina, a trans-

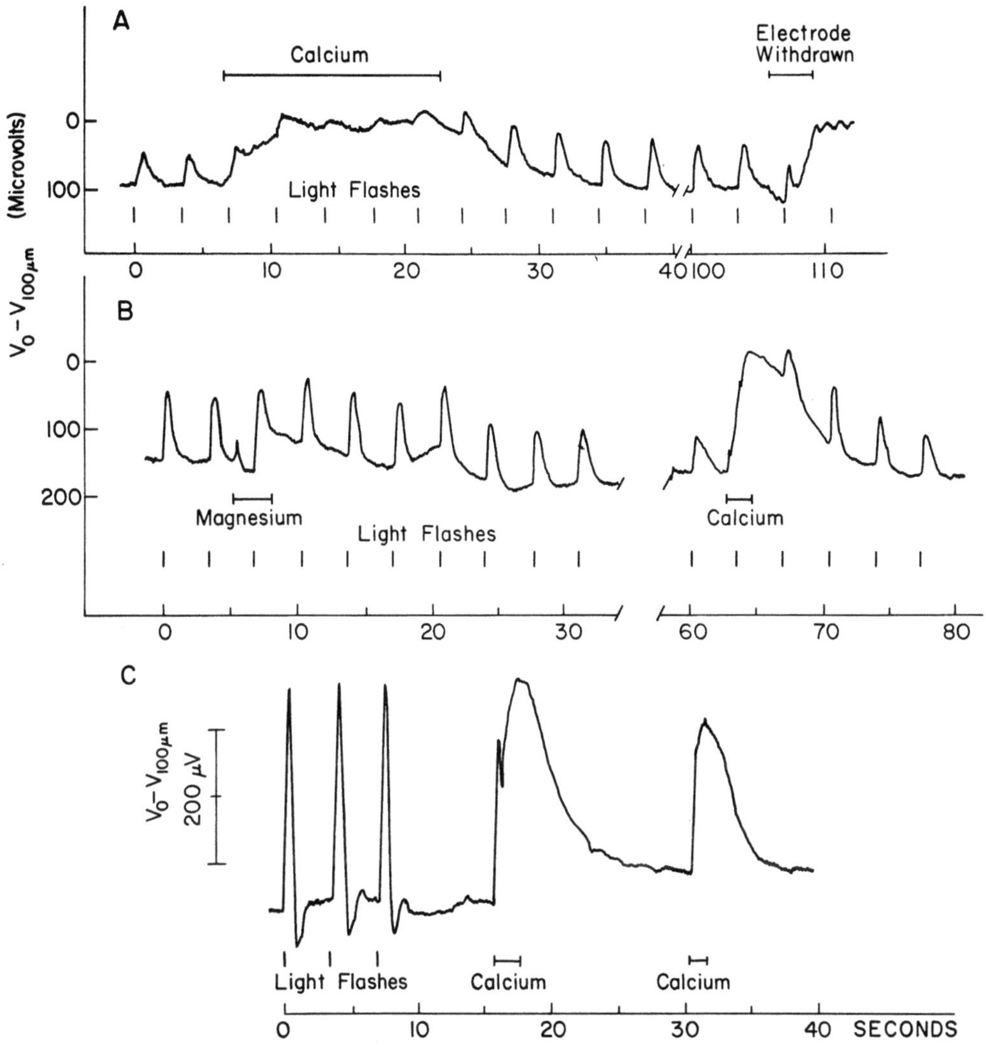

Fig. 4. Effects of externally applied Ca^{++} and Mg^{++} on dark voltage and photovoltage responses of rods of isolated rat retina. Retinas maintained in steady stream of physiological solution. Ca^{++} or Mg^{++} ejected from a large pipette at retinal surface near site where voltages were recorded with microelectrodes. Flowing stream changed solution in chamber in about 3 sec.

A and B: Background $[Ca^{++}]$ 1.36 mM. Applied $[Ca^{++}]$ or $[Mg^{++}]$ was 20mM in Ringer. Size of dark voltage indicated in A by withdrawing recording microelectrode. Both dark voltage and light responses (flashes at vertical lines) were abolished transiently by Ca^{++} but not by Mg^{++}.

C: Large photovoltage responses with overshoot when background $[Ca^{++}]$ is reduced to 10^{-5} M. Responses to applied Ca^{++} are also proportionately larger than in A and B

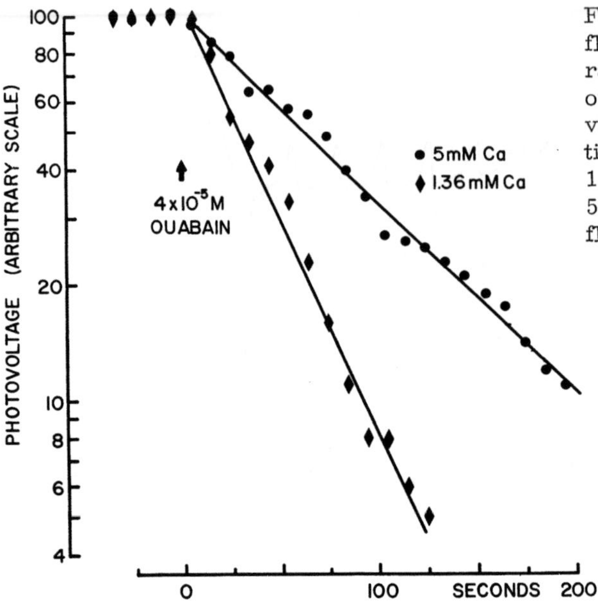

Fig. 5. Effects of [Ca^{++}] in bathing fluid on photovoltage responses of rat retinas poisoned with 40 μ M ouabain. The decline of the photovoltage in 5 mM Ca^{++} is about three times slower than when [Ca^{++}] is 1.36 mM. Stimulus flashes: 2 sec, 560 nm, 30 photons absorbed rod^{-1} $flash^{-1}$

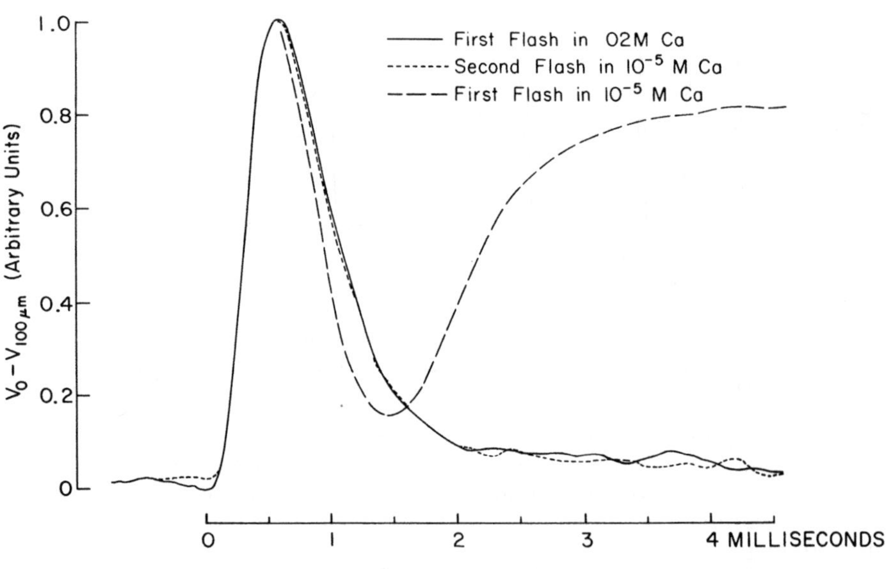

Fig. 6. Shapes of fast photovoltages in response to saturating flashes as functions of light adaptation and external [Ca^{++}]. Stimuli: 100 μsec flashes containing wavelengths 520 nm energy equivalent to ~2.6 x 10^7 photons absorbed rod^{-1} $flash^{-1}$ All responses scaled to unit peak height. First flash in 10^{-5} M Ca^{++} produces a narrow fast photovoltage (FPV) wave followed by a physiological shutdown of the dark current which is complete by ~3.5 msec (dashed curve). Zero dark voltage is at about 0.82 on ordinate scale. — FPV response from a dark-adapted retina in 20 mMCa^{++} has shape identical with that produced by a second flash to a retina light-adapted in 10^{-5} MCa^{++} by a saturating flash delivered 10 sec previously

Fig. 7. Change in the amplitude-flash energy relation of rods exposed to low [Ca^{++}]. Stimuli: 560 nm, 2 μ sec

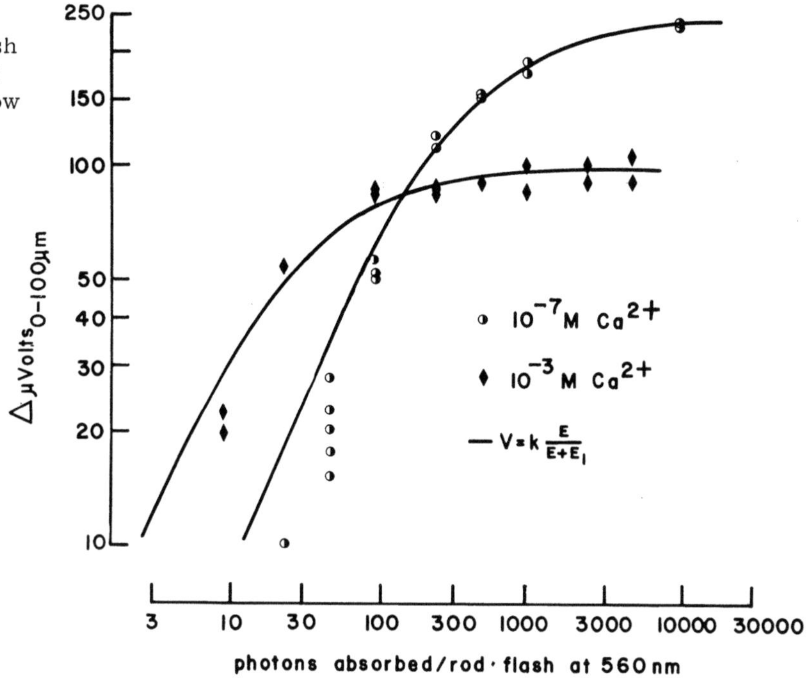

receptor layer voltage waveform like that of Fig. 6 (dashed curve) is seen. There is initially a fast photovoltage (FPV) which is distinctly narrower than that usually seen in the light-adapted retina or in a retina exposed to 20 mM Ca^{++}. Following this, V_D declines rapidly to zero (corresponding to a value of 0.82 on the arbitrary ordinate scale). A second flash of the same intensity now gives a FPV of the usual wide R_2 type (dotted curve). The anomalous shape of the first FPV response can be prevented by exposing the retina to steady illumination equivalent to 1000 photons absorbed rod^{-1} sec^{-1}, an intensity which produces a steady physiological response just sufficient to suppress V_D. This difference in waveform of the FPV between dark and light adapted retinas is even more striking in retinas bathed in lower $a_{Ca^{++}}$ Ringer (10^{-7} M) for several minutes. Conversely, if the external $a_{Ca^{++}}$ is increased to 20 mM, the shape of the FPV response to the first flash is of normal (R_2) width (continuous curve) and is not widened further by light adaptation. The reason for the distortion of FPV responses at low $a_{Ca^{++}}$ will be considered below.

Desensitization of Rods by Ca^{++} Depletion

When rods are exposed to $a_{Ca^{++}} < 10^{-7}$ M, V_D initially becomes very large, sometimes exceeding the normal value of 100-200 μV by fivefold. After a minute or so the V_D then decline to about 100-200 μV. During the first minutes of contact with such low-Ca^{++} solutions when V_D is very large, the relation between test flash intensity and physiological photovoltage response remains the same as that of a normal retina in solutions with $a_{Ca^{++}}$ = 1.36 mM. The light responses simply increase in proportion to V_D (Fig. 7, diamonds), With the elapse of 1-2 min, however, more and more light is required in the test flash to produce half-suppression of V_D and the amplitude: flash energy curve is shifted about 1 log unit towards higher flash energies (Fig. 7, half-filled circles). If $a_{Ca^{++}}$ is returned to its usual 1.36 mM level, the amplitude:energy curve returns to normal within 1 min. Longer

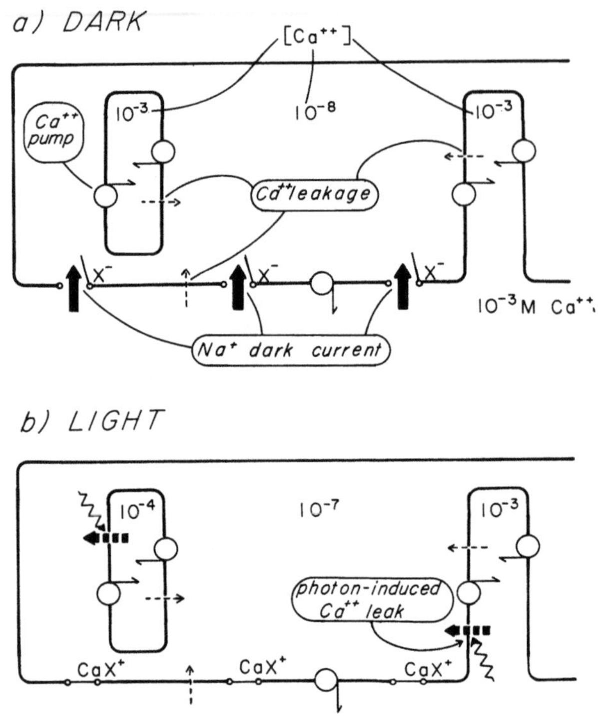

Fig. 8. Hypothetical scheme for the control of the dark current by the activity of free Ca^{++} in the cytoplasm of rods and cones.
a) Heavy arrows indicate Na^{+} fluxes through "D channels" in dark.
b) Heavy broken arrows indicate Ca^{++} fluxes through light-activated "L channels". From Hagins (4)

treatment of the retina with a solution whose $a_{Ca^{++}}$ is less than 10^{-8} M will desensitize the rods so much that flashes causing more than 10^{4} photons to be absorbed per rod may be needed to half-suppress V_D.

A Suggested Mechanism for Control of the Dark Current

The marked effects of external Ca^{++} activity on the properties of the rod photocurrent and dark current suggest a model of the excitatory process shown in Fig.8. The central idea is that I_D enters the outer segments through channels ("D channels") which are specifically permeable to Na^{+} and which are associated with sites on the internal surface of the plasma membrane where Ca^{++} ions may be bound. If a D channel has a Ca^{++} ion bound to it, its Na^{+} permeability is presumed to be blocked. Ca^{++} is presumed to be regulated at the same low level in the rod cytoplasm as it is in muscle fibers (15) or in nerve axoplasm (16,17) by electrolyte pumps. The action of an absorbed photon is then to create a chennel (an "L channel") in either the plasma membrane or one of the disc membranes through which Ca^{++} can reach the cytoplasm and thus block the D channels. Re-accumulation of Ca^{++} in the discs following closure of the channel will then restore the original dark-adapted state. The similarity of the proposed scheme to that operating in sceletal muscle is obvious.

Although we know of no experimental results directly in conflict with the Ca^{++}model, direct attempts to show light-stimulated release of 45Ca from rod discs have not been successful either in our laboratory (unpublished results) or elsewhere (18,19). However, it is not to be expected that light-induced Ca^{++} fluxes will be very large.

The small size of the cytoplasmic space in the outer segment of a rat rod ($<20 \times 10^{-15}$ liter) coupled with the usually low $a_{Ca^{++}}$ in most cytoplasm ($\sim 10^{-6} - 10^{-8}$ M) implies that no more than 100 - 10,000 free calcium ions may be present at any instant in dark adaptation. If an absorbed photon is to reduce I_D by at least 3%, no more than 3 - 300 Ca^{++} ions would need to be released by it. Brighter flashes would release more Ca^{++}, of course, but there is no assurance that the proportionality between photons absorbed and Ca^{++} released will hold above the level of 1 photon absorbed per disc, or 3,000 to 300,000 Ca^{++} ions altogether. Now, our own measurements of 45 Ca binding by isolated frog rod outer segments from a solution containing 100 μ M total Ca^{++} and an $a_{Ca^{++}}$ of 10^{-7} M indicate at least $3 \cdot 10^6$ Ca^{++} ions bound per rod. These bound Ca^{++} ions exchange off against unlabeled Ca^{++} with a time constant of 1-2 min. Against such a background of isotopic exchange, a light-induced release of, say, 30,000 extra Ca^{++} ions would be hard to detect in the usual type of tracer experiment.

Some Speculative Calculations

Although the Ca^{++} hypothesis is little more than a suggestion at present, it raises some interesting quantitative questions. For instance, how many D sites might there be in an outer segment? From the measured size of I_D ($\sim$ 70 pA rod^{-1}) (5) and the likely driving force of 20-50 mV for Na^+ movements, one can estimate that the total g_{Na} need not exceed $\sim$3000 p mho in darkness. We do not know the specific conductance of a D channel but we can make a guess as to its maximum value based on the following argument. Let us assume (a) that in physiological solutions containing 10^{-3} M $a_{Ca^{++}}$ about 20% of the D channels are open (since low Ca^{++} solutions can increase I_D by at least fivefold), and (b) that the fluctuations in total D channel conductance g_{Na} due to random binding and release of Ca^{++} ions is less than the change in g_{Na} produced by an absorbed photon (otherwise rods could not detect photons reliably). If so, the relative standard deviation σ of g_{Na} can be derived from the binomial probability distribution to be

$$\sigma = \frac{1}{\sqrt{N}} \sqrt{\frac{1 - a}{a}}$$

where a is the mean proportion of D sites which have Ca^{++} bound at any moment and N is the total number of D sites. Since σ must be less than .03 in order to be less than the variation due to a photon response, N must be > 3600 for $a \leq 0.2$. If this is so, the conductance of a single D channel need not be more than 1 p mho. This is a low value in comparison with the Na channels of squid photoreceptors (20,4) and squid axons (21) and with channels which have been observed in artifical lipid bilayers (22). If there are indeed more than 3000 D channels in each outer segment, the calcium ions released by a single absorbed photon can be estimated in a second way, the result being similar to the previous estimate. For 3% of the sites to be obstructed by a photon, at least 100 Ca^{++} ions would be required.

The Ca^{++} hypothesis requires that light create new ionic channels in the disc membranes (in rods) and in the plasma membrane (cones). Is there any other evidence for such channels? Perhaps the curious distortions of the FPV wave of retinal rods in low Ca^{++} solutions represents an inward current through Ca^{++} channels in the rod envelope. As is shown elsewhere (13), the R_2 charge displacement in rat rods occurs throughout the outer segment, thus suggesting that rhodopsin is a constituent of the plasma membrane there. Now, the distorted first FPV waveform in low Ca^{++} solutions can be analyzed into two components, one the standard R_2 waveform and the other a somewhat slower inward positive current (the "L current") which, like

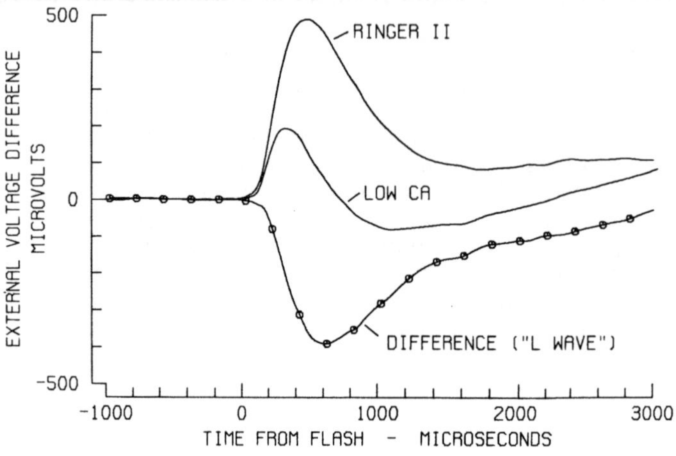

Fig. 9. Analysis of the fast photovoltage transient of the dark adapted rat retina in 10^{-7} M Ca^{++} into two components. The average of 7 responses from different retinas in Ringer II (1.36 mM Ca^{++}) is substracted from the complex biphasic wave obtained in low Ca^{++} Ringer (average of 7 responses from different retinas) to yield a difference "L wave" curve. The R_2 components of the FPV's in light adapted retinas are identical in size and shape in the two solutions. If this is also true for dark adapted rods, the "difference" curve represents an additional component added to the R_2 component in low Ca^{++} solutions to give the distorted FPV waveform to first. This component is a positive inward current like the dark current but its amplitude only saturates with flashes which activate every rhodopsin chromophore. Since the A-wave contribution to both curves was small and appeared after 2 msec, it was not considered in the analysis

---

R_2, is proportional to the number of rhodopsin chromophores activated by the flash (Fig. 9). Since the L current can only be detected when there is a large inward Na^+ ion gradient across the envelope of rod outer segment, it is possible that it is a Na^+ current through channels which normally admit Ca^{++}. If so, the L channels, like the D channels must have a permeability to Na^+ which is reduced by high $a_{Ca^{++}}$ in the rod cytoplasm.

We hope that the near future will bring experiments which will test the calcium hypothesis more severely than do the present indirect ones. If so, new methods will probably be needed for measuring and controlling the activity of free Ca^{++} in the cytoplasm of small cells by non-optical means, and for determining the ionic composition of photoreceptors.

References

1. COHEN, A.I.: New evidence supporting the linkage to extracellular space of outer segment saccules of frog cones but not rods. J.Cell Biol. 37, 424-444 (1968).
2. COHEN, A.I.: Further studies on the question of the patency of saccules in outer segments of vertebrate photoreceptors. Vision Res. 10, 445-453 (1970).
3. PENN, R.D., W.A. HAGINS: Kinetics of the photocurrent in retinal rods. Biophys.J. 12, 1073-1094 (1972).
4. HAGINS, W.A.: The visual process: excitatory mechanisms in the primary receptor cells. Ann.Rev.Biophys.Bioeng. 1, 131-158 (1972).
5. HAGINS, W.A., R.D. PENN, S. YOSHIKAMI: Dark current and photocurrent in retinal rods. Biophys.J. 10, 380-412 (1970).
6. BORTOFF, A., A.L. NORTON: Simultaneous recordings of photoreceptor potentials and the P III component of the ERG. Vision Res. 5, 527-533 (1965).

7. TOMITA, T.: Electrophysiological study of the mechanism subserving color coding in the fish retina. Cold Spring Harbor Symp. Quant. Biol. 30, 559-566 (1965).
8. TOYODA, J., H. NOSAKI, T. TOMITA: Light-induced resistance changes in single photoreceptors of Necturus and Gekko. Vision Res. 9, 453-463 (1969).
9. BAYLOR, D., M.G.F. FUORTES: Electrical responses of single cones in the retina of the turtle. J. Physiol. (Lond.) 207, 77-92 (1970).
10. SILLMANN, A.J., H. ITO, T. TOMITA: Studies on the mass receptor potential of isolated frog retina. II. On the basis of the ionic mechanism. Vision Res. 9, 1443-1451 (1969).
11. YOSHIKAMI, S., W.A. HAGINS: Ionic basis of dark current and photocurrent in retinal rods. Abstr. 14th Ann. Meet. Biophys. Soc. WPM-13 (1970).
12. HAGINS, W.A., H. RÜPPEL: Fast photoelectric effects and the properties of the vertebrate photoreceptors as electric cables. Fed. Proc. 30, 64-68 (1971).
13. RÜPPEL, H., W.A. HAGINS: Spatial origin of the fast photovoltage in retinal cells. This volume, pp. 257-261.
14. KORENBROT, J., R.A. CONE: Dark ionic flux and the effects of light in isolated rod outer segments. J. Gen. Physiol. 60, 20-45 (1972).
15. LUXORO, M., E. YAÑEZ: Permeability of the giant axon of Dosidicus gigas to calcium ions. J. Gen. Physiol. 51, 115-122S (1968).
16. HODGKIN, A.L., R.D. KEYNES: Movement of labelled calcium in squid giant axon. J. Physiol. (Lond.) 138, 253-281 (1957).
17. BLAUSTEIN, M.P., A.L. HODGKIN: The effect of cyanide on the efflux of calcium from squid axons. J. Physiol. (Lond.) 200, 497-527 (1969).
18. BOWNDS, D., A. GORDON-WALKER, A. GAIDE-HUGUENIN, W. ROBINSON: Characterization and analysis of frog photoreceptor membranes. J. Gen. Physiol. 58, 225-237 (1971).
19. NEUFELD, A.H., W.H. MILLER, M.W. BITENSKY: Calcium binding to retinal rod disk membranes. Biochem. Biophys. Acta 266, 67-71 (1972).
20. HAGINS, W.A.: Electrical signs of information flow in photoreceptors. Cold Spring Harbor Symp. Quant. Biol. 30, 403-418 (1965).
21. HILLE, B.: Ion channels in nerve membranes. Prog. Biophys. Mol. Biol. 21, 1-32 (1970).
22. LATORRE, R., G. EHRENSTEIN, H. LECAR: Ion transport through excitability-inducing material (EIM) channels in lipid bilayer membranes. J. Gen. Physiol. 60, 72-85 (1972).

Discussion

S.L. Bonting: You conclude that the rod sacs serve to keep the intracellular Ca^{++} low in the dark. Do you propose that light releases the Ca^{++} which closes the Na-channels from the sacs, or do you also assume extracellular Ca to be involved?

S. Yoshikami: The calcium released by light into the intracellular space could come from rod disc compartment and/or from extracellular space surrounding the cell. The calcium source would depend on whether the visual pigment in the disc membrane or in the plasma membrane absorbed the incident photon. In the former case Ca^{++} would come from the disc space (the pre-dominant source in the case of rods) and in the latter it would come from the extracellular space.

Spatial Origin of the Fast Photovoltage in Retinal Rods

H. Rüppel* and W. A. Hagins
National Institutes of Health, Bethesda, MD 20014/USA

Fast photoelectric effects, of which the "early receptor potential" (1) is the first to be discovered, are found in many types of retinas (2,3,4; for reviews). There is now little doubt that the currents which cause fast photovoltages (FPV's) originate from changes in the distribution of charges in the photopigment molecules themselves, but the electrical arrangements which convert these into externally observable voltages are not yet fully understood. This paper reports some experiments designed to explore the electrical circuitry underlying FPV's of vertebrate photoreceptors. Our aim is to locate the sources of the charge displacements, to estimate their absolute magnitudes on a molecular basis and to determine the cable constants of the rods of isolated rat retinas. The experimental procedure is to compare the FPV's recorded with extracellular microelectrodes from various depths in the receptor layer of the retinas with theoretical solutions of the cable equation for non-uniform cells with the anatomical dimensions of rat rods. The experimental techniques are described previously (5).

Spatial Origin of the Charge Displacements

In attempting to explain the origin of the FPV, Brindley and Gardner-Medwin (6) suggested that the sacs of rod outer segments might have rhodopsin molecules distributed asymmetrically so that photolysis would produce eddy currents like those shown in Fig. 1a. In the long photoreceptors of squids, however, it appears that the sites of the charge displacements are the plasma membranes themselves (7)(Fig. 1b). Fig. 2 shows that the two arrangements predict quite different patterns of voltages in the spaces between photoreceptors. The waveforms for each model are computed for extracellular electrodes at positions A,B,C located as shown in Fig. 1. The cable constants of the rods (which are not critical for the present argument) are assumed to be those for cells with the dimensions of rat rods with cell membrane capacitances of $1\ \mu F\ cm^{-2}$, cytoplasmic resistivities of $200\ \Omega$ cm and membrane time constants of 1 msec. For this example, the charge displacement currents in the cell membranes produced by a short intense flash of light at $t = 0$ are assumed to be of the form

$$A \exp(-t/350\ \mu sec).$$

The transmembrane charge displacement is assumed to be one electronic charge per rhodopsin molecule, the rod envelope being supposed to contain the same density of rhodopsin as the discs.

There are two major differences between the predictions of the two models of Fig. 1. First, because of the impedance of the rod envelope and the shunting effect of the cytoplasm, a monophasic charge displacement in asymmetric discs (Fig. 1a) will produce an external voltage transient which is as large between 0 and 20 μm as it is

*Present Address: Max Volmer Institut, Technische Universität Berlin, 1 Berlin 12, Germany.

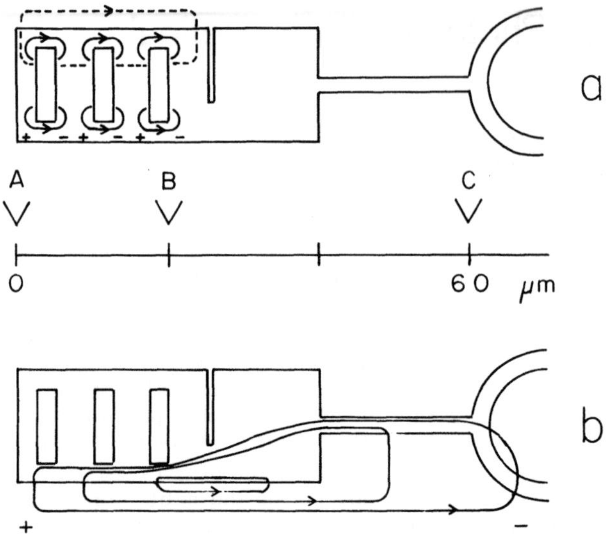

Fig. 1. The models for the generation of fast photoelectric effects in vertebrate rods from charge displacements in rhodopsin molecules. a) Charge displacement in asymmetric discs. b) Charge displacement across envelope membrane enclosing outer segment

between 20 and 60 μm (Fig. 2a). A displacement in the plasma membrane itself, however, yields a monophasic wave which is largest between 20 and 60 μm (Fig. 2b). Second, a charge displacement of a given total size yields a much larger FPV if the current source is in the plasma membrane of the outer segment (Fig. 2b) than if it is within isolated internal discs (Fig. 2a).

Since experimentally observed FPV's resemble those predicted for charge displacements in the rod envelope, it seems likely that the model of Fig. 1b is a good first approximation to the actual situation in rods and is worth further study. The model of Fig. 1a predicts such weak electrical effects (s. Fig. 2a) that it can be rejected.

Spatial Distribution of the Charge Displacement

At 37°C the R_2 component dominates the entire FPV waveform. Measured across the entire rod layer, or transretinally as is usually done, the waveform is a monophasic outer-segment-positive voltage transient. The same statement is true if the FPV is recorded with differential pairs of microelectrodes at several points along the rods (Fig. 3). In particular, the fast photocurrent is outward positive all along the outer segments. Evidently, the entire plasma membrane covering the outer segments is the seat of the R_2 charge displacement and not just a small region at the junction between inner and outer segments. In the latter case, computer solutions indicate that the voltage waveform between electrodes at 0 and 20 μ m would have been reversed in sign from that actually seen. The waveforms often showed a small slow negative component in addition to the main R_2 wave (e. g. in the 0-10 μm curve). This may be a trace of the "L-wave" (8) produced from the incompletely light-adapted rods. A separate study of the FPV at 8°C also indicates that the R_1 component also originates along the entire envelope of the outer segments.

Cable Constants of Rat Rods

Since the R_2 develops with a delay of more than 100 μsec, Cone and Cobbs (9) have suggested that it might be produced by the decay of metarhodopsin I, a process

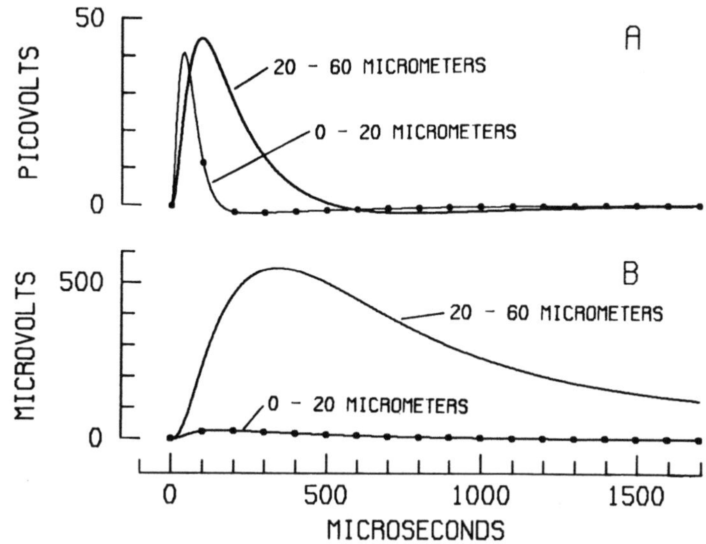

Fig. 2. Computed differential voltage transients in the receptor layer of the isolated rat retina for the two models of Fig. 1. Charge displacement is assumed to be exponential with a 350 μsec time constant and to consist of transfer of one electronic charge (B) across rod envelope membrane or (A) across distal half of each internal rod sac for each rhodopsin chromophore in it. Envelope and disc membranes supposed to bear $6 \cdot 10^{11}$ chromophores cm^{-2}. Electrodes placed at rod tips (0 μ m), near inner-outer segment junction (20 μ m) and in rod fiber layer (60 μ m)

---

which has an exponential time constant of ~350 μsec at 37°C in live retinas (Hagins and Rüppel, unpublished). On this assumption, the cable constants of rat rods can be estimated by fitting the FPV transients observed with electrodes at 0, 20, and 60 μm in the rod layer (Fig. 1, A, B, C) with solutions of the cable equation for a cell with adjustable values of membrane capacitance, cytoplasmic resistivity and membrane time-constants. Table I shows the values obtained by this procedure.

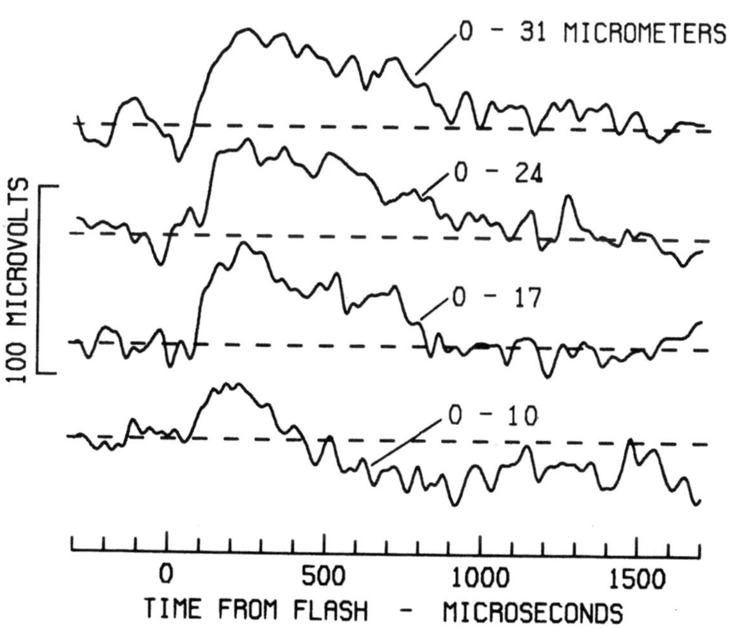

Fig. 3. Experimentally observed voltage transients recorded between differential pairs of microelectrodes, one at rod tips (0 μ m) and one at 10, 17, 24, or 31 μm depth in rod layer. Each trace is the average of 3 responses from different pieces of retina. Flash exposure ~$3 \cdot 10^7$ photons absorbed rod^{-1} $flash^{-1}$ (λ = 520 nm)

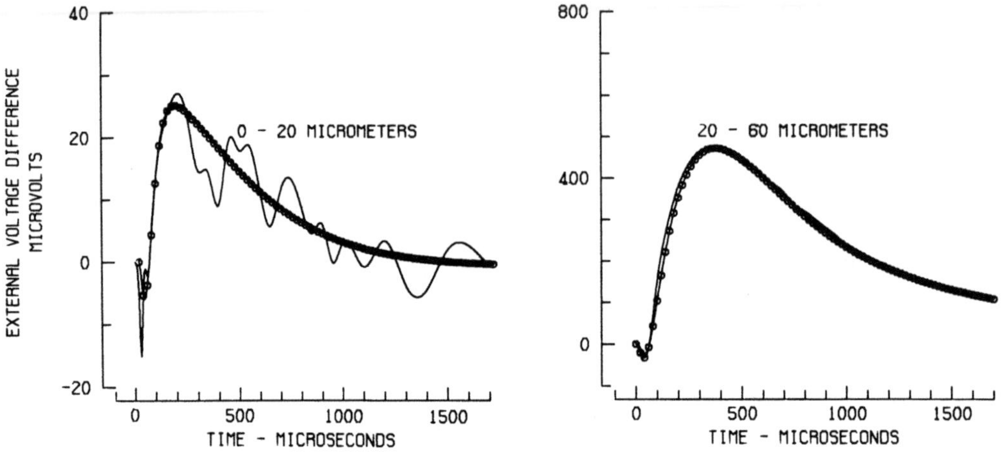

Fig. 4. Fit of experimentally observed and computed fast photovoltages. Observed curves (continuous lines) are averages of 12 responses from different retina blocks recorded with electrodes at 0, 20, and 60 μ m depths. Flashes as in Fig. 3. Computed curves (circles) are based on cable constants of Table I and assumptions that R_1 charge displacement is instantaneous and that causing R_2 has a time-constant of 350 μ sec. Internal longitudinal resistance, membrane capacitance, and membrane conductance of outer segment adjusted to give a best fit. Ordinates of each graph are voltage differences between two electrodes placed at depths shown

---

The fits of observed to computed FPV's are shown in Fig. 4. Not shown in Table I is the relative intensity of the charge displacement per unit length of outer segment. This figure can be determined with only low accuracy, but the best fits are obtained if more than 50% of the total displacement is produced at depths of 20-25 μ m, that is, at the junction between inner and outer segments.

Table I. Estimated cable constants of a typical rat rod

Region	Boundaries+ μm from rod tips	Diameter+ μm	Cytoplasmic Resistivity Ω cm	Membrane Capacitance $\mu F\ cm^{-2}$	Membrane time-constant msec
Outer segment	0 - 24.5	1.7*	150±50	1.0±0.3	10. ±2.
I-O junction	24.5 - 25	1.7§	150±50	25.0±5	10. ±2.
Inner segment	25 - 45	1.7	200±70	1.0±0.2	1.0±0.2
Rod fiber	45 - 60	0.5	200±70	1.0±0.2	1.0±0.2
Nucleus	60 - 70	10.	~200	~1.0	~1.0
Synaptic process	70 - 100	0.5	~200	~1.0	~1.0

+ From measurements of electron micrographs (5)

* 90% of cross-section obstructed by discs

§ 10 discs assumed to be infolded from and attached electrically to plasma membrane

Absolute Size of the Charge Displacement

The relative sizes of the R_1 and R_2 displacements can be estimated to be about 1:9 by comparing the FPV's at 8° and $37^{o}C$. Allowing for the R_1 contributions at $37^{o}C$, the total charge displacement in R_2 required to fit the observed FPV's of isolated rat retinas is $2\pm0.5 \cdot 10^5$ electronic charges per rod. If it is assumed that the density of rhodopsin molecules on the envelope is the same as that on the discs, at least 3% of the $3 \cdot 10^7$ chromophores could lie on the envelope, and each would produce a net charge movement of about (0.5 electronic charge · 70 Å). This figure is similar to that reported previously (2). While it is small relative to the size of ionic currents in photoreceptors (5) it suggests a substantial change in the structure of a metarhodopsin I molecule as it is converted to metarhodopsin II, in situ, a conclusion supported by some x-ray structural evidence (10) but not by optical studies of molecular asymmetry in rhodopsin (11).

The computations also predict that the change in membrane potential due to the FPV at the peak of the R_2 current in rat rods will be about 2-3 mV. This voltage displacement is of the same order of size as that observed directly in salamander and lizard photoreceptors by Murakami and Pak (12).

Further refinements of the present experiments are possible. We hope that future work will lead to more accurate and detailed information about the electrical structure of rods.

References

1. BROWN, K.T., M. MURAKAMI: A new receptor potential of the monkey retina with no detectable latency. Nature (Lond.) 201, 626-628 (1964).
2. HAGINS, W.A., H. RÜPPEL: Fast photoelectric effects and the properties of the vertebrate photoreceptors as electric cables. Fed.Proc. 30, 64-68 (1971).
3. CONE, R.A., W.L. PAK: The early receptor potential. In: Handbook of Sensory Physiology, Vol. I, ed. W.R. LOEWENSTEIN, pp. 345-365. Springer, Berlin-Heidelberg-New York (1971).
4. ARDEN, G.B.: The excitation of photoreceptors. Progr.Biophys.Mol.Biol. 19, 373-421 (1969).
5. HAGINS, W.A., R.D. PENN, S. YOSHIKAMI: Dark current and photocurrent in retinal rods. Biophys.J. 10, 380-412 (1970).
6. BRINDLEY, G., A.R. GARDNER-MEDWIN: The origin of the early receptor potential of the retina. J.Physiol.(Lond.) 182, 185-194 (1966).
7. HAGINS, W.A., R.E. MACGAUCHY: Membrane origin of the fast photovoltage of the squid retina. Science (Wash.) 159, 213-215 (1968).
8. YOSHIKAMI, S., W.A. HAGINS: Control of the dark current in vertebrate rods and cones. This volume, pp. 245-255.
9. CONE, R.A., W.H. COBBS: Rhodopsin cycle in the living eye of the rat. Nature (Lond.) 221, 820-822 (1969).
10. BLAISIE, J.K.: The location of photopigment molecules in the cross section of frog retinal receptor disk membranes. Biophys.J. 12, 191-204 (1972).
11. SHICHI, H., M. LEWIS, F. IRREVERRE, A.F. STONE: Biochemistry of visual pigments. J.Biol.Chem. 244, 529-536 (1969).
12. MURAKAMI, M., W.L. PAK: Intracellularly recorded early receptor potential of the vertebrate photoreceptors. Vision Res. 10, 965-975 (1970).

Photovoltages and Dark Voltage Gradients Across Receptor Layer of Dark-Adapted Rat Retina

G. B. Arden

Department of Visual Science, Institute of Ophthalmology, London WC1 H9QS/Great Britain

Saline filled micropipette electrodes have been used to measure photovoltages and dark voltages across the receptor layers of isolated, incubated rat retinas. The general techniques employed were similar to those previously described. Unless specified experiments were carried out in low calcium Ringers (1). Responses to flashes of light (30 quanta absorbed/rod·flash) were recorded between two pipettes, one stationary at the receptor tips, and another moved across the receptor layer. The electrode movements were made relatively abruptly by hand, or smoothly and slowly with a motorised drive. The DC drift was $\sim 1.0\ \mu V/min$.

Electrode marking experiments showed that small photovoltages could be recorded from an electrode at the receptor tips. The interstitial radial resistance in this region was ca. 120 ohm·cm, and increased in the inner limb and nuclear regions. The photocurrents, determined by the subtraction of averaged records, and scaled for resistance changes, rose along the length of the outer limbs, and fell in the inner limb zone. Further data processing on the averaged records showed that the source current in the outer limb and sink in the inner limb consisted of "fast P III" only (2). The size of the source current is adequate to account for the photovoltage of up to 0.8 mV across the receptor layer. Dark voltage gradients across the same layer never exceeded 0.3 mV. Thus the photovoltage always exceeds the dark voltage.

Since such a finding is incompatible with the hypothesis that the rod receptor potential is due to a reduction in a radial dark current, attempts were made to discover any reason why the dark voltage gradient should be spuriously small. In particular, experiments were performed to discover if any depolarisation deep in the retina produced current flows which "backed off" the sodium current entering the outer limbs. None was found in normal conditions. When sodium was reduced to ca. 4 mM, photovoltages were recorded across the receptor layers which consisted of both fast and slow PIII. These photovoltages appeared although the dark voltage gradient was almost entirely absent. The photovoltages could be as large as in high sodium, but since the radial interstitial resistance was increased, the photocurrents were smaller. It appears that a part of the photocurrent may be due to the reduction in a radial sodium dark current, but part is produced by some other mechanism.

Experiments have also been performed on retinas in which PIII has been isolated either by the use of sodium glutamate or by reducing the phosphate in the medium. Such media contained up to 1.4 mM calcium and no plasma. In both these circumstances it is possible to recognize both fast and slow PIII. The photovoltages were nearly as big as in EDTA and significantly larger than the dark voltage. When TRIS is substituted for sodium in such media, the photoresponses stop, and the dark voltage gradient vanishes. Thus photoresponses occur in the absence of sodium only in the presence of a chelating agent, but the photovoltage exceeds the dark voltage in both low calcium and in normal Ringer.

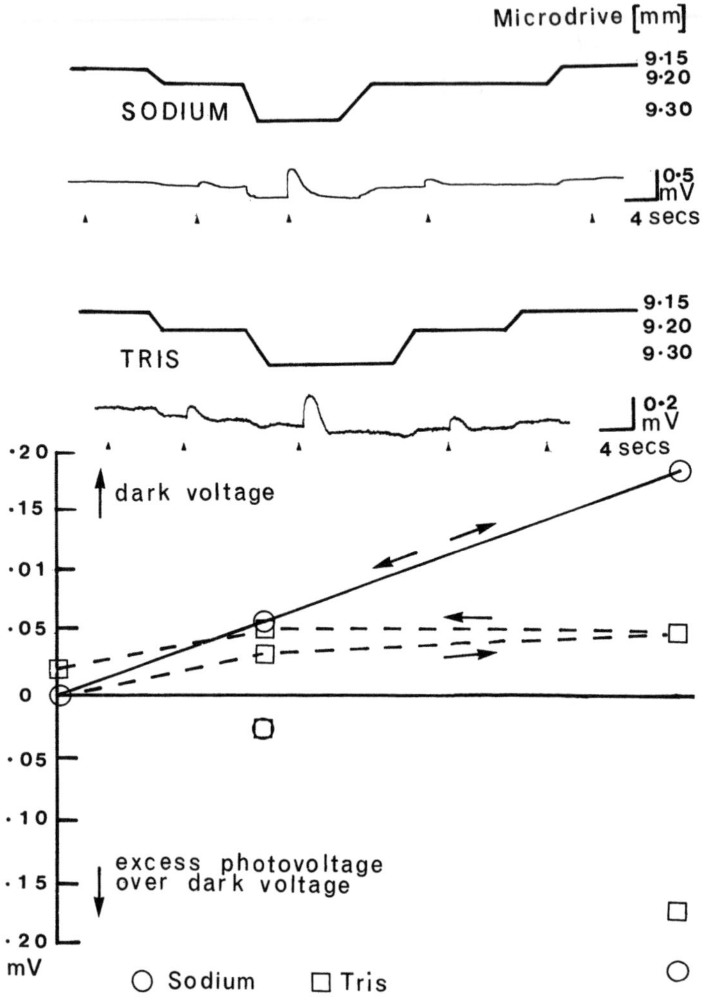

Fig. 1. Upper half: Strip chart record of voltages recorded between two microelectrodes, one fixed at receptor tips, and other moved as indicated across the receptor layer. Records read left to right. Upward deflection indicates negativity at mobile electrode. Stimuli, at arrowheads, 15 msec flashes, 30 quanta$_{502\ nm}$ absorbed/rod/flash. Upper record at beginning of experiment, lower at end, some 90 minutes later. In sodium, insertion of electrode causes downward displacement of trace, and withdrawal, the reverse, indicating that a current flows into a sink in the outer limbs. The photovoltage exceeds the dark voltage. In TRIS (note change of voltage calibration) the electrode motions are associated with smaller changes of voltage, which are not sustained. Microdrive readings are given to the right, and remain unchanged during the experiments which involved 9 changes of incubating medium. The microelectrode penetrates the retina at approximately 45° and thus the retinal depth at which the largest responses obtained is ca. 75 μm.

Lower half: Average of results of experiment (5 changes of solution from sodium to TRIS). Abscissa represents results at (left to right) 9.15, 9.2, 9.30 mm. It can be seen that dark voltage gradients are practically identical for both penetrations and withdrawals of the electrode (arrows), and that the dark voltage is much reduced by substitution of TRIS for sodium. Under all conditions, the photovoltage exceeds the dark voltage. Because interstitial radial resistance of the fluid between the receptors is higher in TRIS than sodium, the photocurrent in TRIS is relatively smaller than appears on the graph. However, some of the responses in tris were recorded some time after the solution changes, when the response was less than maximal. Thus the graph suggests that the photoresponses are due in part to a reduction in a sodium dark current, but also to another mechanism.

Sodium solution contained NaCl, 125 mM; $KHCO_3$, 2.5 mM; citric acid, 1.0 mM; Glucose, 10 mM; $MgSO_4$, 1 mM; sodium glutamate, 5 mM; EDTA 0.18 mM; Tris base, buffered to pH 7.8 with H_3PO_4, 71 mM. In TRIS solution, NaCl replaced by 134 mM TRIS-Cl. All solutions aerated with 95% O_2, 5% CO_2. 3% reconstituted human plasma added. T = 30°C, pH 7.5

Arden and Ernst (1) concluded that small photoresponses continued after the retina had been poisoned with Ouabain, providing a chelating agent was present. The responses then appeared to depend entirely upon the transmembrane sodium gradient. Microelectrode experiments confirm this finding: the small photoresponses (which consist entirely of "fast PIII") are smaller than the radial dark voltage. However, Ouabain causes depolarization deeper in the retina so that the size of the sodium dark current cannot be appreciated unless ion substitution experiments are performed.

Thus it appears that there may be two mechanisms of production of photoreceptor responses, one of which is the reduction of sodium dark current. The other appears to be sensitive to Ouabain but the experiments do not provide any clear indication of how it operates.

References

1. ARDEN, G.B., W. ERNST: A comparison of the behaviour to ions of the PIII component of the pigeon cone and rat rod. J.Physiol. (London) 220, 479-497 (1972).
2. ERNST, W., G.B. ARDEN: Separation of two PIII components in the rat electroretinogram by a flicker method. Vision Res. 12, 1759-1761 (1972).

In vitro Physiology of Frog Photoreceptor Membranes

D. Bownds, J. Dawes, and J. Miller

Laboratory of Molecular Biology and Department of Zoology, University of Wisconsin, Madison, WI 53706/USA

Characterization of the molecular mechanisms underlying visual excitation and adaptation might be greatly simplified if purified suspensions of visual receptor outer segments could be made to function in vitro. The roles of cofactors, regulating molecules, and pharmacologically active agents might then be studied with a precision impossible to obtain in living and intact visual receptor cells. We would like to summarize experiments which demonstrate that purified Bullfrog (Rana catesbeiana) photoreceptor membranes can perform excitation and adaption in vitro in a cycle which utilizes ATP and cyclic AMP. A light dependent phosphorylation of rhodopsin (1,2), possibly a part of dark adaptation mechanisms, will also be discussed.

In Vitro Activity of Frog Photoreceptor Membranes

The activity cycle which isolated frog rod outer segments can be made to perform in vitro is shown in Fig. 1. A frog rod outer segment contains approximately $3x10^9$ rhodopsin molecules. When 100 to 200 of these are struck by light a maximal sodium conductance (g_{Na}) decrease is observed. This is termed "excitation" in Fig. 1. Sodium conductance rapidly returns to normal when illumination is halted. Levels of illumination which bleach 0.1 to 1% of the rhodopsin present cause a maximal conductance change which persists indefinitely in vitro after cessation of illumination. This is termed "inactivation". Thus after 1% of the rhodopsin has been bleached, further illumination has no effect. The excitation and inactivation steps were first observed in vitro by Korenbrot and Cone, who studied intact and freshly broken off outer segments (3). We have used a simple modification of their technique (see below) to extend their findings to aged and purified outer segment suspensions, and also to demonstrate that inactivated outer segments will recover their original sensitivity (i.e., dark adapt) if supplemented with cyclic AMP and ATP.

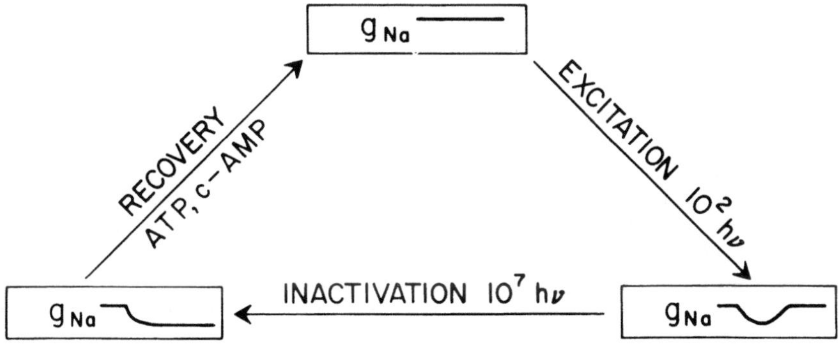

Fig. 1. Activity of frog outer segments in vitro

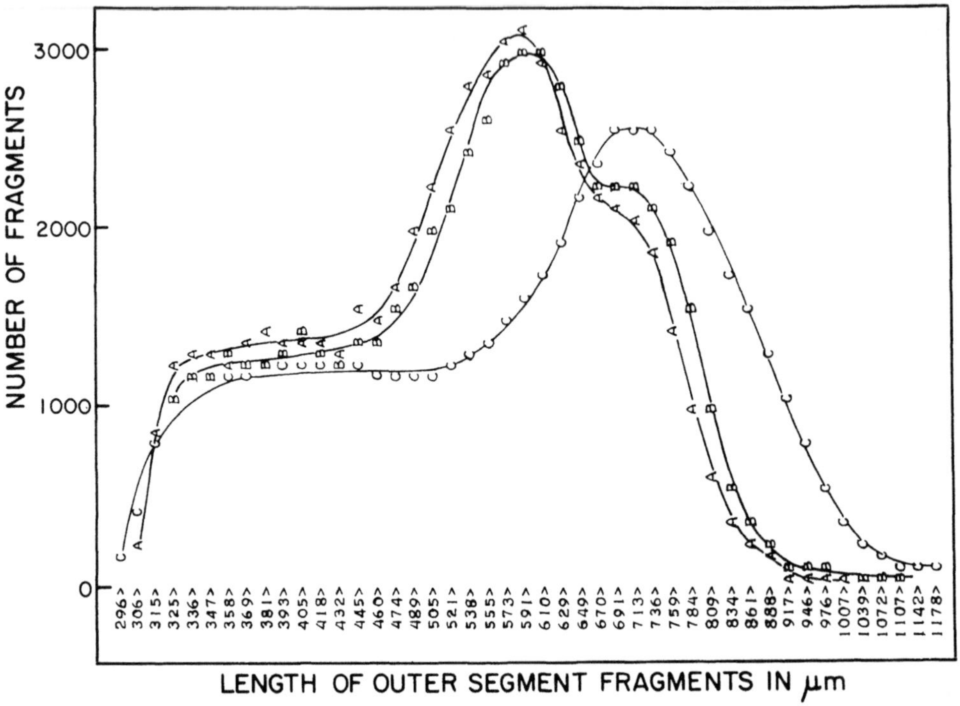

Fig. 2. Response of bullfrog outer segment fragments to hyperosmotic shock. A dark adapted retina was shaken in Ringer's solution. Curve C shows the size distribution of fragments in this suspension. Portions of the suspension were made 0.3 M in NaCl in the light or dark and then sampled on the particle size analyzer. Dark outer segment fragments (curve B) recover partially from the hyperosmotic shock while illuminated fragments (curve A) do not

---

The in vitro assay for the light induced sodium conductance change makes use of an osmotic perturbation. Upon adding NaCl (0.3 to 0.5 M) to the Ringer's solution in which dark outer segments and their fragments are suspended, the fragments first contract as water leaves, and then re-expand within 30 seconds as NaCl enters. These volume changes are reflected mainly as changes in length. If outer segments are illuminated before or during the hyperosmotic shock, their sodium conductance decreases, NaCl cannot enter, and volume recovery does not occur (3). Volume recovery in the dark is virtually complete using fresh and intact outer segments, but is much less dramatic in aged and purified suspensions (Fig. 2). However, the ion specificity and quantum sensitivity of the recovery process is the same for each type of preparation.

The data in Fig. 2 is generated by a Coulter-type counter (Particle Data, Elmhurst, Illinois) interfaced with a PDP-8M computer. Each rod outer segment passing through a 120 μm diameter orifice causes a conductance change in the electrolyte solution which is roughly proportional to its volume. The volume distribution is accumulated and stored by the computer. The data reported in this paper was obtained either with this system or with a simpler apparatus, the Coulter Counter Model B with the model J plotter.

Rod outer segments gently shaken from a retina maintain excitability for many hours even after Ficoll gradient purification (4). If these outer segments are illu-

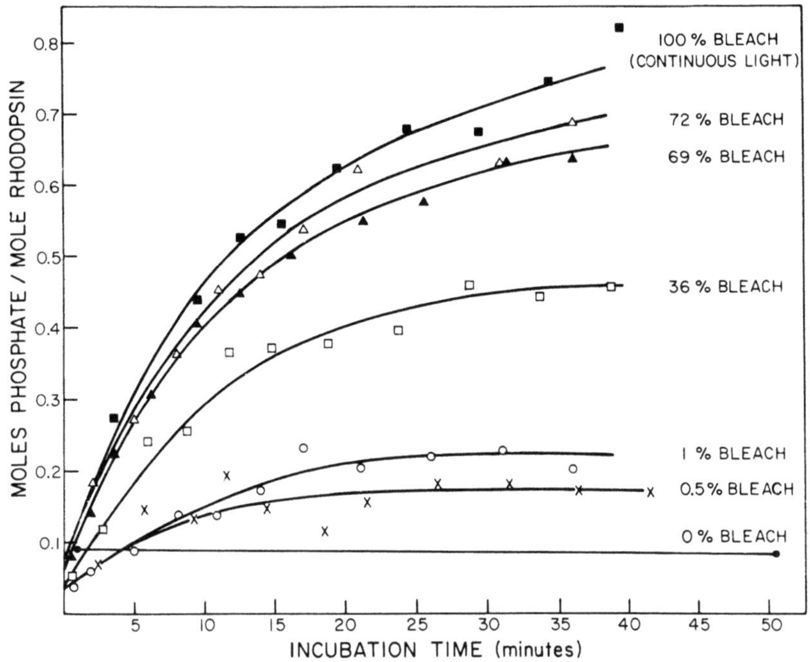

Fig. 3. Phosphorylation initiated by different rhodopsin bleaches. A single preparation of frozen and thawed frog rod outer segments in Ringer's solution (4) was used in this experiment. Portions of that preparation containing 1 mM ATP-γ-^{32}P were irradiated with light (580 nm) to bleach the indicated amount of rhodopsin and then incubated in the dark (1). For the continuous curve, 72% of the rhodopsin was initially bleached, and the sample was then incubated in continuous light

---

minated with 100-300 quanta/rod· sec during osmotic shrinking in 0.5 M NaCl, they are excited and no longer re-expand. As Korenbrot and Cone have noted, re-expansion of dark outer segments after hyperosmotic shock is observed in 0.5 M NaCl but not in 0.5 M KCl, 0.33 M Na_2SO_4, or 0.33 M K_2SO_4. While purified outer segments remain active for many hours in Ringer's solution containing 2 mM Mg^{++} or 2 mM Mg^{++} and 2 mM Ca^{++}, the lifetime of excitability is reduced to one or two hours at 20°C if Ca^{++} is substituted for Mg^{++}. Increasing the Ca^{++} concentration to 5 mM results in a rapid loss of activity. Sodium cyanide (1 mM), theophylline (0.1 mM), and AMP (5 mM) abolish excitability after 15-30 min at room temperature. ATP, ADP, β, γ-methylene ATP, dibutyryl cyclic AMP, and cyclic AMP do not influence excitability after 15-30 min incubations. Iodoacetamide (1 mM), N-ethyl-maleimide (1 mM), strophantinin (5 x 10^{-5} M), helibriginen (5 x 10^{-5} M), dinitrophenol (1 mM), and 1 mM ethylene-glycol-bis (β-aminoethyl ether) N,N'-tetra-acetic acid (EGTA) are also without effect. Thus the excitation mechanism is either resistant or not accessible to sulfhydryl reagents, inhibitors of sodium and potassium transport, and inhibitors of ATP production.

When 0.1 to 1% of the rhodopsin present in suspended outer segments is bleached, the high sodium conductance characteristic of the dark state is lost and cannot be restored even by prolonged incubation in the dark (Fig. 1). After this inactivation further illumination has no effect. We have found, in 36 separate experiments, that

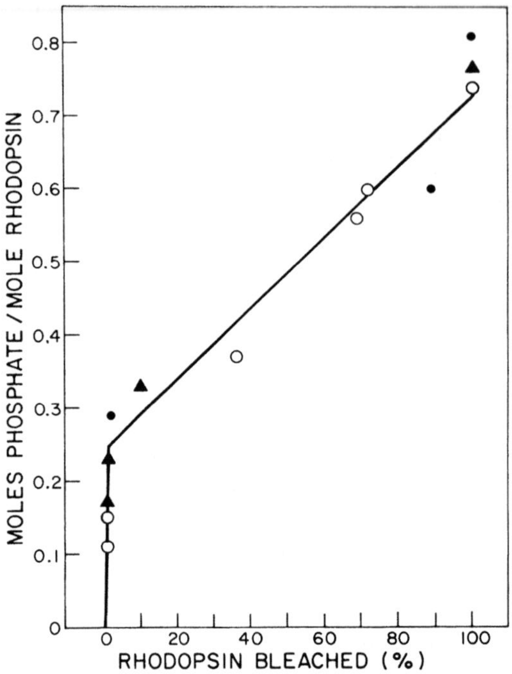

Fig. 4. Relationship between maximum phosphorylation and rhodopsin bleaching. In three seperate experiments similar to that of Fig. 3 the maximum light dependent phosphorylation after different rhodopsin bleaches was measured. The background incorporation in the dark was substracted from incorporation following illumination to give the points shown

excitability can be restored in the dark if bleached outer segments are supplemented with ATP and cyclic AMP or dibutyryl cyclic AMP. If 10% of the rhodopsin in outer segments prepared by Ficoll flotation (4) is bleached, excitability can be restored by 10 mM ATP. After 10-40% bleaches, dark adaptation is observed only if 10^{-6} M cyclic AMP or dibutyryl cyclic AMP is also added to the suspension. Supplemented outer segments can recover their original quantum sensitivity, responding to 100-300 quanta/outer segment. sec, within fifteen minutes after 25-35% of their rhodopsin is bleached. During this process the bleached rhodopsin does not regain its 500 nm absorption; rhodopsin regeneration does not occur. Recovery requires the presence of 10^{-6} M cyclic AMP or dibutyryl cyclic AMP while 10^{-7} is ineffective. Iodoacetate (1 mM), N-ethylmaleamide (1 mM), and EGTA (1 mM) do not inhibit recovery. Different outer segment preparations vary in the ability to recover excitability after large rhodopsin bleaches, with some recovering after 45% of the rhodopsin is bleached while others fail to recover if more than 25% is bleached. The recovery mechanism decays within one hour at room temperature but is stable for several hours at 0°C.

Light Dependent Phosphorylation of Rhodopsin

One role for ATP that may be relevant to the adaptation processes mentioned above is its participation in a light activated phosphorylation of rhodopsin which can leave 50 rhodopsin molecules phosphorylated for each photon absorbed. We originally reported (1) that incorporation of γ-^{32}P from γ-^{32}ATP, followed in the dark after irradiation, is maximal after approximately 1% of the rhodopsin present has been bleached. Subsequent work has shown that illumination which bleaches more than 1% of the rhodopsin does cause greater incorporation (Figs. 3 and 4). In the experiment shown in figure 3, a rhodospin bleach of more than 80% causes the incorporation of 0.8 moles of phosphate per mole of rhodopsin. Figure 4 illustrates a further point: small bleaches are much more effective than large bleaches. Thus, the "quantum yield" of the reaction (the number of phosphate groups incorporated per effective photon absorbed) is 10-50 if less than 1% of the pigment is bleached but is less than one for bleaches of 10-100%.

The function of the phosphorylation reaction in outer segments is not known. In other systems phosphorylation of proteins has been shown to either activate or suppress their enzymic activity (5). The fact that the phosphorylation of rhodopsin proceeds at approximately the same rate as dark adaptation in vitro and in perfused frog retinas (6) suggests that it might be associated with recovery of sensitivity, perhaps restoring rhodopsin's ability to trigger conductance changes. Rhodopsin phosphorylation is not influenced by cyclic AMP, a regulatory molecule which is involved in many protein kinase reactions (5). However, cyclic AMP may be important in controlling either the extent or rate of dark adaptation, for it must be added for recovery after large rhodopsin bleaches.

Acknowledgement. This work was supported by a grant (EY 00463) from the National Eye Institute and National Institutes of Health Training Grant (GM 01874).

References

1. BOWNDS, D., J. DAWES, J. MILLER, M. STAHLMANN: Phosphorylation of frog photoreceptor membranes induced by light. Nature New Biology 237, 125-127 (1972).
2. KÜHN, H., W.J. DREYER: Light dependent phosphorylation of rhodopsin by ATP. FEBS letters 20, 1-6 (1972).
3. KORENBROT, J.I., R.A. CONE: Dark ionic flux and the effects of light in isolated rod outer segments. J.Gen.Physiol. 60, 20-45 (1972).
4. BOWNDS, D., A. GORDON-WALKER, A.D. GAIDE-HUGUENIN, W.E. ROBINSON: Characterization and analysis of frog photoreceptor membranes. J.Gen. Physiol. 58, 225-237 (1971).
5. HOLZER, H., W. DUNTZE: Metabolic regulation by chemical modification of enzymes. Ann.Rev.Biochem. 40, 345-374 (1971).
6. FRANK, R.N.: Properties of "neural" adaptation in components of the frog electroretinogram. Vision Res. 11, 1113-1123 (1971).
7. FALK, G.B., P. FATT: Rapid hydrogen ion uptake of rod outer segments and rhodopsin solutions on illumination. J.Physiol. (London) 183, 211-224 (1966).

Discussion

C. Baumann: Falk and Fatt (7) have shown that hydrogen ions enter ROS through the puncture at the cilium. May it be that something similar could help ATP and AMP to get into the cell?

D. Bownds: Yes, perhaps the resealed portion of the outer segment is leaky to these molecules. We assume that the break at the cilium does heal, because these outer segment fragments do behave as osmometers.

M. W. Bitensky: By anology with other systems, in response to your remark that cyclic AMP might be working at any locus, in almost all known examples in vertebrates it functions from within the cell.

D. Bownds: We think it likely that the site of cyclic AMP action is inside the outer segment. However, the experiments reported here do not tell us where it is acting.

M. W. Bitensky: To be more certain of cyclic AMP specificity it is important to compare its effects with those of other closely related nucleotides. Have you made such comparisons and at what concentrations?

D. Bownds: No, we have not.

A. I. Cohen: Adding cyclic AMP may not be valid if it acts on the inside of the membrane and the ROS plasma membrane is intact. Also, regarding ATP effects - how does ATP get through an intact plasma membrane?

D. Bownds: It is curious that molecules that normally don't get into cells influence this system. ATP must be able to reach the disc membranes because the light activated phosphorylation reaction can leave all of the rhodopsin in the disc membranes phosphorylated. We know that the plasma membranes of these outer segments are very leaky to sodium ions, unlike normal nerve cell membranes. Perhaps they are also more permeable to nucleotides. With regard to cyclic AMP addition, we have also added the dibutyryl derivative, which is thought to pass more easily across membranes.

H. Kühn*: With frog ROS you got 50 molecules of rhodopsin phosphorylated when only one molecule has absorbed light. This indicates that non-bleached rhodopsin is phosphorylated. - We have worked on phosphorylation of cattle rhodopsin (2) and found a different efficiency of light. In our case, the phosphorylation extent was about linear, which means that not more rhodopsin molecules can be phosphorylated than have been bleached. - Light must influence some partner in this reaction: Either the substrate (rhodopsin) or the enzyme (kinase). In our case, the substrate rhodopsin needs to be bleached in order to be available to phosphorylation. The enzyme on the other hand works independent of light: it is able to phosphorylate other substrates like histones and protamines in a mixture with ROS suspension in the dark as well as in the light, whereas no rhodopsin is phosphorylated in the dark. - Do you assume "your" kinase to be stimulated directly by light? And how can the differences be explained, which were found in your frog and our cattle ROS systems?

D. Bownds: I think that the differences between the frog and cattle data might be easily resolved. I would suggest that you are simply missing the first part of the curve shown in the last figure. Because cattle retinas obtained from a slaughterhouse are frequently bleached to some extent, you would not see the "high quantum yield" portion of the curve. We can obtain frog outer segments with a full rhodopsin complement and thus accurately measure the effect of very small bleaches. We do not yet understand the molecular mechanism which permits single photon absorption to result in the phosphorylation of as many as 50 unbleached rhodopsin molecules. Current experiments are trying to elucidate this mechanism.

---

* Max-Planck-Institut für Medizinische Forschung, Heidelberg, W.-Germany; as visitor.

W.T. Mason: Have you examined the incorporation of PO_4 into other cyclic nucleotides?

D. Bownds: No.

W. Sickel: It may be gratifying to mention that several of the effects you found in your experimental set up can be seen in the perfused-retina preparation, provided the experimental changes are small; but then the retina will show you the effects with considerable resolution. As to the effects which one would expect to be exerted by Ca^{++} but were seen by you exerted by other ions: this need not contradict the Ca^{++}-story, because on the Chatelier's principle there should be interaction. I wonder if your observation were of the "right" direction, particularly with H^+.

D. Bownds: The perfused retina preparation is indeed a useful one for some purposes. We would have found it very difficult to study either the phosphorylation reaction or the cyclic AMP and ATP effects on dark adaptation in the perfused retina.

The Internal Transmitter Model for Visual Excitation: Some Quantitative Implications

Richard A. Cone
The Thomas C. Jenkins Department of Biophysics, Johns Hopkins University, Baltimore, MD 21218/USA

Fuortes and his co-workers (1-3) have suggested that rhodopsin, in both vertebrate and invertebrate photoreceptors, may excite the cell by releasing a transmitter substance which diffuses internally to other regions of the receptor membrane and there modulates ionic conductance. This suggestion has been based primarily on theoretical models which accurately describe the amplitude and time-course of the photoresponse. More recently, Hagins and his co-workers have reported some persuasive evidence which implies vertebrate receptors may employ an internal transmitter (4,5), and Yoshikami and Hagins (6,7) have proposed that the transmitter is Ca^{++}.

In this report I wish to begin by outlining some of the strongest available evidence that rhodopsin must indeed excite both vertebrate and invertebrate receptors by releasing an internal transmitter. Next, a semi-quantitative analysis of the internal transmitter model will be considered. The aim of this analysis is to determine the number of transmitter particles which a photoactivated pigment molecule (Rh*) must release to produce an adequately reliable signal. The analysis also predicts quantitatively some characteristics of the Na-channels which the postulated transmitter particles must regulate.

Evidence that a Transmitter Mediates Excitation

Among invertebrate receptors, the ventral photoreceptors of Limulus provide perhaps the clearest evidence that a transmitter must be employed. In these receptors, voltage-clamp observations of quantum-bumps reveal that a single Rh* can initiate a 1-5 nA current through the membrane (8,9). Since the current is carried by Na ions, the driving voltage is in the range of 70 mV. Hence the conductance increase must be ~50,000 p mho. If all the photocurrent flows through a single pore, and if the conductivity within the pore is comparable to that of the surrounding saline, the diameter of the pore must be nearly 100 Å. Since this is considerably larger than the diameter of a rhodopsin molecule, and since a pore with such a large diameter could not have the required ion selectivity, rhodopsin cannot, by itself, form the pore and must instead somehow initiate the opening of numerous Na-selective channels elsewhere in the membrane. Moreover, a rhodopsin molecule which happens to be located near the tip of a microvillus encounters a further difficulty: the diameter of a villus is so small (~700 Å) that the conductance of its interior will permit no more than ~1 nA to flow into the cell even if it were possible for the Rh* molecule to sever the tip from the villus (10). Hence, a rhodopsin molecule located near the tip must not only initiate the opening of numerous Na-channels, but some of these must be located at least as far away as the other end of the villus (~1 μm). As will be discussed, the internal transmitter model easily satisfies both these conditions.

In vertebrate rods and cones several characteristics reveal the necessity for an internal transmitter. In a turtle cone, for example, a single Rh* can transiently reduce the conductance of the plasma membrane by about 1 part in 10^3 (3). Since there are about 10^8 pigment molecules electrically in parallel in the plasma membrane, each molecule, if it acts alone, could reduce the conductance by no more than its share of the membrane area, or no more than about 1 part in 10^8. Since the observed conductance decrease is some 10^5 times greater than this, the pigment molecule cannot act alone, but must instead act upon at least 1 in 10^3 of all the light-regulated Na-channels in the membrane. If the Na-channels are distributed uniformly throughout the pigment-containing membrane, given the total area of the membrane, many of the Na-channels blocked by an activated pigment molecule must be at least 1 μm away on the membrane surface.

When this argument is applied to rods (11) it can again be seen that even within one disc, a single rhodopsin molecule, acting alone, is incapable of producing the required conductance decrease. Moreover, in rods the evidence is now quite convincing that most of the discs are free-floating: they are structurally (12), electrically (13), and osmotically (14) uncoupled from the plasma membrane, and it is the Na conductance of the plasma membrane which is regulated by light (13,14). Hence, in rods the need for an internal transmitter is particularly evident: an Rh* situated in a free-floating disc must somehow regulate at least 1 in 10^3 of the Na-channels located in the plasma membrane.

Quantitative Characteristics of the Internal Transmitter Model

In *Limulus* photoreceptors, the ~50,000 pmho conductance increase which occurs during a quantum-bump is about 100 times larger than the 450 pmho conductance of the "diffusion limited" Na-channels in squid axon (20). Thus it is likely that ~100 or more Na-channels open during a quantum-bump. Therefore, if each transmitter particle released by rhodopsin opens a Na-channel, one Rh* must release ~100 particles or more. (On an entirely different basis, Borsellino and Fuortes estimate at least 25 particles must be released (2)). Given the time course of a quantum-bump, ions or small molecules could diffuse several micrometers from the microvillus in which they are released, and could therefore act on Na-channels distributed over an area encompassing at least 1000 microvilli. Thus, even if there is less than 1 Na-channel per villus, the transmitter could reach a sufficient number of channels to account for the photocurrent. The time-course of the quantum-bump also indicates that rhodopsin must release the transmitter particles in less than about 0.1 sec. Hence, while active, a *Limulus* rhodopsin molecule must release the postulated transmitter particles at a rate of ~1000 per sec or more.

Similar quantitative estimates are more difficult to make in the case of vertebrate rods and cones, but the semi-quantitative analysis which follows does serve to suggest the order of magnitude of the number of transmitter particles which must be released. Sufficient data are now available to apply this analysis to four different receptors: rat rods (4,13,16), frog rods (16,17), turtle cones (3,22), and pigeon cones (18,24). In all four receptors the excitatory mechanism appears to be the same, but the values of some of the parameters differ considerably, as shown in Table I.

Conductance-Intensity Relationship

In all four receptors, at the peak of the photoresponse, the Na conductance, G, of the plasma membrane of the outer segment appears to satisfy the following relationship:

Table I. Measured parameters of outer segments of vertebrate photoreceptor cells

Dimensions of outer segment	Frog rod	Rat rod	Turtle cone	Pigeon cone
Length (μm)	50	24	45	~40
Radius (μm)	3	0.85	0.6	0.5
A = area of plasma membrane (μm^2)	1300	150	3600	~2200
V = aqueous volume (μm^3)	700	27	24	~18
Photoresponse for 1 Rh*/outer segment				
T_r = risetime (msec)	~250	100	50	70
T = time to peak (msec)	500	200	90	130
f = fractional decrease in dark current	$\frac{1}{150}$	$\frac{1}{30}$	$\sim\frac{1}{1000}$	$\sim\frac{1}{1000}$
G_d = Na conductance in darkness (p mho)	~17,000	~700		

(Temperature: ~20°C for frog and turtle, 30°C for rat and pigeon. G_d calculated assuming a driving force of 30 mV.)

$$G = \frac{G_d}{1 + fI} \qquad 1$$

where G_d is the Na conductance in the dark, I is the number of pigment molecules (Rh*) activated by a stimulus flash in a single outer segment, and f is the fraction by which 1 Rh* reduces the dark current. Baylor and Fuortes (3) have suggested that this nonlinear relationship between G and I can be explained with the internal transmitter model using an appropriate relationship between transmitter concentration and flash intensity. It turns out that if the amount of transmitter released is assumed to be linearly proportional to I, then equation 1 can only be satisfied by a condition which implies that the transmitter binds reversibly to the Na-channels, and that the binding and release reactions are sufficiently rapid to permit at least a near-equilibrium condition at the peak of the photoresponse. This can be seen as follows:

Let C = concentration of free transmitter particles in extradiscal cytoplasm
K = equilibrium constant for reversible binding of transmitter to the Na-channel (K = concentration at which $\frac{1}{2}$ of the channels bind transmitter = 1/affinity.)
N = total number of Na-channels in the plasma membrane of the outer segment
n_o = number of open channels
n_c = number of closed channels
[ ] = concentration
g = Na conductance of an open channel
G = total Na conductance of the open channels

Then $N = n_o + n_c$, and $G = g \cdot n_o$. If a channel is blocked whenever a transmitter particle is bound to it, and if equilibrium binding is achieved, then

$$\frac{C\,[n_o]}{[n_c]} = K$$

Together these relations yield

$$G = \frac{g \cdot N}{1 + \frac{C}{K}}$$

Hence if $C = IfK$, equation 1 is satisfied. Thus the observed nonlinear relationship between G and I can be satisfied by the model if each Rh* releases a quantity of transmitter which produces an increase in transmitter concentration of

$$\Delta C = fK \qquad 2$$

(In darkness, regardless of the resting concentration of transmitter, C_d, equation 1 is still satisfied as long as C increases linearly with I, and if C_d is small compared to K, $\Delta C \approx fK$ still holds true.)

Diffusion Distances

If the viscosity of the cytoplasm is nearly that of water (0.01 poise), and if the transmitter is an ion or small molecule with a diffusion coefficient $D \sim 10^{-5}$ cm^2/sec, then most of the transmitter should diffuse to and contact the plasma membrane of a rod within 1-10 msec after being released by an Rh*. This is a short time compared to the rise-times of near-threshold photoresponses. The lamellar arrangement of the discs slightly speeds this transverse diffusion but the discs greatly impede diffusion along the length of the rod since they act as a stack of baffles which occlude 95-98% of the cross-sectional area. The extent of longitudinal diffusion is an important consideration for the internal transmitter model since two conditions must be satisfied: 1) the postulated transmitter must be capable of reaching a large enough area of plasma membrane to be able to reduce its conductance by ~1% (Table I), and 2) since Hagins et al. (4) have shown that in rat rods the dark-current is blocked only in the illuminated portion of the rod, the longitudinal spread of transmitter by diffusion must be considerably less than the length of the rod. Only approximate calculations can be done at this time but they indicate that during the photoresponse in frog and rat rods the transmitter could diffuse a few micrometers longitudinally, and hence reach about 1-10% of the plasma membrane. Thus the internal transmitter model appears capable of satisfying both conditions.

Minimum Signal Requirements

Let S be the number of transmitter particles released by an activated pigment molecule. The following five conditions lead to an estimate of the minimum value for S:

a) S must produce $\Delta C = fK$. Thus

$$S = \alpha \Delta C L V = \alpha f K L V \qquad 3$$

where L is Avogadros's number, V is the aqueous volume of the outer segment, and α is an empirical "loss" factor introduced to account for partial uptake and/or degradation of the transmitter prior to the peak of the response. (The effective concentration at the membrane surface may be somewhat altered by surface charge effects. To account for such an effect an additional correction factor would have to be applied to the volume.)

b) If a Na-channel is blocked whenever a transmitter particle is bound to it, then the number of particles released must exceed the increase in the number of Na-channels which bind transmitter at the peak of the response. Thus

$$S = \alpha \beta f N \qquad 4$$

where β is an empirical "unbound excess" factor introduced to account for the likelihood that only a fraction of the available transmitter binds to the channels at any given time.

c) In darkness, it is reasonable to expect that most Na-channels will be open to permit a wide dynamic range for the photoresponse. Thus the resting concentration of transmitter, C_d, should be small compared to K: $C_d = \gamma K, \gamma \leq 1$, where γ is an empirical "resting concentration" factor which relates the resting concentration to the equilibrium constant of the Na-channels. In isolated frog rods, G increases by about 20% in low Ca solutions (16). Thus it appears that in darkness perhaps $\frac{1}{5}$ of the Na-channels are normally closed, and hence that $\gamma \approx \frac{1}{5}$.

d) The limiting source of noise in vertebrate photoreceptors is unknown, but it is clear that noise inherent in the internal transmitter model must not exceed the noise level observed psychophysically. In the model, the primary noise source appears to be the random closing of the Na-channels. If each closing can be considered a discrete event, then, during a response time of the photoreceptor, the noise level from this source is effectively set by the fluctuations in the total number of channel closings which may occur during this interval. From kinetic theory it can be shown that for $C<K$ the average frequency of closings for a single channel is equal to σz, where σ is the effective cross-section for a binding collision and z is the collision frequency per unit area of membrane. At $20^{\circ}C$, for ions and small molecules with a molecular weight of about 100, $z = w \cdot C$, where $w \approx 3 \cdot 10^8$ collisions/sec Å^2 for a one molar solution. The average number of channel closings which will occur in the dark during a response time T is then $N \sigma z T = N \sigma w C_d T$. During this interval of time, the transmitter released by 1 Rh* must increase, on average, the total number of channel closings by significantly more than the random fluctuations which occur in the dark. Hence:

$$N \sigma w \Delta C T = \varepsilon \, (N \sigma w C_d T)^{\frac{1}{2}} \qquad 5$$

where ε is an empirical "signal/noise" factor which determines the reliability of the transmitter signal. A lower limit for ε can be estimated from the "dark light" observed psychophysically in human rods and cones (19). If the reasonable assumption is made that somewhere in the receptor and/or retina a high frequency filter limits the signal bandwidth to $\sim 1/T$ (see ref. 4), calculations indicate that the signal must exceed the noise by about 3 standard deviations in rods and be comparable to the noise in cones ($\varepsilon \approx 3$ for rods, $\varepsilon \approx \frac{1}{2}$ for cones). It is important to note that even if random fluctuation in the number of channel closings is not the primary source of noise in receptors, only a small increase in the value of ε is needed to render this "transmitter noise" entirely negligible compared to the observed dark light. Thus the calculated value of S will not be much affected if dark light arises from another source, such as the thermal activation of rhodopsin (19).

e) The collision cross section σ for a binding collision between a transmitter particle and a Na-channel, although not known, can be estimated within adequate limits

Table II. Calculated parameters

	S	ν	K	N	N/A	g (p mho)
Frog rod	370	1400	$0.4 \cdot 10^{-7}$	6000	5	~2
Rat rod	120	1200	$0.7 \cdot 10^{-7}$	400	3	~2
Turtle cone	25	500	$6 \cdot 10^{-7}$	3000	1	
Pigeon cone	20	300	$7 \cdot 10^{-7}$	2500	1	

since, as will be shown below, both S and K vary only as the square root of $(1/\sigma)$. The cross-sectional area of a selective binding site cannot much exceed the cross-section of the transmitter particle. Thus for ions and small molecules an upper limit for σ is about 10-30 $Å^2$. This is almost certainly an overestimate. For example, saxitoxin (15), which reversibly binds to and blocks Na-channels in a manner analogous to the postulated transmitter particle, has an equilibrium constant ($\sim 10^{-9}$) and dissociation time ($\sim$ 1-3 sec) which indicate that σ for this reaction is $\sim 1\ Å^2$.

On solving the above equations for S, K, and N, the results are:

$$S = \alpha\varepsilon\left(\frac{\beta\gamma LV}{\sigma wT}\right)^{\frac{1}{2}} \qquad K = \frac{S}{\alpha f LV} = \frac{\varepsilon}{f}\left(\frac{\beta\gamma}{LV\sigma wT}\right)^{\frac{1}{2}} \qquad N = \frac{S}{\alpha\beta f}$$

Table II shows calculated values for S, K, and N using the experimentally observed parameters shown in Table I. In addition, Table II includes an estimate of the rate at which rhodopsin must release transmitter particles during its active phase, $\nu = S/T_r$, and an estimate of the number of Na-channels/μm^2 in the plasma membrane, N/A. For the calculations, ε, σ and the empirical factors α, β, γ were chosen as follows:

α (loss) = 3
β (unbound excess) = 3
γ (resting concentration) = $\frac{1}{5}$
ε (signal/noise) = 3 for rods, $\frac{1}{2}$ for cones
σ (collision cross section) = 1 $Å^2$

The values for α and ε were chosen to yield a minimum reasonable value for S.

It can be seen in this table that, except for N, rods and cones fall into separate groups: apparently differences between rods and cones are more significant than differences between warm and cold blooded animals. Significantly, rat rods yield essentially the same calculated values as frog rods despite the fact that rat rods compare much more closely with cones in volume (see Table I).

A surprising result of the analysis is the calculated value for g, the Na conductance of a single channel. When the dark conductance G_d of both frog and rat rods is divided by the number of channels, N, the result in both cases is ~2 p mho. Since neither the magnitude of the dark current nor the magnitude of g were considered in the analysis this result was unforseen, yet it yields a reasonable value. The Na-channels of frog and rat rods are known to be unusually selective, accepting Na ions over potassium ions by at least a 100:1 ratio (16). Hence they can be expected to have a somewhat smaller conductance than, for example, the "diffusion limited" Na-channels in frog axons which have a conductance of about 100 p mho (20). The

Na-channels in squid photoreceptors appear to have a conductance of about 20 pmho (21). Thus, since the analysis yields a reasonable value for g there is some reason to believe that the other calculated values may at least be within about an order of magnitude of the actual values.

It is interesting that K, the equilibrium constant for the binding of transmitter particles by the Na-channels, falls in the range 10^{-8}-10^{-6}. If Ca is in fact the transmitter, this would appear to be an appropriate range since in muscle, where Ca is known to act as an internal transmitter, it operates in this same range (23,24).

In Table II, the calculated value for the number of particles released, S, ranges between 20 and 370. Since the empirical factors were chosen to yield a reasonable minimum value for S the results clearly indicate that a single photoactivated rhodopsin molecule is likely to release at least 10-100 particles. Therefore, if equilibrium binding occurs, the possibility that rhodopsin releases only a single particle appears unlikely. However, if the evidence for equilibrium binding is put aside, it is possible that a single particle can produce a signal with sufficient speed and reliability if the following conditions are met: a) The particle must bind "irreversibly" to the channel for the duration of the photoresponse. b) There must be 1/f Na-channels. c) Since the particle must collide with the channel in less than the rise-time of the photoresponse, the collision cross section σ must satisfy the condition $\sigma > LV/NwT_r$ which in the case of rods implies σ must exceed 20 Å^2. d) In rods, the rest concentration must be less than 10^{-13} molar. This is so low that it is unlikely for such a transmitter particle to be an ion; it would more likely be a small molecule.

If equilibrium binding does occur, as indicated by the light-dependence of the Na conductance (eq. 1), the analysis suggests rhodopsin releases transmitter particles at about the same rate in both rods and cones. It also suggests that the Na-channels may have about the same properties (K,σ) in both types of receptor. With this in mind, the calculations were repeated, this time permitting the empirical factors to vary within reasonable limits but with the additional constraint that ν, K and σ be the same in all four receptors. The results indicate that with this additional constraint the release rate must be about 3000/sec, and the Na-channels must have an equilibrium constant of about $2 \cdot 10^{-7}$ for $\sigma = 1$ Å^2. In addition, the values of the empirical factors needed to meet this constraint suggest that in rods there may be a slower rate of uptake, a higher resting concentration, and a greater excess of unbound transmitter compared to cones. Interestingly, the calculated number of Na-channels per unit area of plasma membrane turns out to be about the same in both rods and cones ($\sim 1/\mu m^2$). This suggests that the dark current per pigment molecule may be much higher in cones than in rods.

Finally, it may be noted that the calculated release rates for vertebrate rhodopsins ($\sim 10^3$ particles/sec) fall in the same range as that predicted for _Limulus_ rhodopsin, raising the possibility that both _Limulus_ and vertebrate rhodopsins may release transmitter particles by essentially the same mechanism. The required rate of release is sufficiently slow that each of the current models of rhodopsin (enzyme, carrier, pore) deserve experimental consideration.

Acknowledgement. I wish to thank D.C. Petersen and E.E. Lattman for helpful discussions.

References

1. FUORTES, M.G.F., A.L. HODGKIN: Changes in time scale and sensitivity in the ommatidia of Limulus. J.Physiol. 172, 239-263 (1964).
2. BORSELLINO, A., M.G.F. FUORTES: Reponses to single photons in visual cells of Limulus. Proc.IEEE 56, 1024-1032 (1968).
3. BAYLOR, D.A., M.G.F. FUORTES: Electrical responses of single cones in the retina of the turtle. J.Physiol. 207, 77-92 (1970).
4. HAGINS, W.A., R.D. PENN, S. YOSHIKAMI: Dark current and photocurrent in retinal rods. Biophys.J. 10, 380-412 (1970).
5. HAGINS, W.A.: The visual process: excitatory mechanisms in the primary receptor cells. Ann.Rev.Biophys.Bioeng. 1, 131-158 (1972).
6. YOSHIKAMI, S., W.A. HAGINS: Light, calcium, and the photocurrent of rods and cones. Biophys.Soc.Abst. TPM-E16 (1971).
7. YOSHIKAMI, S., W.A. HAGINS: Control of the dark current in vertebrate rods and cones. This volume, pp. 245-255.
8. MILLECCHIA, R., A. MAURO: The ventral photoreceptors of Limulus, III. A voltage-clamp study. J.Gen.Physiol. 54, 331-351 (1969).
9. BROWN, J.E.: personal communication.
10. FEIN, A., F.A. DODGE: personal communication.
11. CONE, R.A.: Rotational diffusion of rhodopsin in the visual receptor membrane. Nature New Biol. 236, 39-43 (1972).
12. COHEN, A.I.: Further studies on the patency of saccules in outer segments of vertebrate photoreceptors. Vision Res. 10, 445-453 (1970).
13. PENN, R.D., W.A. HAGINS: Kinetics of the photocurrent of retinal rods. Biophys.J. 12, 1073-1094 (1972).
14. KORENBROT, J.I., D.T. BROWN, R.A. CONE: Membrane characteristics and osmotic behavior of isolated rod outer segments. J.Cell Biol., in press (1972).
15. HILLE, B.: Quaternary ammonium ions that block the potassium channel of nerves. Biophys.Soc.Abst. WC3, (1967).
16. KORENBROT, J.I., R.A. CONE: Dark ionic flux and the effects of light in isolated rod outer segments. J.Gen.Physiol. 60, 20-45 (1972).
17. TOMITA, T.: Electrical activity of vertebrate photoreceptors. Quart.Rev.Biophys. 3, 179-222 (1970).
18. ARDEN, G.B., W. ERNST: The effect of ions on the photoresponses of pigeon cones. J.Physiol. (Lond.) 211, 311-339 (1970).
19. BARLOW, H.B.: The physical limits of visual discrimination. In: Photophysiology, vol. 2, ed. A.C. GIESE, pp. 163. Academic Press, New York (1964).
20. HILLE, B.: Dynamic characteristics of excitable membranes. Biophys.Soc. Abst. FAM-A5 (1970).
21. HAGINS, W.A.: Electrical signs of information flow in photoreceptors. Cold Spring Harbor Symp.Quant.Biol. 30, 403-418 (1965).
22. LIEBMAN, P.A.: Microspectrophotometry of photoreceptors. In: Handbook of Sensory Physiology, vol. VII/1, ed. H.J.A. DARTNALL, pp. 482-528. Springer, Berlin (1972).
23. REUBEN, J.P., P.W. BRANDT, M. BERMAN: The relation between pCa and steady state tension in skinned crayfish muscle fiber. Fed.Proc. 29, 846 (1970).
24. HELLAM, D.C., R.J. PODOLSKY: Force measurements in skinned muscle fibers. J.Physiol. (Lond.) 200, 807-819 (1969).

VI. Enzymology and Molecular Architecture of the Light Sensitive Membrane

Chemo-Surgical Studies on Outer Segments

Adolph I. Cohen*
Department of Ophthalmology, Washington University, School of Medicine,
St. Louis, MO 63110/USA

Current Views of Outer Segment Structure

The outer segments of vertebrate photoreceptors contain photopigment and interact with light. They are therefore objects of investigation by those interested in the transduction process - i.e. the chain of causal events whereby the capture of quanta of light by photopigment molecules of the outer segment leads to signalling by the rods and cones. This leads to an interest in the molecular architecture of the membranes of the outer segments, since the visual photopigments seem to constitute very substantial fractions of the total protein of the outer segment (ca. 80% in frogs (1)), and the insolubility of the pigments, save in detergents, suggests that they are built into the proteinaceous membranes of the outer segments.

Our laboratory, over some years, has been investigating details of the structure of outer segments, and differences between rods and cones, but before describing some current experiments, it seems appropriate to briefly summarize some of the current data in this area.

The outer segment may be cylindrical or conical and at the ultrastructural level one sees an enveloping cell membrane enclosing a column of flattened saccules or double-membrane discs. A primate rod may have as many as 1000 of these discs, a frog rod some 1800 discs. The rod discs are about 150 Å thick, possess a narrow lumen or less dense interior, and may vary in diameter from 5-6 μm (frogs) to about 2 μm (rats). Furthermore, rod discs lack any physical connections to each other and to the cell membrane, and are spatially separated from neighboring discs within the column of discs. Experiments dealing with the infiltration of barium sulfate into unfixed rods with damaged cell membranes (2) showed that this colloid could not readily penetrate between discs. This suggests that there may be an interdisc gel. Transparent adhesions of the surviving rims of successive fixed discs were observed by Falk and Fatt (3) when discs of osmium-fixed rods were eroded by washing with tris-hydroxymethyl amino methane (TRIS) buffer. This also suggests the presence of interdisc material. An exception to the isolation of rod discs consists of a small number of discs with larger lumina at the very base of the rod outer segment. The membranes of these exceptional rod discs are continuous with the cell membrane as if these were forming by ingrowth or invagination of the latter.

In cones of most species, most if not all of the discs of the outer segment still retain physical connections with the cell membrane and, in appropriate longitud-

*This work was supported by grant EY-00258-09 from the National Eye Institute. It was performed during the tenure of career development award (5-K03-EY03170-08) from the National Eye Institute.

nal sections, give the appearance of forming by a pleating of this membrane. Thus each disc has a lumen physically connected with extracellular space and this has been verified by exposing fixed (2,4) or living cones (4,5) to tracer materials. However, in cones of mammals, while the inner third of the outer segment reveals discs connected to the cell membrane and open to the exterior, such discs are not readily apparent in the outer two-thirds of the outer segments.

Turning to the molecular architecture of the discs, it is known from the electron microscopic study by Clark and Branton (6) of replicas of freeze-fractured surfaces of cross-sections of fresh, glycerol treated rods that the plane of fracture of the rod discs passed through the disc membranes in a manner akin to that seen with other biological membranes. It now seems to be established that the favored plane of freeze-cleavage is an internal plane of the membrane corresponding to a supposed hydrophobic interior of the membrane. The appearance of the replica in the electron microscope in somewhat unique, at least quantitatively, in having a cobblestone appearance with the tightly packed "stones" having dimensions of 200-250 Å.

X-ray diffraction data by Blaurock and Wilkins (7) are consistent with the periodicity data for the frequency, spacing, and dimensions of saccules in electron micrographs of rod outer segments of frogs. X-ray diffraction data used in a different approach by Blasie et al. (8,9) on isolated frog disc membranes treated with anti-rhodopsin antibody and other proteins, suggests that there are particles of 40-50 Å dimensions in square array in the membrane, and their electron micrographs of negatively stained disc membranes seem to show square arrays of particles of about 40 Å dimensions. However, even granting certain problems in determining true particle dimensions in freeze-etching replicas and in negative-stained preparations, there seems to be a dimensional disparity between the Clark and Branton "cobblestones" and the particles of Blasie et al.

That the chromophores of the photopigment molecules are parallel to the electric vector of light passing axially through the outer segments and are thus parallel to the plane of the discs has been established by studies on the efficacy of bleaching with various orientations of plane polarized light. It has also recently been independently established by P. Brown (10) and by R. Cone (11) that the molecules of photopigment are free to rotate within the plane of the disc, thus confirming a prediction of Hagins and Jennings (12).

The question is then raised as to whether the rhodopsin molecules are also free to move non-rotationally within the disc membrane. Recently Singer and Nicolson (13) have proposed a "fluid mosaic" model for certain cell membranes, including the rod disc membrane. These investigators argue that several aspects of the data of Blasie _et al._ (8,9) indicate a lack of order of rhodopsin in the membrane. In a fluid mosaic model of the rod disc, the rhodopsin molecules may randomly move about the disc membrane through thermal agitation. A possible boundary to the fluid area might be the rim of the rod disc. Its non-fluidity is indicated by its resistance to deformation when saccules swell (14), or its resistance to disc erosion when the osmium-fixed disc is exposed to TRIS buffer (3). Rhodopsin molecules might possess a lipophilic keel dipping into the interior of the membrane. Thus they could randomly diffuse laterally, and rotate, while maintaining some vertical orientation. Such diffusability of visual pigment might explain a difference between rods and cones observed by R. Young (15,16). Young injected radioactive amino acids into frogs and monkeys, fixed animals at different intervals, and then carried out radioautographic studies on sections of the retina. He found that radioactive protein in rods was largely incorporated into discs at the base of the outer segment. With time, radioactive rod discs were displaced apically by new unlabeled discs forming at the base of the outer segment. Eventually groups of labeled discs were displaced to and separated from the apex of the outer segment and were ingested and digested by cells of the pigment epithelium (17). In addition, a relatively minor fraction of labeled protein in the rod outer segment had a diffuse distribution and also turned over. In cones, however, all the labeling of the outer segment was diffuse. The fluid mosaic hypothesis suggests that a contributory factor underlying the absence of banding in cones could lie in the fact that the discs are topologically continuous and lack a rim where they join the plasma membrane. There might be a path by which protein inserted into any disc membrane of cones could diffuse intramembranally from disc to disc and diffuse the radioactivity. Against this hypothesis however, is the realization that something must stabilize the membranes of the cone outer segment so as to preserve its shape and the progressive decrease in the diameter of the cone discs. Moreover, the diffuse labeling of the outer segments of cones seems to parallel a weaker diffuse labeling of rods which exists in addition to the band-like labeling (18). However, some of the diffuse labeling of rod outer segments might be due to diffusing molecules in the plasma membrane.

What is achieved by the isolation of the rod discs? What is the significance of the continuous production and destruction of rod discs? If thermal effects of light permanently damage pigment protein, is there some unique character in the structure of the rod disc which precludes molecular replacement in situ in a manner akin to that seen in cones? It thus becomes desirable to explore more closely the question of the molecular assemblage of the photoreceptor discs.

Experiments and Discussion

Our laboratory has begun a sequence of experiments designed to study the behavior of the plasma and disc membranes of the outer segments of frog rods and cones after (1) extractions with solvents, (2) physical damage, or (3) exposure to enzymes which attack membrane components.

In the first of these experiments, retinas were detached and then plunged directly into ammoniacal acetone solutions (12 µl of 28% w/v ammonia per 100 ml of 10% water in acetone). This was followed by exposure to 1% OsO in carbon tetrachloride for 1 hr., 1 hr. in toluene, and then embedment in epoxy resin. Sections were stained with lead. In another group of experiments, the same sequence was followed except that fixation for one hour in 2.5% glutaraldehyde in 0.1 M PO_4 buffer preceded the plunge into ammoniacal acetone. According to Fleischer et al. (19), ammoniacal acetone extracts >95% of membrane phospholipids when applied to fresh membranes.

The material prefixed in glutaraldehyde has some portion of its lipid content preserved despite subsequent exposure to the ammoniacal acetone by cross-linking groups such as the amino group in phosphatidyl ethanolamine to amino groups of protein. However, the generally good preservation of form serves to illustrate the appearance provided by the unconventional application of osmium tetroxide in carbon tetrachloride after dehydration(Fig. 1). This leads to the critical experiment whose results are illustrated in Figs. 2 and 3, where no glutaraldehyde exposure has preceded extraction with ammoniacal acetone. In the latter instance, the acetone extraction causes major distortions of the form of the outer segment

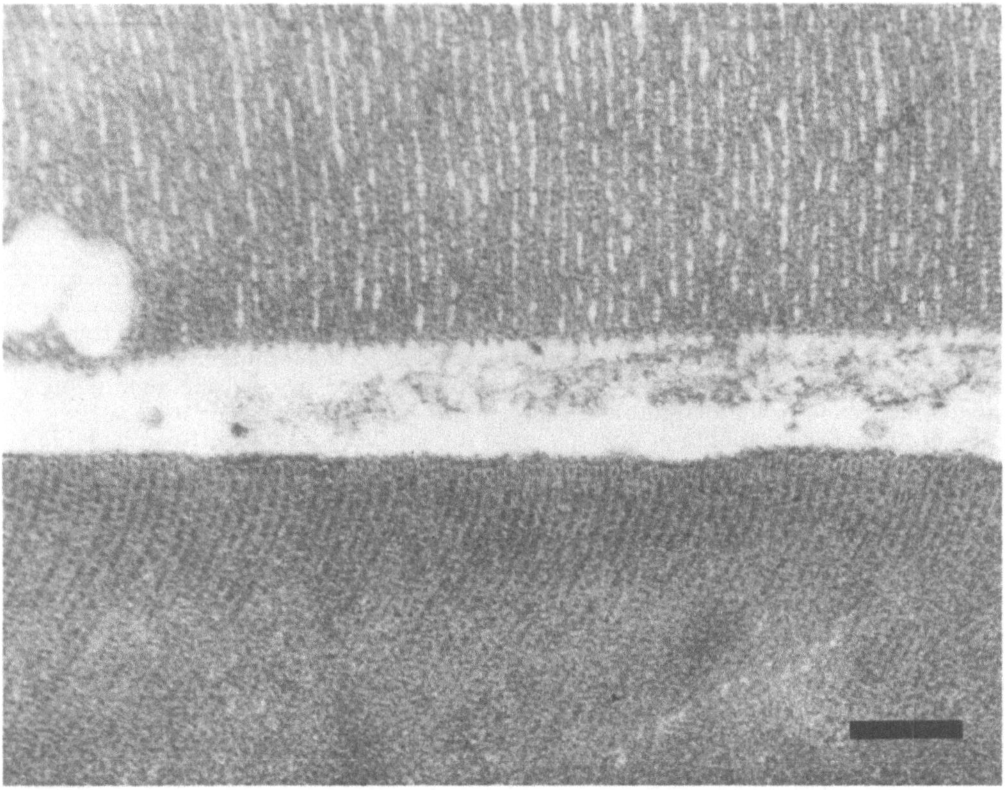

Fig. 1. Outer segments from glutaraldehyde fixed frog retina extracted with ammoniacal acetone. Cone above, rod below. Bar equals 155 nm

but one can yet discern what seem to be white membranes invaginating in cones and turning at the disc edges in rods. The white membranes correspond to the hydrophobic or supposed lipidic interior of each of the two membranous walls of the discs. The inner, dark surface line of the osmicated membrane of the disc becomes optically confluent with a dark staining matrix within the disc and the outer dark lines become optically confluent with a dark staining interdisc matrix. The discs seem somewhat compacted except in the case of cones in the retinas pretreated with glutaraldehyde.

The ammoniacal acetone treatment strips most of the water and most of the phospholipid from the membranes and the removal of water blocks hydrolysis. While this treatment takes a finite time during which molecular rearrangements could have occured, it is clearly possible to obtain a membrane with a trilaminate appearance in the absence of most phospholipid. Removal of water by acetone treatment may denature proteins, but it is not likely to cross-link them. The remaining sheets of protein are clearly spaced by some matrix which could be protein or polysaccharide.

Nilsson (20) has published high resolution pictures of membranes of rod discs in which a globular substructure was suggested by partitioning dense spurs from the surface dense lines across the trilaminate membrane. Robertson (21) has criticized this appearance as meaningfully indicative of membrane structure on the grounds that optical artifacts may be involved. As all structures are in focus in the full

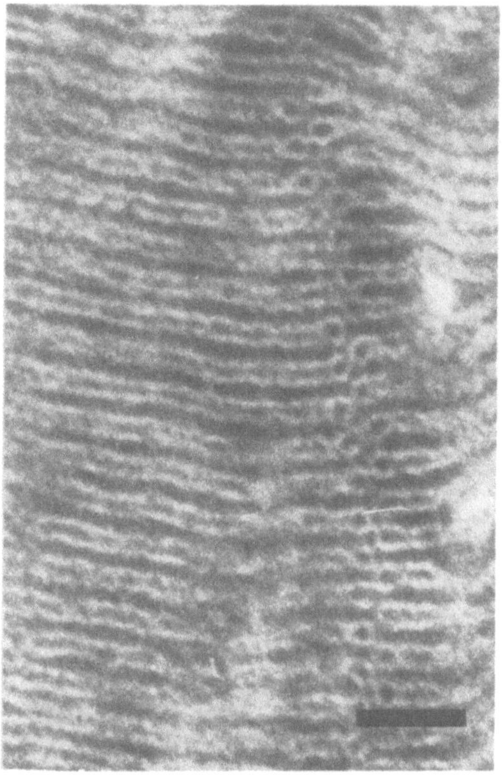

Fig. 2. Rod outer segment from fresh frog retina extracted with ammoniacal acetone. Bar equals 50 nm

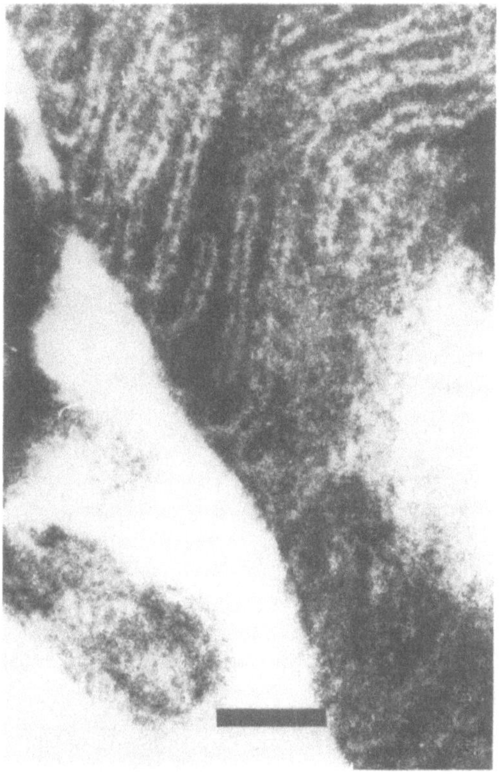

Fig. 3. Cone outer segment from fresh frog retina extracted with ammoniacal acetone. Bar equals 65 nm

thickness of the usual sections employed in electron microscopy (ca. 500 Å to 800 Å), the photograph of a membrane which is perpendicular to the plane of a section is a projection onto the film plane of all densities in the height of the membrane. These could produce the globular appearance.

While Robertson's cautions are worth bearing in mind, the maintenance of a gap between the two dark surface lines in the lipid extracted membrane requires some explanation beyond a simplistic view of the membrane as having protein surfaces and a lipid core. It is a separate and most difficult question as to how this or any detailed appearance in fixed and processed membranes relates to the situation in life. It therefore seems more useful to study grosser effects of agents on unfixed membranes.

Isolated, detached frog retinas were given 15 seconds of exposure to ultrasonic vibrations under a standard geometric arrangement. The retinas were in ice-cold control (physiological) or experimental solutions and the vessels containing them surrounded by an ice-water mixture. Following sonication they were incubated at 20°C for various periods and then processed for electron microscopy. After this brief sonication, many outer segments were broken free at their ciliary connection and lost. Of the remaining attached outer segments, many showed extensive gaps in their plasma membranes and some were truncated, having apparently been

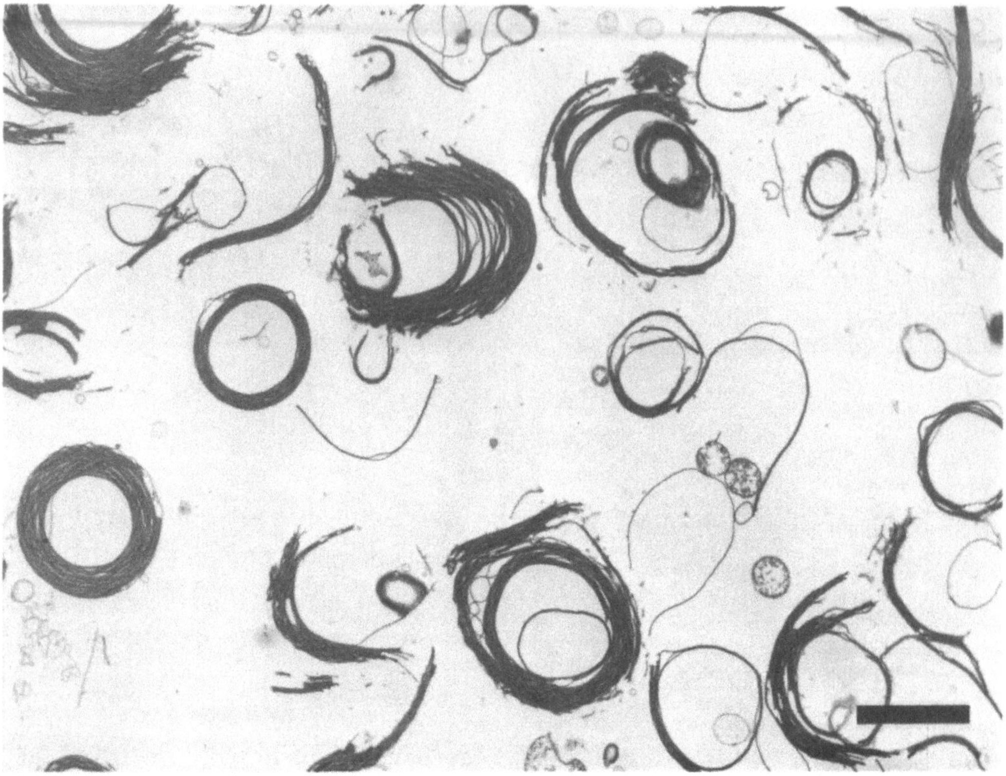

Fig. 4. Disc clumps from sonicated fresh frog retina incubated in Pronase. Bar equals 3.2 μm

broken in their length. Because the discs in the two latter groups were seen to have direct access to the ambient medium such outer segments were termed "open". Normal attached outer segments were termed "closed". There was no difference in the impression of the frequency or extent of patency of "open" segments after incubations up to 2 hrs. in physiological media, hence no impression that there was significant healing of damaged plasma membranes was obtained. Discs in outer segments extensively lacking plasma membranes remained adherent to one another and did not float off. A variable number of both closed and open outer segments, after the longest incubations (2 hrs) exhibited an osmiophilic material which accumulated in droplets along the discs. However, most discs in outer segments of retinas incubated for 90 minutes in physiological media seemed quite normal.

After a 40 minute (20°C) exposure of a sonicated retina to a 0.5% concentration of Pronase in physiological salt solution, many outer segments entirely lacked a plasma membrane, but others were intact. The column of discs in open outer segments briefly exposed to Pronase seemed undisturbed, but after longer exposures (>30 min.) the discs separated into clumps (Fig. 4), and relatively rarely, a single disc could be observed. The lobular subdivisions of the discs became clearly separated, suggesting that it was confinement by an intact plasma membrane that held the lobe margins in close approximation. The clumps of discs tended to remain with the retina when the incubation did not exceed an hour, perhaps being adherent to some sort of gel, but with longer incubations they tended to wash away. Such an applica-

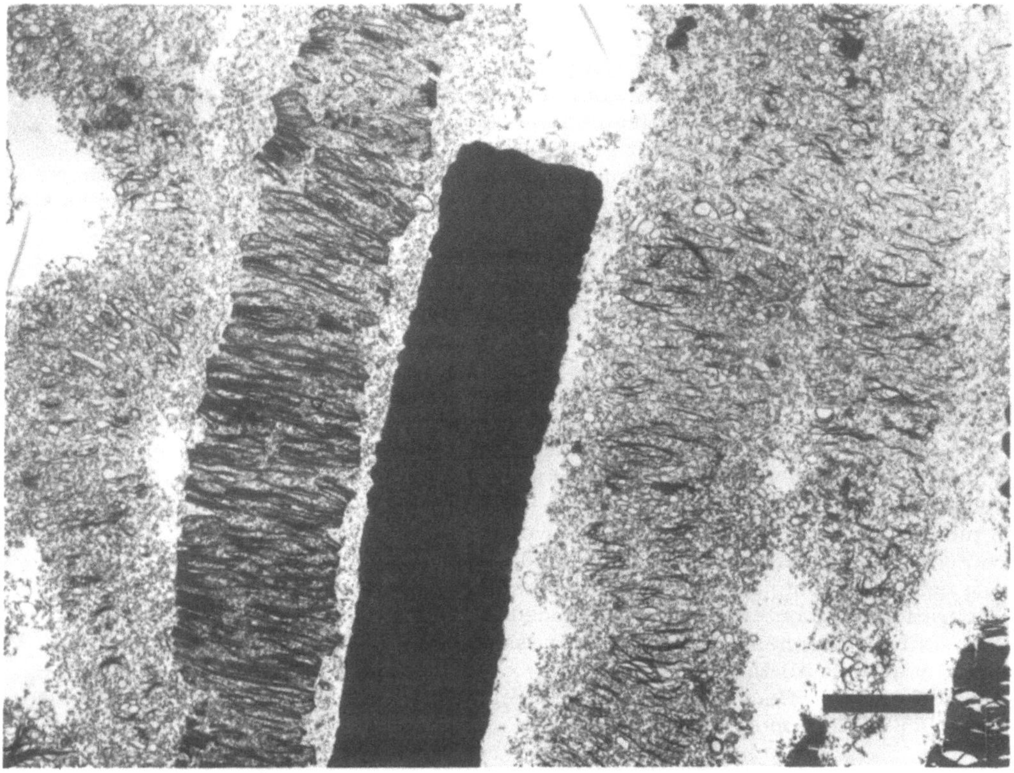

Fig. 5. Intact and disintegrating outer segments from fresh frog retina exposed to phospholipase A. Bar equals 4.5 μm

tion of Pronase or other proteolytic enzymes may lead to a gentle means of isolating intact rod discs. After Pronase exposure for two hours at 20°C the retina itself broke down as a tissue, but by centrifugation a pellet could be recovered for fixation and microscopy and many double-membrane discs lacking enclosing plasma membranes were identified in the fixed pellet.

Because Pronase is known to be capable of almost completely hydrolyzing many proteins to amino acids and dipeptides, the relative resistance of the plasma membrane to the action of these enzymes in closed outer segments and the resistance of the discs in open segments suggests that the proteins in these membranes are in a form which largely denies access of the enzymes to the linkages they attack. However, the rapid destruction of damaged plasma membranes suggests that sensitive groups may be exposed at the edges of any tear in these membranes.

A situation where each disc was freed from its neighbors was not achieved, even with two hours of incubation. The inter-disc distance is too great to seriously consider other mechanisms than some type of physical connection between the discs or a "sticky" and relatively Pronase resistant gel which binds them together. Since Nilsson (22) reported that no membranous inter-disc connections were seen in serial sections of tadpole rods, a gel seems the more likely candidate.

In a recent study of the swelling of isolated outer segments of frog rods in hypotonic solutions by Heller et al. (23), these authors found no intact plasma membranes by electron microscopy and support their view that it is the discs which are swelling by illustrating several stages of progressive enlargement of the intradiscal space. However, these illustrations (their Figs. 7a,b,c,d) also show enlargements of the inter-disc space that greatly exceed those of the intra-disc space. This suggests that the inter-disc substance may also swell in hypotonic solutions, and that the relative contribution to the total swelling of open outer segments by discs and inter-disc materials remains to be determined. In a related study from our laboratory (24) when whole receptor cells with intact plasma membranes were exposed to hypotonic salt or sucrose solutions, the outer segments became spherical but the saccules within them showed highly variable expansion whatsoever.

Studies with lipolytic agents are also in progress. Sonicated retinas were exposed to a 0.1 mg/ml concentration of phospholipase A in a physiological medium. This agent is known to produce lysolethicins by removing fatty acid from lecithin and to be capable of rapid hydrolysis of phosphotidyl ethanolamine (25). Some outer segments immediately begin to lengthen and curl while the dimensions of others were unaffected (Fig. 5). The discs in the lengthening outer segments break down into tubules and vesicular elements. However, these do not wash away, but are apparently held as a part of a mass of gelled material and/or tangled microtubules. These outer segments lack a confining plasma membrane, yet have the general configuration of an outer segment expanded in length and somewhat in diameter. The basis for the cohesiveness of this mass is being investigated. Phospholipase A eventually attacks all the outer segments, even those with intact plasma membranes, but the rapid effect is on open outer segments. The basis for the lengthening could be an imbibition of water.

Falk and Fatt (26) have also described the appearance of disordered structure at the light microscopic level in isolated outer segments of frog rods which were subjected to 10-minute exposures at different temperatures in the range 38° to 52°C. The disordering increased with temperature and is described as either a curling or an elongation with cross-striations appearing. Bleaching made the segments more sensitive to thermal disordering. Thus an exposure at 41.5°C gave 20% disordered outer segments in bleaching preparations. No electron microscopic observation on the material is mentioned, but it would be desirable to know whether the outer segments were closed or had been opened by isolation damage and whether changes in disc structure were involved in the phenomena. The observation of Krinsky (27) that more lipid can be extracted from bleached outer segments suggests that bleaching produces a changed relationship involving some lipid in the membrane structure. It is not clear whether this changed relationship predisposes towards thermal damage or is a consequence of thermal damage by light. Falk and Fatt point out that Hubbard (28) has observed a greater thermal stability of rhodopsin as compared to opsin.

References

1. BOWNDS, D., A. GORDON-WALKER, A. GAIDE-HUGUENIN, W. ROBINSON: Characterization and analysis of frog photoreceptor Membranes. J.Gen. Physiol. 58, 225-237 (1971)
2. COHEN, A.I.: New evidence supporting the linkage to extracellular space of outer segment saccules of frog cones but not rods. J.Cell Biol. 37, 424-444 (1968)

3. FALK, G., P. FATT: Distinctive properties of the lamellar and disk-edge structures of the rod outer segment. J. Ultrastruct. Res. 28, 41-60 (1969).
4. COHEN, A.I.: Further studies on the question of the patency of saccules in outer segments of vertebrate photoreceptors. Vision Res. 10, 445-453 (1970).
5. LATIES, A.M., P. LIEBMAN: Cones of living amphibian eye: Selective staining. Science 168, 1475-1476 (1970).
6. CLARK, A.W., D. BRANTON: Fracture faces of frozen outer segment from the guinea pig retina. Zeitschr. Zellforsch. 91, 586-603 (1968).
7. BLAUROCK, A.E., M.H.F. WILKINS: Structure of frog photoreceptor membranes. Nature (Lond.) 223, 906-909 (1969).
8. BLASIE, J.K., M.M. DEWEY, A.E. BLAUROCK, C.R. WORTHINGTON: Electron microscope and low angle x-ray diffraction studies on outer segment membranes from the retina of the frog. J. Mol. Biol. 14, 143-152 (1965).
9. BLASIE, J.K., C.R. WORTHINGTON, M.M. DEWEY: Molecular localization of frog retinal receptor photopigment by electron microscopy and low angle x-ray diffraction. J. Mol. Biol. 39, 407-416 (1969).
10. BROWN, P.K.: Rhodopsin rotates in the visual receptor membrane. Nature New Biology 236, 35-38 (1972).
11. CONE, R.A.: Rotational diffusion of rhodopsin in the visual receptor membrane. Nature New Biology 236, 39-43 (1972).
12. HAGINS, W.A., W.H. JENNINGS: Radiationless migration of electronic excitation in retinal rods. Faraday Society Discussions 27, 180-189 (1959).
13. SINGER, S.J., G.L. NICOLSON: The fluid mosaic model of the structure of cell membranes. Science 175, 720-731 (1972).
14. DE ROBERTIS, E., A. LASANSKY: Ultrastructure and chemical organization of photoreceptors. In: The Structure of the Eye, ed. G.K. SMELSER, pp. 29-49. Academic Press, New York (1961).
15. YOUNG, R.W.: A difference between rods and cones in the renewal of outer segment protein. Invest. Ophth. 8, 222-231 (1969).
16. YOUNG, R.W.: The renewal of rod and cone outer segments in the Rhesus monkey. J. Cell Biol. 49, 303-318 (1971).
17. YOUNG, R.W., D. BOK: Participation of the retinal pigment epithelium in the rod outer segment renewal process. J. Cell Biol. 42, 392-403 (1969).
18. BOK, D., R.W. YOUNG: The renewal of diffusely distributed protein in the outer segments of rods and cones. Vision Res. 12, 161-168 (1972).
19. FLEISCHER, S., B. FLEISCHER, W. STOECKENIUS: Fine structure of lipid-depleted mitochondria. J. Cell Biol. 32, 193-208 (1967).
20. NILSSON, S.E.G.: A globular substructure of the retinal receptor outer segment membranes and some other cell membranes in the tadpole. Nature (Lond.) 202, 509-510 (1964).
21. ROBERTSON, J.D.: Granulo-fibrillar and globular structure in unit membranes. Ann. N.Y. Acad. Sci. 137, 421-440 (1966).
22. NILSSON, S.E.G.: Receptor cell outer segment development and ultrastructure of the disc membrane in the retina of the tadpole. J. Ultrastruc. Res. 11, 581-620 (1964).
23. HELLER, J., T.J. OSTWALD, D. BOK: The osmotic behavior of rod photoreceptor outer segment discs. J. Cell Biol. 48, 633-649 (1971).
24. COHEN, I.A.: Electron microscopic observations on form changes in photoreceptor outer segment and their saccules in response to osmotic stress. J. Cell Biol. 48, 547-565 (1971).
25. DAWSON, R.M.C.: The nature of the interaction between protein and lipid during the formation of lipoprotein membranes. In: Biological Membranes; Physical Fact and Function, ed. D. CHAPMAN, pp. 203-232. Academic Press, New York (1968).

26. FALK, G., P. FATT: Physical changes induced by light in the rod outer segment of vertebrates. In: Handbook of Sensory Physiology, Vol. VII/1, ed. H.J.A. DARTNALL, pp.200-244, Springer, Berlin-Heidelberg-New York (1972).
27. KRINSKY, N.I.: The lipoprotein nature of rhodopsin. Arch.Ophth. 60, 688-694 (1958).
28. HUBBARD, R.: The thermal stability of rhodopsin and opsin. J.Gen.Physiol. 42, 259-280 (1958).

Discussion

S.L. Bonting: You ascribe the effect of cobra venom to its phospholipase activity. But cobra venom would also contain other enzymes, like proteases. And noting the dispersion of the matrix material (which is not phospholipid in nature), could this be a proteolytic effect primarily?

A.I. Cohen: Since pronase, a most powerful group of proteolytic enzymes, does not give the effect, it is most unlikely to be primarily a proteolytic effect. Conceivably it could be a combined effect of phospholipase A and a proteolytic enzyme, but surely the simplest hypothesis is to attribute the effect to lysolecithin formed by phospholipase action.

W. T. Mason: I would like to comment that in my unpublished studies of the deformation behavior of pronase-treated disc photoreceptor membranes, I observed that the external structural protein of the disc membrane appeared markedly disrupted and in some experiments digested to a large extent. While I did not study the action of pronase on "tears" in the membrane or by thin section techniques, it does appear that pronase may be capable of exerting some effect of the interdisc matrix or on the disc membrane protein itself.

A.I. Cohen: Since I am not aware of what is meant by "deformation behavior" or "disrupted membrane" in the absence of thin sectioning, it is difficult to reply. One should note that experimental embryologists routinely use pronase in low-calcium media with mild mechanical forces to disassociate cells. These remain fully viable and reassociable in media containing normal calcium levels. Pronase can remove some glycopeptides and perhaps surface coat proteins during the disassociation procedures.

W. Sickel: You touched on the electrical implications of the different membrane arrangements in rods and cones, which to me appear electrically parallel in cones and in-series in rods, i.e. opposite to the morphological aspect. Would you care to comment on this point?

A.I. Cohen: The isolation of rod discs from the plasma membrane as opposed to the maintained confluence of many cone discs with this membrane seems to correlate with the minimal contribution which rods make to fast photovoltages (ERP). I have also noted (24) that when most discs are confluent with the plasma membrane they greatly increase the total cell surface. This doubtlessly has some electrophysiological consequences, such as increasing the total capacitance, etc.

Ultrastructure of the Photoreceptors of the Bovine Retina*

W. T. Mason
Department of Chemistry, Case Western Reserve University, Cleveland, OH 44106/USA

The bovine retina is the tissue studied in greatest detail by the vision biochemist but from the standpoint of gross morphology and ultrastructure it has been neglected. From the present observations of the bovine retina, it appears that this visual tissue is similar in some respects to the primate retina (1,2,7) but unique in many respects. In addition, the possible existence of cone photoreceptor structures in the cattle retina has been considered but never investigated in any detail. We have therefore undertaken to study the ultrastructure of the bovine retina by electron microscopy.

Eyes are enucleated from freshly slaughtered cattle, fixed intact with 3% glutaraldehyde in Millonig's buffer for $3\frac{1}{2}$ hours, and post fixed in 1% osmium tetroxide for 1 hour. The eye was then sliced to expose the retina; the retinal and scleral tissue was dehydrated in ethanol and infiltrated with propylene oxide for 30 min. The tissue was then bathed in 50% propylene oxide/50% Epon for 12 hours and pure Epon for 2 hours. After embedding, thin sections were cut with an LKB microtome and stained with 1% uranyl acetate followed by 0.25% alkaline lead citrate in methanol. Electron microscopy was performed on a Hitachi 11-UB instrument.

In Fig. 1 a region of the outer segment area is shown which illustrates the relative sizes of the rods and cones in the bovine retina. Rod and cone outer segments (ROS and COS) were seen to be 2 μm in diameter, the COS being 3-4 μm in length and the ROS 7-10 μm in length.

Several distinctions were noted between the rod and cone outer segments, namely the disc thickness, intradisc spacing and the number of discs per segment. In bovine cones, a disc thickness of 375 Å with an intradisc spacing of 225 Å was observed. Rod disc membranes, on the other hand, were 250 Å thick with an intradisc spacing of 75-100 Å. The appearance of saccules in bovine cones was observed to consist of two main types: one exhibiting a uniformly dense pigmentation and highly osmiophilic and the other showing a large intradisc space with low osmiophilicity. The presence of a terminal loop was noted on the end of the rod disc membranes; cone saccules showed this structure only in those discs which exhibited high osmiophilicitiy in the intradisc space. In cone saccules of high intradisc osmiophilicity the terminal ends were continuous and showed no end loop. ROS contained on the average approximately 450 disc membranes per segment while COS were observed to contain 100-120 discs.

The discs of the ROS were of a uniform osmiophilic nature; however, the cone saccules varied in this respect in that a "banding" effect was observed. This difference was seen to arise from the fact that the intradisc space of the disc membranes exhibits differential affinity for the stain and distinctly different spacing thicknesses although no explanation for this is available.

* Supported by grant EY-00471 from the National Eye Institute, the National Institute of Health.

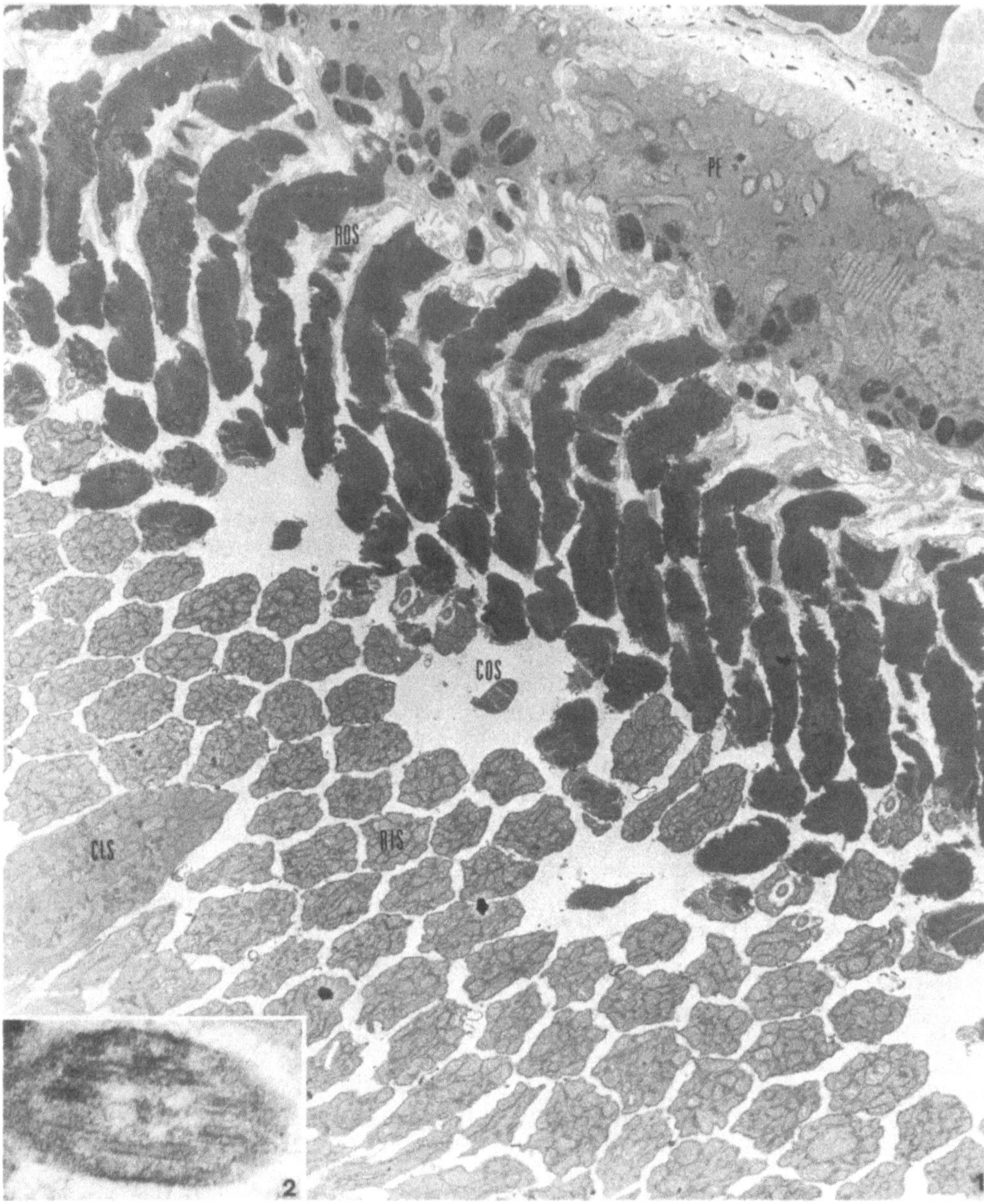

Fig. 1. Electron micrograph of an oblique longitudinal section through the bovine retina; note relative rod (ROS) and cone (COS) outer segment size and cytoplasmic pocket surrounding the COS. Glutaraldehyde-osmium fixation with uranium-lead stained. Magnification x 14,000

Fig. 2. Tangential section through the connecting cilium of a rod inner and outer segment; note the nine peripheral doublets with a central osmiophilic structure. Glutaraldehyde-osmium fixation with lead citrate/uranyl acetate staining. Magnification x 80,000

In the rods and cones of the central retinal area studied here, a plasma membrane was observed to continuously surround the outer segment with no tactile junctures; this is especially significant because it has been postulated that the cone saccules are invaginations of the plasma membrane (3-7). In the bovine retina it appeared that the morphogenetic origin of the cone saccules is a complex vesicular system located at the base of the outer segments. However, vesicles of a similar nature were observed at several additional sites along the entire length of the COS, thereby suggesting that the COS possesses multiple sites from which disc material may originate.

The ROS showed no vesicular structures of the type exhibited by the COS with the exception of several small vesicles at the base of the segment. In ROS, the disc membranes appeared to originate as the result of fusion of the smaller vesicles at the base of the segment. This vesicular material of disc-like appearance at the outer segment base was 500-700 A in diameter. Without exception these discs did not appear to arise from the plasma membrane, but rather appeared to originate in the region adjacent to the juncture of the ciliary stalk with the base of the outer segment. This observation is countered by Young's data (6) which suggests that disc material in the frog <u>Rana</u> <u>catesbiana</u> arises from invaginations of the plasma membrane surrounding the disc.

Another interesting observation of the bovine retina is that the COS are isolated from ROS by a large region of low densitiy cytoplasm 7-10 μm in diameter. The cones are located vitreally to the ROS but sclerally to the rod inner segments. This cytoplasmic region therefore separates the COS from both the rod inner and outer segments. In the bovine retina, the pigment epithelium was observed to envelop a number of lysosomal and melanosomal bodies. These bodies were noted in addition to numerous other bodies in the pigment epithelium and appeared to participate in an extensive and specific digestive cycle of ROS material, although this question may only be fully resolved by autoradiographic studies. The COS were not seen to make contact with the pigment epithelium.

In longitudinal sections of the bovine retina, the cilium structure in cones and rods is identical to that observed in other vertebrates (7-11). The cilium in the bovine retina is 1-1.5 μm in length with a second centriole pair in the inner region of the cilium and a connecting axial centriole (basal body) perpendicular to the long axis of the cilium and located vitreally to the termination of the second centriole pair in the inner segment. Numerous fibrillar processes were seen to extend as much as 0.2 μm into the inner segment of the cilium. The bovine retina is unique in that cross sections of the receptors at the level of the cilium reveal nine peripheral microtubles with a central structure (Fig. 2). This configuration is unusual when compared to the 9+0 configuration normally obtained in vertebrates (4), i.e., nine peripheral doublets lacking a central doublet core.

A marked difference of size between inner segments of rods (RIS) and cones (CIS) in the cattle retina was also apparent (Fig. 1). CIS were 3-4 times the diameter and 10 times the total volume of RIS. From observations of longitudinal thin sections of bovine retina, CIS were estimated to contain 700 mitochondria and RIS about 50 mitochondria. The mitochondria are oriented parallel to the axis of the photoreceptor and are densely concentrated in the scleral portion of the structure. In the vitreal portion of the inner segment are numerous granular endoplasmic reticulum. Villous processes were observed to extend from the base of the inner segments and make various contacts with the outer nuclear layer. The appearance of these processes and their contacts with the outer nuclei suggest that the large numbers of endoplasmic reticulum present in the inner segment may synthesize outer nuclear material as well as outer segment material. This would imply that protein newly synthesized in the inner segment may flow bidirectionally within the segment.

Glial cells were noted to predominate in the inner segment region and the presence of cone pedicles was noted toward the vitreal side of the retina in the synaptic region. The numerous processes of the glial cells in this area were noted to be diverse in size and fibrillar in character. In the region of the external limiting membrane they appeared to diversify so as to sheath the inner segments at this level.

Extensive contacts between α- and β-type cells were noted in the outer plexiform nuclear layer, as well as synaptic junctions between horizontal cells and α and β cells in this same region. Furthermore, in the inner plexiform layer extensive bipolar cell types were noted.

In summary, in this study of the bovine retina, most notable was the detection of cone outer segments in the outer receptor layer, a unique microtubular process in the connecting cilium, and a number of highly specialized neural relationships in the plexiform neural layers.

References

1. DOWLING, J.E., B.B. BOYCOTT: Organization of the primate retina: Electron microscopy. Proc.Roy.Soc.London, B 166, 80-111 (1966).
2. DOWLING, J.E.: Foveal Receptors of the monkey retina: Fine structure. Science 147, 57-59 (1965).
3. SJÖSTRAND, F.S.: Fine structure of cytoplasm: The organization of membranous layers. Rev.Mod.Phys. 31, 301-318 (1959).
4. MOODY, M.F., J.D. ROBERTSON: The fine structure of some retinal photoreceptors. J.Biophys.Biochem.Cytol. 7, 87-97 (1960).
5. TOKOYASU, K., E. YAMADA: The fine structure of the retina studied with the electron microscope. IV. Morphogenesis of outer segments of retinal rods. J.Biophys.Biochem.Cytol. 6, 225-230 (1959).
6. YOUNG, R.: The Organization of vertebrate photoreceptor cells. In: The Retina, pp. 177-204. Univ. of California Press, Los Angeles (1969).
7. COHEN, A.I.: The fine structure of the visual receptors of the pigeon. Exp. Eye Res. 2, 88-97 (1963).
8. SJÖSTRAND, F.S.: The ultrastructure of the retinal receptors of the vertebrate eye. Ergeb.Biol. 21, 128-160 (1959).
9. COHEN, A.I.: The ultrastructure of the rods of the mouse retina. Am.J.Anat. 107, 23-48 (1960).
10. BROWN, P.K., I.R. GIBBONS, G. WALD: The visual cells and visual pigment of the mudpuppy, Necturus. J.Cell Biol. 19, 79-106 (1963).
11. HOLMBERG, K.: Ultrastructure of the hagfish retina. Z.Zellforsch. 111, 519-538 (1970).

Microspectrophotometry of Visual Receptors

P. A. Liebman
Department of Anatomy, School of Medicine, University of Pennsylvania,
Philadelphia, PA 19104/USA

During this brief talk, I would like to try to give an updated view, since my Handbook review (1) written in 1969 of the kinds of things I believe microspectrophotometry (MSP) of single vertebrate receptor outer segments has contributed to our fund of knowledge of visual pigments and photoreceptor structure and where I anticipate future understanding will be gained by further application of this technique.

Background

Following the pioneering application of bile salt extraction methodology to retinas by König and by Köttgen and Abelsdorf, refinements in technique and explorations of vertebrate pigments from the late 1930's until present can be attributed in bulk to Wald and to Crescitelli in America and to Dartnall in England together with their many students and collaborators (for references cf. 1). From the early work came at first the generalizations that there were two basic types of visual pigment prosthetic group derived from vitamins A_1 and A_2 in particular stereo-isomeric form which combined with a kind of visual protein to form two visual pigments respectively, rhodopsin and porphyropsin. Wald early found evidence for another pigment which he named iodopsin and later showed to be dependent on vitamin A_1 (retinal) and a separate visual protein. In addition, Wald was able to generate a new (non-visual) pigment by combining his bleached iodopsin preparation with the aldehyde of vitamin A_2 (dehydroretinal) to yield cyanopsin, a pigment that has not to this day been extracted from a retina. These discoveries led Wald to the compact generalization of two chromophores + two proteins = four visual pigments. The absorption spectra alone of these pigments has proved extremely useful over the years especially to electrophysiologists and psychophysicists of the visual system.

Meanwhile both Dartnall and Crescitelli et al. had been gathering a considerable diversity of visual pigment data on a great number of species showing significant variation from the theme of a 500 nm A_1 pigment and a 522 nm A_2 pigment - notably in the fishes (Munz, Schwanzara and Beatty and Dartnall, Lythgoe and Bridges) and in geckos (Crescitelli) until the range of λ_{max}'s soon reached from ca. 470-528 for A_1 and from 513-543 for A_2 rod-pigments. These pigments are all presumed quite reasonably to originate in rods. The generalization to be made from this work is that there seem to be many rod pigment proteins, whose chromophores span a considerable range of λ_{max} and these may be related to the particular environment of the animal of origin. The several examples of 500 and 522 nm pigments at first discovered may reflect simply the feature of common photic environment of the animals from which obtained. The recent data showing λ_{max} diversity have brought into question the hypothesis that there is a "class" of pigments that may be called rhodopsins or what you like that have a spectroscopic biochemical or morphological feature in common. From the above and other work, it is not now clear that the limited class hypothesis is justified. But certainly no knowledgeable visual pigmentologist would stop with the rod pigments for there are a number of other known and putative pigments that must be acknowledged as falling at least into classes distinct from that of rod pigments, the phenomenon of colour vision providing one of the more persuasive pieces of circumstantial evidence pointing in this direction for as long as men have known about molecules. It is of course generally known that this last mentioned area of investigation has been the one most frustrating to extraction technique and it is at this point that my story and that of MSP begins.

In 1962-1963 after gaining some perspective with work on single frog rod spectroscopy (2), I began, as did Marks in MacNichol's laboratory, at about the same time to get successful spectral recordings from single cones using a new breed of home-made microspectrophotometer. P. Brown was starting similar work with his adaptation of a commercial recording spectrophotometer in Wald's laboratory. In the intervening years, even better instruments have been constructed on similar principles - minimizing light exposure and noise sources and maximizing instrument sensitivity. Although the most exciting initial phase of this work was the confirmation of trichromatic theory by finding three separate cone pigments each in goldfish, monkey and man, many other provocative and less anticipated discoveries have been made by MSP.

Role of P 620 in Vertebrates

Without attempting to keep things in chronology of discovery the first such discovery I shall mention is now proof of Wald's hypothesis that pigments like the synthetic cyanopsin mediate cone vision in some vertebrates. In 1962-1963, Entine and I (3) as well as Marks (4) found a 620 nm pigment in goldfish cones. In 1964 Entine and I (5) found a 620 nm pigment in tadpole (*Rana pipiens*) principle cones and this corresponded to an iodopsin-like 575 nm pigment we found in the homologous frog principal cones. In 1965 Granda and I (6) discovered the 620 nm cone of the swamp turtle. In 1969-1970 one of my students discovered spontaneous displacement in the dark of goldfish 620 nm cone pigment to 575 nm in the presence of 11-cis retinal (R. Kanter, unpublished). This was confirmed in my own experiments and together with the frog-tadpole results (frog A_1, tadpole A_2 retina; 5) constitutes the most direct evidence on chromophore identity and duplicates Wald's results in nature. In 1971,I found 620 nm principal cones in both sunfish (*Lepomis microlophus*) and channel catfish (*Ictalurus punctatus*) (unpublished), and in the past several weeks Dartnall and I have found the same in tench (*Tinca tinca*). Thus, P 620_2 or cyanopsin has now been found in six species from three vertebrate classes and has been shown to be related to a P 575_1 by simple change of prosthetic group.

In our frog-tadpole investigation, Entine and I (5) also discovered that the accessory cones of *Rana pipiens* contain respectively rhodopsin (P 502_1), and porphyropsin (P 527_2). Granda and I (6) later found "rhodopsin" or "porphyropsin" containing cones in two species of turtle and found further that some of the oil-drop containing cones harboured these pigments. More recently I have found accessory cones of sunfish and channel catfish to contain P 537_2 or P 543_2 respectively, like the rods of each and also that all cones of the bullhead catfish (*Ictalurus melas*) contain P 543_2 identical to that of the rods (Fig. 1). Finally, in the last few weeks, Dartnall and I have demonstrated "rod pigment"-containing accessory cones in tench retina. Thus, the new principle that "rod pigment" is not confined to rods but also occurs in accessory cones seems now reasonably well established through MSP.

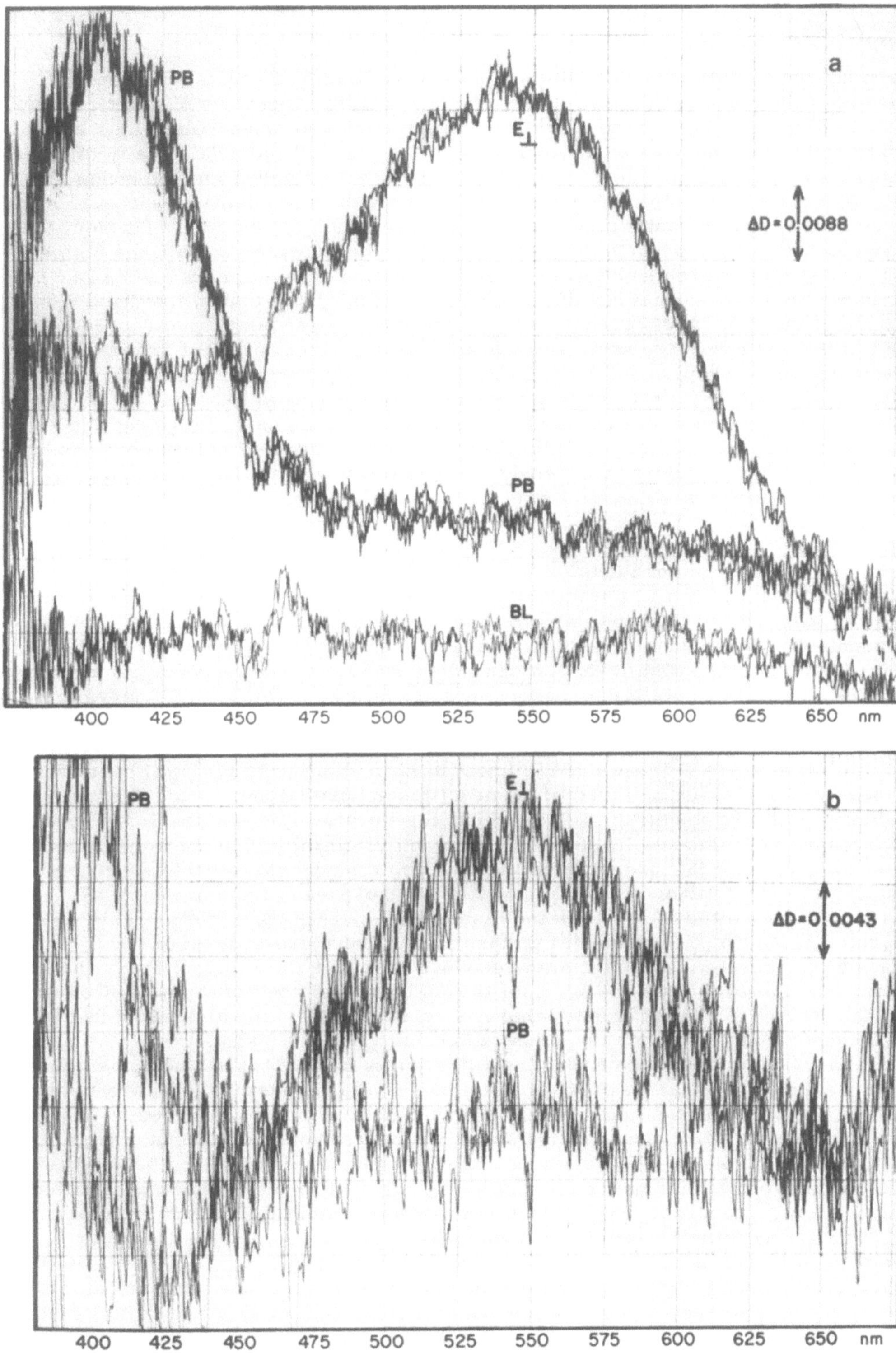

Fig. 1. Spectra recorded from single rod (a) and accessory cone (b) of catfish (Ictalurus melas). $E_\perp$ - Electric vector perpendicular to rod long axis, PB - postbleach spectrum, BL - baseline

Trichromaticity

We are not yet certain of the physiologic role of rod-pigment-containing cones. However, if they turn out to have the electrophysiologic properties of other receptors of cone morphology, we are led to the realization that trichromaticity may be mediated through somewhat different receptor and pigment morphologies in different animals. Thus, in the primates, man and monkey, only single rods and cones are known to occur and in these a rod pigment and three cone pigments of separate identity occur. In goldfish, as in man, there are three cone pigments distinct from the rod pigment; but these occur mostly in "twin cone" pairs which contain the pigment pairs, red-green or green-blue. In frog, the situation changes more radically. The blue (cone?) pigment is in a rod of morphology distinct from that of the "real" (red) rod. One of the cone types (accessory) contains a "rod" pigment (spectrum identical to that of the red rod in the same animal) and the third (red) pigment is in a cone paired with the "rod"-pigment cone. In colored oil droplet containing photoreceptors like those of turtles (and presumably of birds, too) there is yet another twist, for not only are there three cone pigments, one of which is also a rod pigment and which occurs as well in oil droplet containing cones, but the intensely colored oil droplets sufficiently modify individual cone spectral sensitivity to yield five distinct cone-receptoral sensitivities - i.e. no longer trichromatic but pentachromatic:

Concept of Visual Pigment Class

All of you will know of the violet absorbing receptors of frog (green rods), P 432_1 and tadpole P 438_2 based on a common visual protein. In the violet absorbing region we also know about goldfish P 455_2, Pseudemys P 450_2, Chelonia P 440_1, and primate P 445_1. These would seem reasonably to constitute a "class" of blue-violet absorbing pigments.

In goldfish, there are P 522_2, P 537_2, and P 620_2 in addition to P 455_2 (3,4) so there are clearly four classes of visual pigment in this species alone. In addition there is a cone pigment of Necturus, P 575_2 which regenerates P 540_1 with 11-cis retinal. This later pair of pigments (based on a common protein) would seem to represent a class (λ_{max}) separate from any of the four goldfish. Goldfish P 537_2 regenerates P 510_1 with 11-cis retinal. The λ_{max} of a number of these pigments, in A_1-A_2 pairs have been plotted in Fig. 2 in the fashion of Dartnall and Lythgoe. This figure illustrates at least two fundamental truths. First, there are at least five λ_{max} classes of vertebrate visual pigment. Second, the longer the λ_{max} of the A_1 based pigment, the longer will be the λ_{max} of the A_2 based member of the pair. Conversely, for short λ_{max} A_1 pigments, there is relatively speaking almost no change in λ_{max} caused by switching to the A_2 based chromophore. Finally, the A_1-A_2 pair plot is quite linear. When the first generalization is fortified with the pigments of invertebrates we see that the P 345_1 of Hamdorf et al. (7) easily suggests a sixth class and the presumptive tench P 680_1 of Naka and Rushton (8), a seventh. However, if continuing work uncovers the same diversity among cone pigments in species of diverse habitat as already found for rod pigments, it will soon be a justifiable question of whether the concept of visual pigment (protein) class is justified at all. We can, of course, expect some perspective on this in the next ten years of MSP. The second generalization concerning A_1 versus A_2 λ_{max} positions bears some thought by theoreticians, and some synthetic work by organic chemists so that appropriately variant chromophores can be used to test hypotheses of mechanism. I have already described a possible mechanism of this effect as a result of discussions P. Blatz (University of Wyoming, Laramie), i.e. that of π cloud decoupling

Fig. 2. Each point represents a single visual pigment protein in combination with retinal (ordinate) or 3-dehydroretinal (abscissa) 1, 2, 3, 4, 5, denote "classes" of visual pigment represented. Data obtained by MSP of single receptors from • frog, ▲ turtle, ■ goldfish, o catfish, △ mudpuppy in naturally occuring or in vitro regenerated pairs

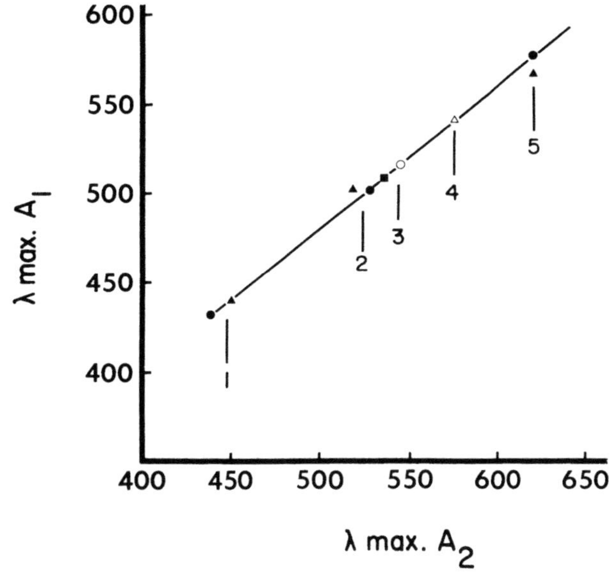

through twist around, or charge localization at the $C_6 - C_7$ bond of the retinal-Schiff base of visual pigments (1).

I have been asked a number of times about photoproducts in single outer segments after bleaching. On the time scale of our measurements, we see a 387 nm peak in A_1 based rods and a 405 - 410 nm peak in A_2 based rods. In single and principal cones, we see little or no photoproduct. This I speculate may be a corollary of the infolded nature of the pigment containing plasma membrane of this cone, providing much aqueous contact for diffusion of products away from the outer segment in contrast to the pinched-off, included lamellae of rods which retain products for some time. It is of some interest to find a rod-like photoproduct in accessory cones (Fig. 1b).

Molecular Orientation and Stoichiometry

It is generally known that rod visual pigment is dichroic having 4.5 - 5 times more absorption in the plane of the outer segment lamellae than perpendicular to them. This has proved true of all cells (rods and cones) so far examined. In addition, it is clear from records like Fig. 1 that the violet absorbing photoproducts are dichroic in the same sense as the original pigment. My work (2) on frog rods suggests that MI and MII intermediates are highly dichroic and work that extended that of Denton confirms the large positive dichroism of vitamin A, both intrinsic and extrinsic in origin, using both fluorescence and absorption measurements (9). Extrinsic retinal does not appear to orient in rod outer segments though the 11-cis form will support visual pigment regeneration as described above.

It has been possible for us to record from single rod outer segments through the ultraviolet spectrum. This reveals the protein band (280 nm) and the even larger 235 nm band. We have as yet not positively identified the origin of the 235 nm band. Neither the 235 nm nor the 282 nm band is dichroic. Using the known extinction coefficient for rhodopsin, we can calculate rhodopsin concentration in frog rods and this is 2.2 mM, corresponding to $3 \cdot 10^9$ molecules per outer segment or $7.5 \cdot 10^5$ molecules per lamellar unit membrane. For a 6 μm diameter rod this gives an average spacing of about 62 Å between rhodopsins, in very close agreement with results of X-ray diffraction (10). We have found all visual cell outer segment (ROS) pigment densities to fall in the range 0.013-0.018 (lower value likely low due to light leakage around cones) per micron pathlength, so for comparable extinction coefficients, all visual pigments will have similar average spacing in the membranes. After correction for α band dichroism, a comparison of ROS D 280/500 nm ratio with that of the purest rhodopsin extracts shows that about 25-30% of the 280 nm band is unaccounted for as visual pigment. Thus, ROS protein is mostly but not all visual pigment. ROS membranes are known to be highly unsaturated in fatty acids of their phospholipids. Part of the 235 nm band of ROS may be due to conjugated unsaturations in these lipids. If so, these would constitute about 10% of the total fatty acids based on 235 nm extinction. From the hydrocarbon chain spacing of the phospholipids of ROS membranes (X-ray) it can then be calculated that there are approximately 100 phospholipid molecules per rhodopsin in ROS membranes and this is in good agreement with phospholipid-rhodopsin analysis of extracts. It is therefore possible to confirm by MSP that contamination of extracts by non-ROS material is negligible and that highly pure rhodopsin has indeed been prepared. This is also born out by inspection of D 400/500 nm ratios of MSP records for A_1 rods.

Conclusion

In summary, although it has been a hope of those concerned with the identification of visual pigments by extraction methods that techniques would be forthcoming that would allow all the visual pigments of a retina to be discovered, such work has met with very limited success. Visual pigments have yet to be fractionated from extracts so that their spectra could be studied separately with the same accuracy as that enjoyed when a single preponderant pigment is measured. Although something has been learned in a number of cases where there was more than one dominant pigment present (e.g. by differential bleaching etc.), with one exception, no cone pigments have been detected by extraction methodology. Finally, we can be fairly confident that extraction methodology alone will never allow pigments to be identified with their photoreceptors of origin. It has been and will continue to be the unique role of MSP to provide answers to such questions.

References

1. LIEBMAN, P.A.: Microspectrophotometry of photoreceptors. In: Handbook of Sensory Physiology, Vol. VII/1, pp. 481-528, ed. H.J.A. DARTNALL. Springer, Heidelberg (1972).
2. LIEBMAN, P.A.: In situ microspectrophotometric studies on the pigments of single retinal rods. Biophys. J. 2, 161-178 (1962).
3. LIEBMAN, P.A., G. ENTINE: Sensitive low-light-level microspectrophotometer - detection of photosensitive pigments of retinal cones. J.Opt.Soc.Am. 54, 1451-1459 (1964).
4. MARKS, W.B.: Visual pigments of single goldfish cones. J.Physiol. 178, 14-32 (1965).
5. LIEBMAN, P.A., G. ENTINE: Visual pigments of frog and tadpole. Vision Res. 8, 761-775 (1968).
6. LIEBMAN, P.A., A.M. GRANDA: Microspectrophotometric measurements of visual pigments in two species of turtle, Pseudemys scripta and Chelonia mydas. Vision Res. 11, 105-114 (1971).

7. HAMDORF, K., J. SCHWEMER, M. GOGALA: Insect visual pigment sensitive to ultraviolet light. Nature (Lond.) 231, 458-459 (1971).
8. NAKA, K.I., W.A.H. RUSHTON: S-potentials from colour units in the retina of fish (Cyprinidae). J. Physiol. (Lond.) 185, 536-555 (1966).
9. LIEBMAN, P.A.: Microspectrophotometry of retinal cells. Ann. N.Y. Acad. Sci. 157, 250-264 (1969).
10. BLASIE, J.K., C.R. WORTHINGTON, M.M. DEWEY: Molecular localization of frog retinal receptor photopigment diffraction. J. Mol. Biol. 39, 407-416 (1969).

Dichroism in Rods during Bleaching

C. M. Kemp
Department of Visual Science, Institute of Ophthalmology, London WC1 H9QS/Great Britain

The long-lived intermediates in the rhodopsin bleaching sequence have been linked with the state of adaptation of isolated retinae (1, 2, 3, 4). Correlations of this type can only be made when the kinetics of the relevant dark reactions are established. Since spectrophotometric data for the isolated retina, on which kinetic analyses are normally based, are obtained using light passing axially through the rod outer segments (ROS), the concentrations of the intermediates at a given time after bleaching the retina can only be calculated if the orientation of each species with respect to the ROS axis is known. While the transverse orientation (i.e. perpendicular to the ROS long axis) of rhodopsin is well established (5, 6, 7, 8), the relative angles of the chromophores of the decay products are known less certainly. Denton's measurements (6) in the near u.v. imply that metarhodopsin II (MII) is transversely orientated, and he deduced from polarized fluorescence data that retinol is aligned axially. In contrast, Wald et al. concluded that retinol lies in the same plane as rhodopsin (8). The present work seeks to establish the relative orientations of MII, metarhodopsin 465 (MIII), retinal and retinol in frog (Rana pipiens) and albino rat retinae. The technique used follows that of Denton (6) closely: the linear dichroism of rhodopsin and the long-lived products in the aligned rods on the edge of a folded retina were examined on a microspectrophotometer (MSP).

Methods

The experiments were carried out on a Shimadzu MPS-50L multipurpose spectrophotometer equipped with the double-beam MSP attachment. Light from a tungsten or Xenon source was monochromated, chopped, and the analyzing beam passed through a field limiting diaphragm (d=20 μm), a polarizer (Polaroid, type HNP'B), a reflecting condenser and objective (50 x; N.A. ≈ 0.5) and a further field stop (d = 20 μm) onto an end-on photomultiplier. The reference beam passed through air only, to a second photomultiplier. When spectra were recorded in the UV region a visible light absorbing filter (Wratten 18B) was interposed between the objective and the photomultiplier of the MSP.

The isolated dark adapted retina was folded, immersed in bathing medium (Baumann's Ringer (9) was used for frog retinae; for rat, the medium was identical to that used by Ernst and Kemp (3)), sealed with a cover-slip in a chamber of path length ~ 200 μm built on a quartz slide, and the slide mounted on the MSP. Adjustments to the position and focus of the retina were made using light from the field illuminating lamp attenuated by a Wratten 70 filter to remove wavelengths shorter than 640 nm. Bleaches were effected with a flash-bulb (Philips PFB-1) filtered by a Wratten 12 to remove wavelengths less than 490 nm. A single flash removed 60-80% of the rhodopsin in the rods in the centre of the field.

Experiments on frog retinae were carried out at room temperature; those on rat at ~ 30°C.

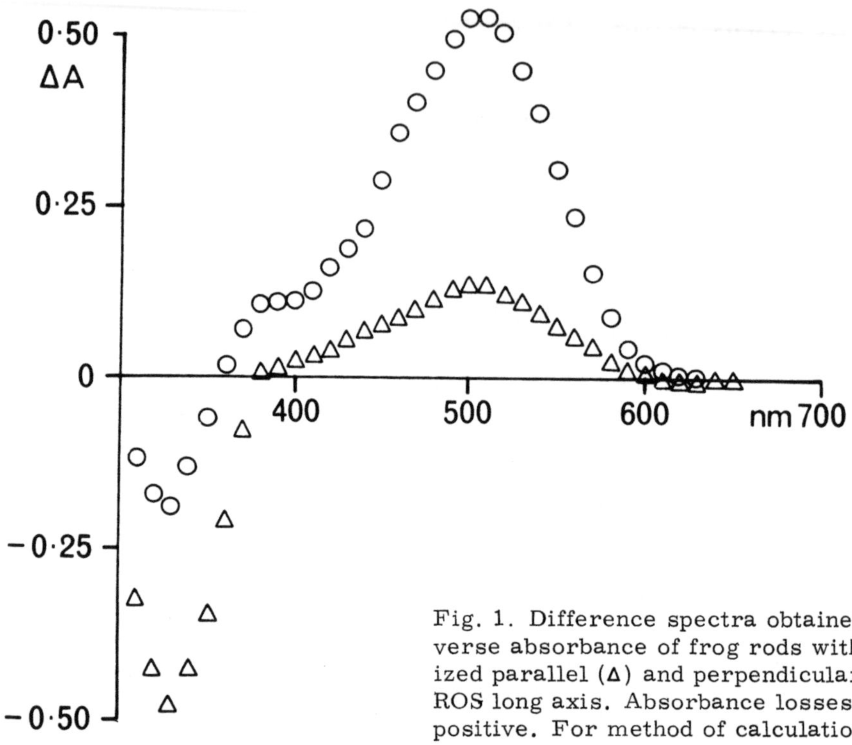

Fig. 1. Difference spectra obtained for transverse absorbance of frog rods with light polarized parallel (Δ) and perpendicular (o) to the ROS long axis. Absorbance losses shown as positive. For method of calculation, see text

Results and Discussion

Difference curves to establish the orientations of rhodopsin and retinol were obtained by subtracting the absorbance at a given wavelength obtained more than 40 min after bleaching from the dark-adapted value for the same wavelength and polarizer setting. Thus $\Delta A_{\perp\lambda}$ is the change in absorbance at wavelength λ for light polarized perpendicularly to the ROS long axis, and $\Delta A_{\parallel\lambda}$ is the axial component. In the example shown in Fig. 1, the loss $\Delta A_{\perp 500}$ is about three times the gain in $\Delta A_{\perp 328}$, while for the changes in $\Delta A_{\parallel}$ the converse is found.

Over several experiments the ratio of the gain $\Delta A_{\parallel 328}$ to the loss $\Delta A_{\perp 502}$ ranged from 0.9 to 1.3. The dichroic ratio at 502 nm ($D_{502} = \Delta A_{\perp} / \Delta A_{\parallel}$) also varied; for most experiments it was between 3 and 3.5. The inverse ratio $\Delta A_{\parallel 328} / \Delta A_{\perp 328}$ ($= D_{328}$) was usually rather lower than D_{502} for a given experiment: in most cases it lay between 2.5 and 3.

Kinetic data for the absorbance changes at 480 nm in the frog ROS following flash illumination are shown in Fig. 2. $\Delta A_{\perp 480}$, the absorbance change for light polarized transverse to the ROS long axis is always greater than the axial component $\Delta A_{\parallel 480}$. For a given experiment the dichroic ratio D_{480} ($= \Delta A_{\parallel 480} / \Delta A_{\parallel 480}$) of this transient absorbance was equal to that of the permanent absorbance losses at $\sim$ 500 nm, within experimental error. Over the time period when D_{480} could be measured with any certainty, it was roughly constant.

In Fig. 3 the mean data for experiments of two types are compared. In one series of experiments the absorbance changes at 390 nm following flash illumination were

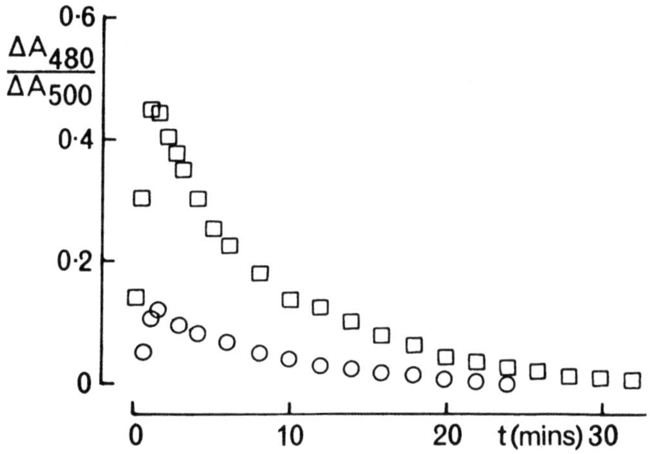

Fig. 2. The changes in absorbance at 480 nm in frog ROS following a flash bleach for light polarized parallel (○) and perpendicular (□) to the long axis. The curves are normalized with respect to the permanent absorbance loss at 500 nm for light polarized perpendicular to the ROS long axis

monitored axially with unpolarized light (○), while in other (Φ) light passed transversely through the ROS and was polarized perpendicularly to the ROS long axis. Within experimental error, the two sets of data are superposable. The curve on which the points lie cannot be described by a single exponential: after an initially rapid decay phase, the absorbance decreases more gradually.

At 328 nm changes in absorbance parallel ($\Delta A_{\parallel 328}$) and perpendicular ($\Delta A_{\perp 328}$) to the ROS long axis are more complex (Fig. 4).

Both $\Delta A_{\perp 328}$ and $\Delta A_{\parallel 328}$ show an immediate increase in absorbance after illumination, but the latter is very small. This component grows rapidly, however, in contrast to $\Delta A_{\perp 328}$, which first decreases and subsequently increases only slowly. Thus the dichroic ratio D_{328} ($= \Delta A_{\parallel 328}/\Delta A_{\perp 328}$) changes very rapidly initially and only approaches constancy after more than 20 min.

The results on rat retinae were less satisfactory: it was difficult to obtain dichroic ratios at 500 nm for dark-adapted retinae much greater than 2 in the present experimental conditions. Moreover, the rat preparations showed an increase in scatter during the course of the experiment, which made data obtained in the UV particularly suspect.

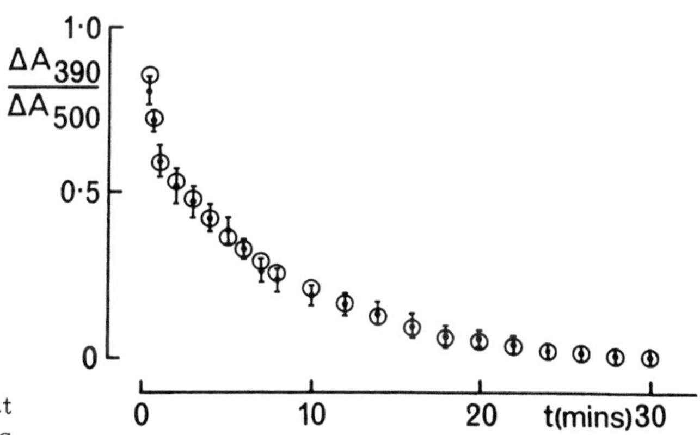

Fig. 3. The changes in absorbance at 390 nm in frog ROS following flash illumination, for axial transmission of unpolarized light (for 3 experiments) (○); and for transverse transmission of light polarized perpendicular to the long axis (mean of 4 experiments ± 1 S.D.) (Φ). Both sets of data have been normalized by reference to the permanent absorbance loss at 500 nm in the same conditions

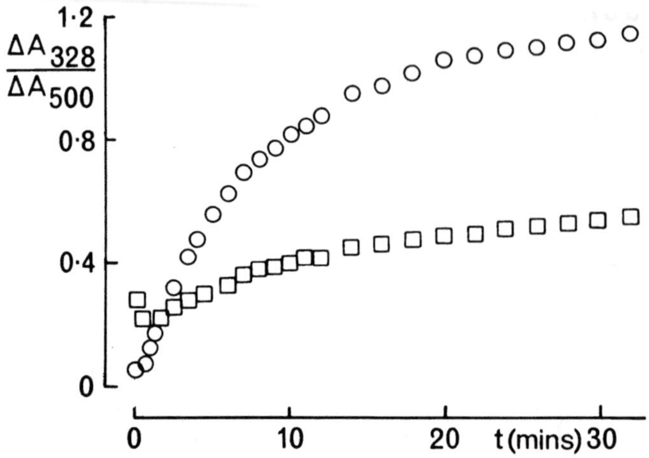

Fig. 4. Normalized absorbance changes in frog ROS at 328 nm following flash bleaching, for light polarized parallel (O) and perpendicular (□) to the long axis

For this reason, absorbance changes could only be followed profitably at 390 nm and 480 nm. At these wavelengths the results obtained were qualitatively similar to those obtained with frog retinae.

Baumann (9) has analyzed the kinetics of the long-lived decay intermediates in the frog retina. He attributes the changes in absorbance at 480 nm to the presence of MIII, those at 380 nm principally to MII and retinal, and those at 328 nm to retinol (and, to a lesser extent, MII and retinal). On this basis, and the assumptions that MII, MIII and retinal have the same orientation in the ROS as does rhodopsin, and that retinol is aligned axially, he showed that only one of the four kinetic schemes he tested was compatible with experiment. The present results on frog ROS justify his assumption about the orientations of the intermediates. Thus the data illustrated in Fig. 2, which reflect the changes in MIII concentration, indicate that this species is aligned in much the same plane as rhodopsin, and those of Fig. 3 are consistent with the idea that all the species absorbing appreciably at 390 nm have similar orientations. In contrast, Figs. 1 and 4 show clearly that retinol, which gives rise to the absorbance at 328 nm, is orientated at a large angle to the rhodopsin chromophore. The behaviour of $\Delta A_{\perp 328}$ in Fig. 4 further indicates that this initial absorbance increase arises from species other than retinol, which are aligned perpendicular to the ROS long axis.

On the basis of the qualitative similarity between the dichroism changes during bleaching for both frog and rat retinae, kinetic analyses of the kind applied to the frog by Baumann (9) have been extended to the rat. Using experimental techniques described elsewhere (3,4), the axial spectral changes in isolated rat retinae at 30°C were measured following intense light exposure. The four kinetic models described by Baumann (9), plus a further one suggested by Cone and Cobbs (10) from ERP measurements on rat were then tested for compatibility with these changes. Only two schemes were consistent with the data and the known rapid formation of retinol in the rat retina (11). The first was the one which Baumann finds applicable to the frog:

$$\text{Rhodopsin} \dashrightarrow \text{MII} \xrightarrow{k_1} \text{MIII} \xrightarrow{k_3} \text{Retinal} \xrightarrow{k_4} \text{Retinol;} \quad (\text{MII} \xrightarrow{k_2} \text{Retinal})$$

where: $k_1 \sim 3 \cdot 10^{-3} sec^{-1}$; $k_2 \sim 1 \cdot 10^{-3} sec^{-1}$; $k_3 \sim 5 \cdot 10^{-4} sec^{-1}$; $k_4 \sim 4 \cdot 10^{-3} sec^{-1}$, and the second was the sequence proposed by Cone and Cobbs, in which MIII decays directly to retinol. Further experimentation in different conditions may eliminate this ambiguity.

A correlation between the disappearance of MIII and the later stages of an adaptation process in the isolated trans-retinal PIII component of the e.r.g. of the perfused rat retina has recently been proposed (3,4). Using these kinetic analyses, preliminary comparisons between the time course of this adaptation (a slow increase in the maximum amplitude elicitable from PIII following extensive rhodopsin bleaching) and the time courses of intermediates other than MIII have been made. The results tentatively suggest that a more extensive correlation may exist between the electro-physiological changes and the formation of retinol, rather than the disappearance of specific intermediates in the decay sequence, but further work is necessary to establish whether or not this is the case.

Acknowledgement. The author wishes to thank the Medical Research Council for the provision of the Shimadzu spectrophotometer.

References

1. BAUMANN, CH., H. SCHEIBNER: The dark adaptation of single units in the isolated frog retina following partial bleaching of rhodopsin. Vision Res. 8, 1127-1138 (1968).
2. DONNER, K.O., T. REUTER: Dark adaptation processes in the rhodopsin rods of the frog's retina. Vision Res. 7, 17-41 (1967).
3. ERNST, W., C.M. KEMP: The role of metarhodopsin III in the recovery of the PIII photoresponse of isolated rat retina after an intense light exposure. In: Advances in Experimental Medicine and Biology, Vol. 24, ed. G.B. ARDEN, pp. 81-86. Plenum Press, New York (1972).
4. ERNST, W., C.M. KEMP: The effects of rhodopsin decomposition on PIII responses of isolated rat retinae. Vision Res. 12, 1937-1946 (1972).
5. SCHMIDT, W.J.: Polarisationsoptische Analyse eines Eiweiss-Lipoid-Systems erläutert am Aussenglied der Sehzellen. Kolloid -Z. 85, 137-148 (1938).
6. DENTON, E.J.: The contributions of the orientated photosensitive and other molecules to the absorption of whole retina. Proc.Roy.Soc. B 150, 78-94 (1959).
7. LIEBMAN, P.A.: In situ microspectrophotometric studies on the pigments of single retinal rods. Biophys.J. 2, 161-178 (1962).
8. WALD, G., P.K. BROWN, I.R. GIBBONS: The problem of visual excitation. J.Opt.Soc.Amer. 53, 20-35 (1963).
9. BAUMANN, CH.: Kinetics of slow thermal reactions during the bleaching of rhodopsin in the perfused frog retina. J.Physiol. (Lond.) 222, 643-663 (1972).
10. CONE, R.A., W.H. COBBS: Rhodopsin cycle in the living eye of the rat. Nature (Lond.) 221, 820-822 (1969).
11. DOWLING, J.E.: Chemistry of visual adaptation in the rat. Nature (Lond.) 188, 114-118 (1960).

Discussion

S.L. Bonting: What is the alternatively acceptable bleaching sequence to the one proposed by Baumann?

C.M. Kemp: The alternative scheme (that of Cone and Cobbs, 10) is

Rhodopsin ---> MII → MIII → Retinol
MII → Retinal → Retinol

S.L. Bonting: If "retinal" is still orientated in the membrane plane and retinol is perpendicular, then this reinforces the suggestion that "retinal" is mostly bound. If free it would, just as retinol, align itself according to phospholipid chains.

C.M. Kemp: One uses the term "retinal" loosely in this context: it is unlikely that it is, in fact, free.

G. Wald: Retinal is probably never free in the retina; and in metarhodopsin II it keeps its orientation as in rhodopsin. But retinol (vitamin A) is of course free. It represents the transport (migratory) form, ready to migrate from the outer segments into the pigment epithelium. Its orientation parallel to the outer segment axis probably means that it enters the lipid layers, lining up among the fatty acid chains of the phospholipid molecules, which lie parallel with the rod axis; and then it migrates in this orientation along the lipid layers.

C.M. Kemp: I agree. It is interesting that the final product of bleaching the frog retina in the presence of hydroxylamine, retinal oxime, is orientated axially rather than transversely (8).

Octopus Rhodopsin *in situ:* Microphotometric Measurements of Orientational and Spectral Changes

U. Täuber
Institut für Tierphysiologie der Ruhr-Universität Bochum, Bochum-Querenburg/W. Germany

The rhabdomeres of the octopus photoreceptor cells are built up by highly ordered microvilli structures, which are arranged perpendicular to the long axis of photoreceptor cells (1). Like rod outer segments, invertebrate photoreceptors show dichroic absorption, which is believed to play a part in the primary mechanism underlying the perception of linear polarized light (2). This paper will present some microphotometric data on the kinetics of reactions induced by unpolarized and polarized light in the photoreceptors of the cephalopod Eledone moschata (Fam. Octopidae).

Methods

Kinetic measurements on cross sections of the retina were performed with a one beam microphotometer (MPM Zeiss), which was used to measure the change in optical transmission perpendicular to the receptor layer at given wavelengths. The retina consists of two populations of receptor cells, the microvilli of which are perpendicular to one another. Fig. 1 demonstrates the orientation of the retina slice under the microscope. The 0°-direction of the incident polarized light is perpendicular to the axis of the photoreceptor cells, whereas the 90°-direction is parallel to it.

The concept of solution photometry leads to the following expression for the light J, falling on the photomultiplier tube:

$$J = J_o \exp(-(E_G + E_R + E_M))$$

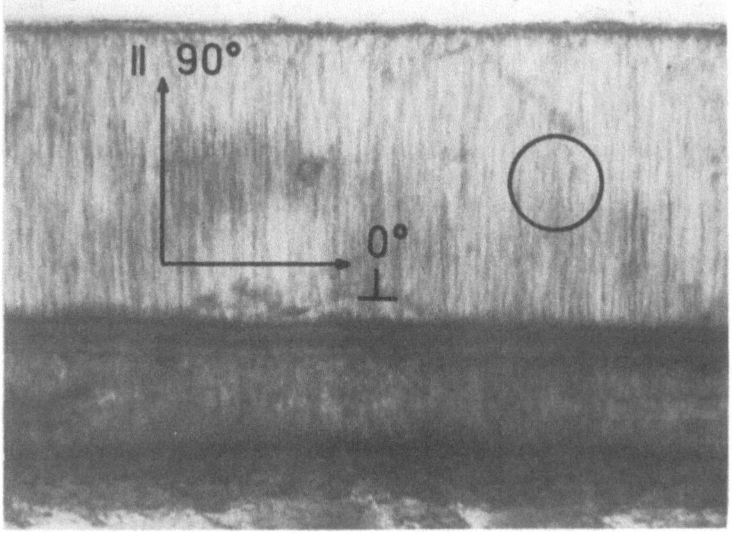

Fig. 1. Retina slice. Circle shows diameter of lightspot, 100 µm

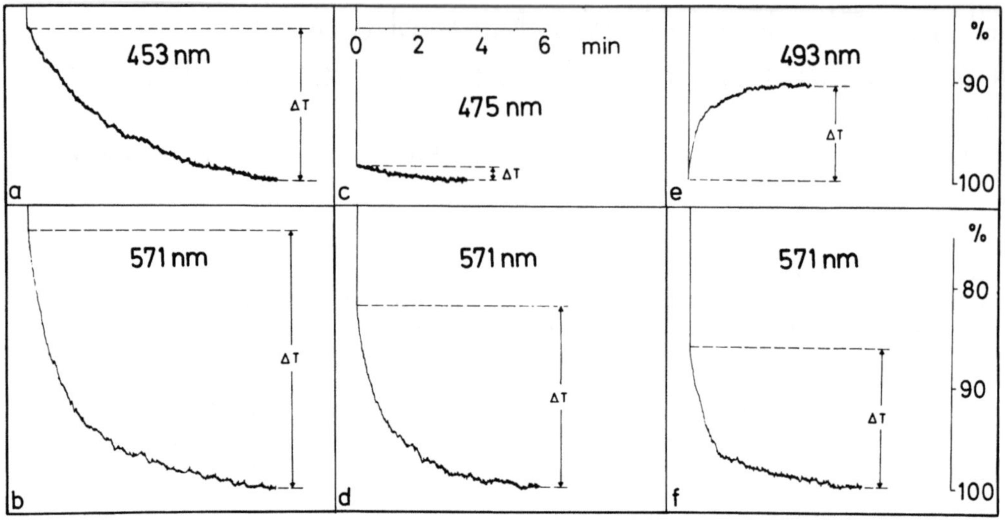

Fig. 2. Original registrations of transmission changes in a retinal slice during illumination with light of different wavelengths. Illumination by 453 nm after dark-adaptation (a), afterwards by 571 nm (b); further illumination by 475 nm (c), afterwards by 571 nm (d), further illumination by 493 nm (e), afterwards by 571 nm (f)

The total optical absorbance is the sum of the background absorbance of the tissue, the absorbance of rhodopsin and, depending on the adaptation state, another absorbance due to metarhodopsin. When the tissue absorbs dichroically, equation 1 is valid for both the transversal and the longitudinal direction. Hence:

$$J_{\parallel} = J_o \exp(-(E_{G_{\parallel}} + E_{R_{\parallel}} + E_{M_{\parallel}})); \; J_{\perp} = J_o \exp(-(E_{G_{\perp}} + E_{R_{\perp}} + E_{M_{\perp}})) \qquad 2$$

In the following measurements it is not the dichroic ratio N which has been determined (i.e. the quotient of the absorbances perpendicular and parallel to the receptor axis) but the dichroism D, derived from equations 2 as follows:

$$D = \ln \frac{J_{\parallel}}{J_{\perp}} = (E_{G_{\perp}} - E_{G_{\parallel}}) + (E_{R_{\perp}} - E_{R_{\parallel}}) + (E_{M_{\perp}} - E_{M_{\parallel}}) \qquad 3$$

Results

When a dark-adapted retinal slice is illuminated by sufficiently intense light of a given wavelength, a photoequilibrium results. This can be seen as a change of absorbance from that of rhodopsin to that of an equilibrium mixture (Fig. 2). Such absorbance changes are plotted as a function of wavelength in Fig. 3a (solid curve). There is a decrease in absorbance at shorter wavelengths and an increase at longer ones with a maximum at about 530 nm and with an isosbestic point at about 480 nm. The amounts of rhodopsin and metarhodopsin in the photoequilibria obtained by monochromatic illumination of the photoreceptor membranes agree very well with the values calculated from solution data (3) (Fig. 3b). Fig. 3a also shows the decrease in absorbance at 571 nm, which results when the equilibrium mixture is further exposed to light of 571 nm, sufficiently intense to reconvert all acid metarhodopsin to rhodopsin (dashed curve). 571 nm was chosen for these measurements, since only metarhodopsin absorbs light of this wavelength to any significant extent. Hence the dashed curve represents the equilibrium amounts of metarhodopsin that are formed by lights of different wavelengths.

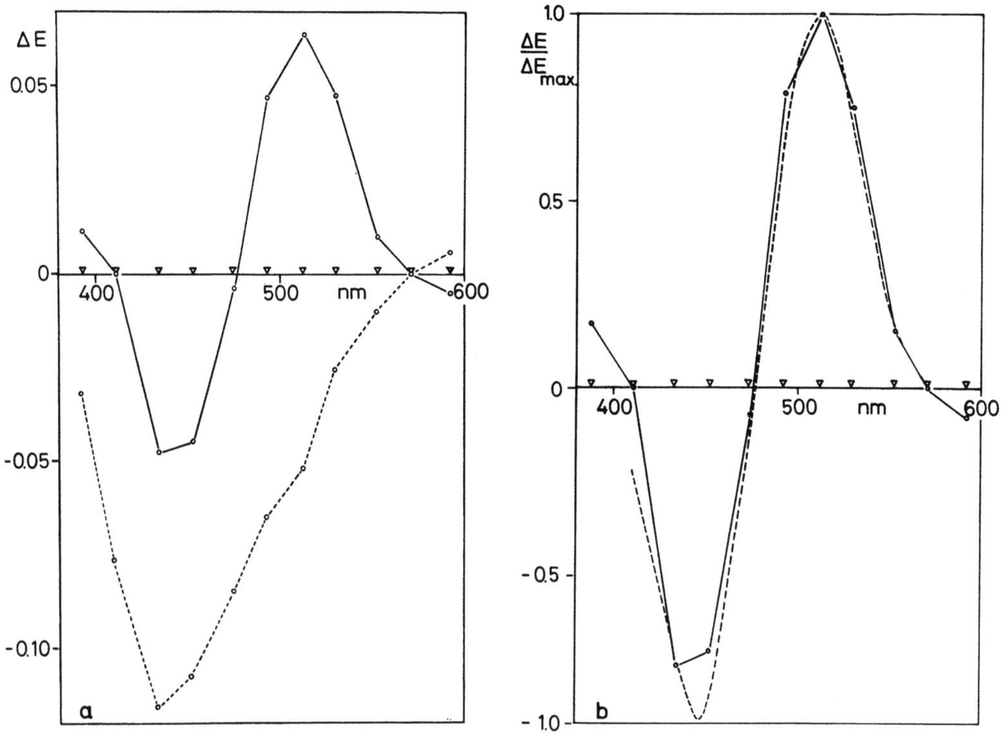

Fig. 3.a. Change of absorbance in a retina slice between dark-adapted state and light adaptation by the indicated wavelengths (solid line). Change in absorbance at 571 nm after a saturating illumination by indicated wavelengths (dashed line). Fresh preparate, pH 6.7. b. Comparison between the normalized spectrum of Fig. 3a (solid line) and a spectrum calculated from measurements on solution (dashed line)

It is known from behavioural experiments (1) that the octopus is able to discriminate the planes of linear polarized light. This ability must have its basis in the dichroic absorption of rhodopsin. Fig. 4a (solid curve) shows the dependence of the dichroism D of a dark-adapted retina slice on wavelength. It agrees very well with the absorption curve of rhodopsin (curve R in Fig. 4b) as would be expected for theoretical reasons.

Illumination by unpolarized light of 453 nm, which produces a certain mixture of rhodopsin and acid metarhodopsin (the absorptance of this mixture calculated from solution data is shown in Fig. 4b, dashed line), changes dichroism, in agreement with the shift of the absorption curve (Fig. 4a, dashed line). All preparations exhibit a negative dichroism at wavelength longer than about 580 nm, where neither rhodopsin nor metarhodopsin absorb ("form" dichroism of the tissue matrix?).

It must be concluded from Fig. 4a, that acid metarhodopsin absorbs dichroically like rhodopsin. For both substances maximal absorption is in a direction parallel to the long axis of the microvilli in the rhabdomeres. The appearance of metarhodopsin can therefore be detected by an increase of dichroism at longer wavelengths. Since the dichroism change parallels the change in absorption, the best wavelength

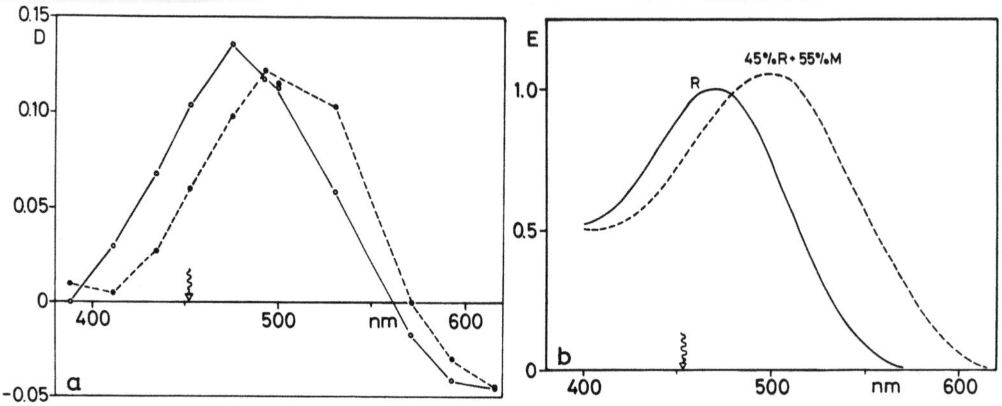

Fig. 4. a. Dichroism D in dependance on wavelength in a retina slice in dark-adapted state (solid line) and after illumination by 453 nm, unpolarized light (dashed line). Fresh preparate, pH 6.0. b. Absorbance spectra of rhodopsin (solid line) and of a mixture of rhodopsin and metarhodopsin obtained after illumination by 453 nm, unpolarized light, calculated from solution data (dashed line)

---

to observe dichroism changes should be 530 nm. Regarding the anisotropic absorption properties of rhodopsin and metarhodopsin, we must assume that the photoequilibria are not only dependent on the wavelength of irradiation but also on the plane of polarization. If the dichroic ratio is the same for rhodopsin and metarhodopsin, the quotient of the concentrations of rhodopsin and metarhodopsin and therefore the dichroism D, should be independent of the plane of the electric vector of the illuminating light and vice versa.

The dichroism D of a dark-adapted retina slice fixed by glutaraldehyde was measured at 388, 475, 530, and 616 nm (Fig. 5, solid line). Illumination with a strong polarized adapting light of 475 nm (15 sec) - the plane of the electric vector perpendicular to the long axis of photoreceptor cells - leads to a change in dichroism owing to the formation of a mixture of rhodopsin and acid metarhodopsin (Fig. 5, dashed curve 0°). After further illumination with light of the same wavelength, but with the plane of the electric vector parallel to the long axis of the receptor cells, an increase of dichroism is observed in the whole wavelength range (Fig. 5, dashed curve 90°).

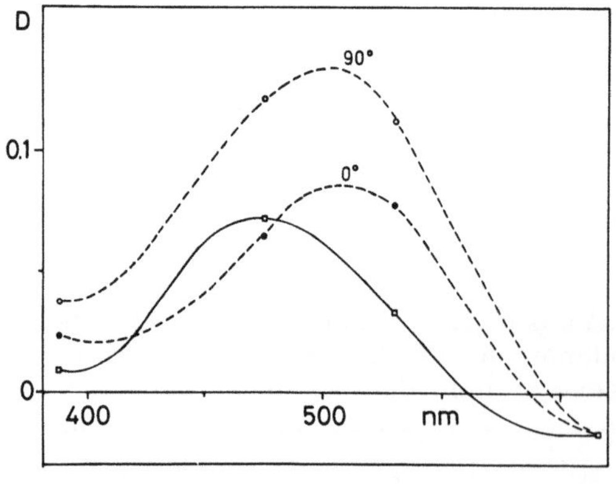

Fig. 5. Dichroism D of a retina slice in dark-adapted state (solid line) and after illumination by 475 nm (isosbestic point) under 0° and under 90°. Preparate fixed in 5% glutaraldehyde, pH 7.0

Discussion

As was shown by Hamdorf et al. (4) photoregeneration of visual pigments plays an important role in invertebrate vision. The agreement between the measured and calculated photoequilibria in the photoreceptor cells shows, that the over all quantum efficiency for the reaction from rhodopsin to metarhodopsin and for the backward reaction is the same. The amount of rhodopsin present in the microvilli membranes is not only determined by the wavelength of illumination, but also by the anisotropic absorption properties of both substances rhodopsin and acid metarhodopsin. Illumination with polarized light of a given wavelength leads to a higher concentration of that component, the absorption of which is less than that of the other component under the chosen angle of polarization.

References

1. MOODY, M.F., J.R. PARRIS: The discrimination of polarized light by Octopus: a behavioural and morphological study. Z.vergl.Physiol. 44, 268-291 (1961).
2. LANGER, H.: Nachweis dichroitischer Absorption des Sehfarbstoffes in den Rhabdomeren des Insektenauges. Z.vergl.Physiol. 51, 258-263 (1965).
3. HAMDORF, K., J. SCHWEMER, U. TÄUBER: Der Sehfarbstoff, die Absorption der Rezeptoren und die spektrale Empfindlichkeit der Retina von Eledone moschata. Z.vergl.Physiol. 60, 375-415 (1968).
4. HAMDORF, K., R. PAULSEN, J. SCHWEMER: Photoregeneration and sensitivity control of photoreceptors of invertebrates. This volume, pp. 155-166.

Studies on the Assembly of Rod Outer Segment Disc Membranes

Michael O. Hall, Scott F. Basinger, and Dean Bok
Jules Stein Eye Institute and Department of Anatomy, UCLA Center for the Health Sciences, Los Angeles, CA 90024/USA

The vertebrate photoreceptor cell outer segment is a stack of many hundreds of closely packed membranous discs, oriented at right angles to the long axis of the cell, and enclosed within the cell membrane. The photosensitive visual pigment molecules are believed to be embedded in these discs.

The membranous discs of the frog rod outer segment (ROS) are comprised of about equal quantities of protein and lipid (Table I). Recent studies (1,2) indicate that approximately 80 per cent of the protein in washed ROS is rhodopsin. The ROS has one of the highest lipid contents of any tissue. The major part (~ 74 per cent) of this lipid is phospholipid (Table I).

In 1967, in a series of elegant autoradiographic studies, Young showed that the ROS discs undergo continual renewal (3). Using biochemical techniques, it was subsequently shown that rhodopsin, an integral component of the ROS disc, is renewed as the discs are renewed (1). Since synthetic enzyme activities are excluded from the ROS, all components of the disc membranes must be synthesized in the inner segment and transported through the connecting cilium to the outer segment. Biochemical and autoradiographic studies have confirmed this idea. Rhodopsin synthesized on the ribosomes of the inner segment migrates through the connecting cilium to the outer segment base, where it is assembled into membranous discs. The newly-formed discs are continually displaced along the outer segment by newer discs, which are formed by repeated invaginations of the cell membranes. In order to maintain the length of the outer segment, discs at the apex of the rods are continually shed from the end of the cell and phagocytized by the cells of the pigment epithelium (PE). In the frog, the time taken to completely renew the outer segment of a red rod is about 8 weeks - or one complete disc every 40 minutes (1,3). In warm blooded animals such as the rat and rhesus monkey, the renewal time is much shorter - 9 days and 12 days respectively (3,4). Thus a considerable portion of the metabolic machinery of the photoreceptor cell must be geared to synthesizing the components of the ROS disc membrane - particularly rhodopsin and phospholipids.

We have studied the synthesis of rhodopsin in the inner segment of the photoreceptor cell, its transport to the outer segment and its assembly into the ROS disc membranes. At the same time we have investigated the synthesis of phospholipids in the retina, and their relationship to rhodopsin transport and incorporation into outer segment discs. Finally, we have tried to localize the site at which vitamin A (as retinal) is complexed with opsin to form the photosensitive rhodopsin molecule.

Rhodopsin

The recent studies of O'Brien *et al.* (5) on the biosynthesis of rhodopsin by the isolated bovine retina, have confirmed and extended the earlier *in vivo* results (1). The time course of rhodopsin synthesis, transport and incorporation into newly synthesized discs in this *in vitro* system closely followed that previously demonstrated

Table I. Phospholipid composition of frog rod outer segments

			Mole % Phosphorus
Total Protein	~50%		
Total Lipid	~50%		
Phospholipid	37%	Phosphatidyl Choline (PC)	45.3 ± .77
		Phosphatidyl Ethanolamine (PE)	37.3 ± .78
		Phosphatidyl Serine (PS)	11.3 ± .52
		Phosphatidyl Inositol (PI)	2.4 ± .43
		Sphingomyelin (Sph.)	1.6 ± .26

in the frog. The transfer of newly synthesized rhodopsin from the ribosomes of the inner segment to the outer segment was demonstrated in vitro in an elegant pulse-chase experiment (5). Current studies in our laboratory have shown that rhodopsin biosynthesis in isolated bovine and frog retinas has very similar metabolic requirements and follows the same time course. Autoradiographic studies using labeled amino acids have shown that radioactive protein migrates from the ribosomes to the Golgi apparatus (6). The reason for this is unknown, although it has been suggested that this might be the site at which sugars and retinal are complexed with opsin to form rhodopsin (1). Recently O'Brien has obtained some evidence that the addition of glucosamine to opsin might occur shortly after the assembly of the polypeptide chain (7). This observation supports the suggestion that the Golgi body may be the site of addition of the carbohydrate moiety of opsin, as it is known to be for some other proteins (8).

In none of the previous in vivo and in vitro studies, has it been clear where retinal is added to opsin to form the photosensitive rhodopsin molecule. This could occur either in the inner segment or in the outer segment, after opsin has been incorporated into newly forming discs. If the latter possibility were correct, the retinal could be added "actively" during new disc synthesis, or "passively" to vacant opsin molecules after a complete disc has been synthesized. This "passive" addition would presumably occur by the same mechanism functioning during the regeneration of bleached visual pigment. In order to investigate this problem, one large group of frogs was dark-adapted for 24 hours; a similar group was kept in constant light for 24 hours. At the end of the adaptation period, the animals were injected intravenously in either light or dark with vitamin A-11, 12-^{3}H-acetate (a gift from F. Hoffman La Roche). The animals were killed at intervals after injection, and the incorporation of ^{3}H- retinal into purified visual pigment was measured. The results are shown in Fig. 1. In animals kept in constant light, some rhodopsin becomes labeled very rapidly. This is presumably due to addition of ^{3}H-retinal to vacant opsin molecules in the partially bleached outer segments. However, in the completely dark-adapted retina, no vacant opsin molecules are present, thus eliminating any possibility of such an addition. Thus, in the dark, the only manner in which ^{3}H-retinal could appear in rhodopsin, would be by disc renewal. As new discs are assembled at the base of the outer segment, ^{3}H-retinal is incorporated into these discs as a part of the rhodopsin molecule.

As shown in Fig. 1, no labeled visual pigment can be isolated from the retinas of dark-adapted frogs prior to 6 hours after injection of the isotope. Since it only takes 1-2 hours for an opsin molecule to be synthesized in the inner segment, and transported to the outer segment, this delay of 6 or more hours probably reflects some

Fig. 1. Effect of light and dark on the incorporation of ^{3}H-vitamin A into rhodopsin

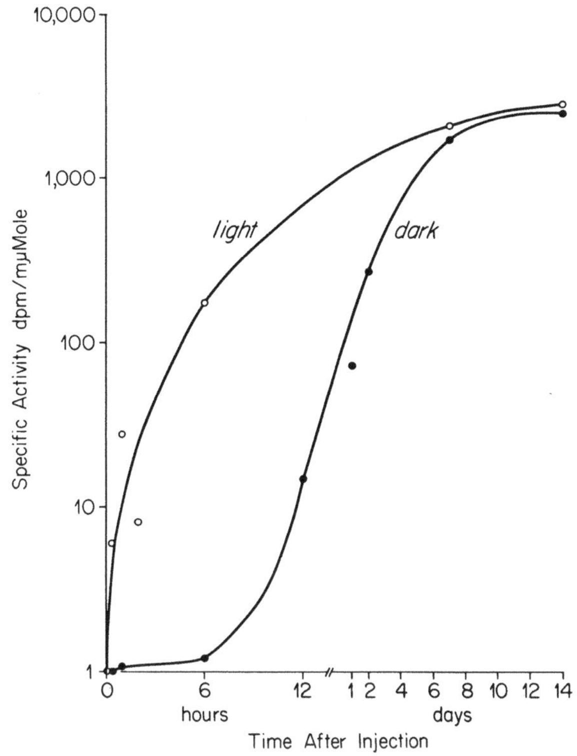

holdup of the retinal by the PE. In in vitro studies, in which the PE is removed prior to incubation of the retina, the time course of incorporation of retinal into rhodopsin closely follows the time course of rhodopsin synthesis. Neither of these experiments localize the site of addition of retinal to the inner or outer segment of the photoreceptor cell. They do however, conclusively show that retinal can be incorporated into rhodopsin under conditions in which light-induced exchange is eliminated. Further evidence comes from studies on the biosynthesis of opsin. When ^{3}H-leucine is injected into frogs (as a precursor of rhodopsin) small amounts of labeled rhodopsin can be extracted from outer segment preparations prior to the autoradiographic appearance of labeled molecules in the outer segment discs (1). Since the outer segment preparations always show some contamination by inner segment fragments, this suggests that the inner segment is the site at which retinal is added to opsin to form the photosensitive rhodopsin molecule. Following its synthesis in the inner segment, rhodopsin would then be assembled into the newly forming discs at the base of the outer segment.

Phospholipids

Lipids comprise about 50 per cent of the dry weight of the retina (Table I). Of this total, about 75 per cent is phospholipid. It is becoming increasingly evident that in addition to their structural role, phospholipids play a vital role in membrane function. For example, it has recently been shown that phospholipids are necessary for rhodopsin regeneration in vitro (9,10). In addition they provide a suitable environment which allows for precise orientation of the amphiphilic rhodopsin molecule (11). The hydrophobic forces between rhodopsin and certain phospholipids are

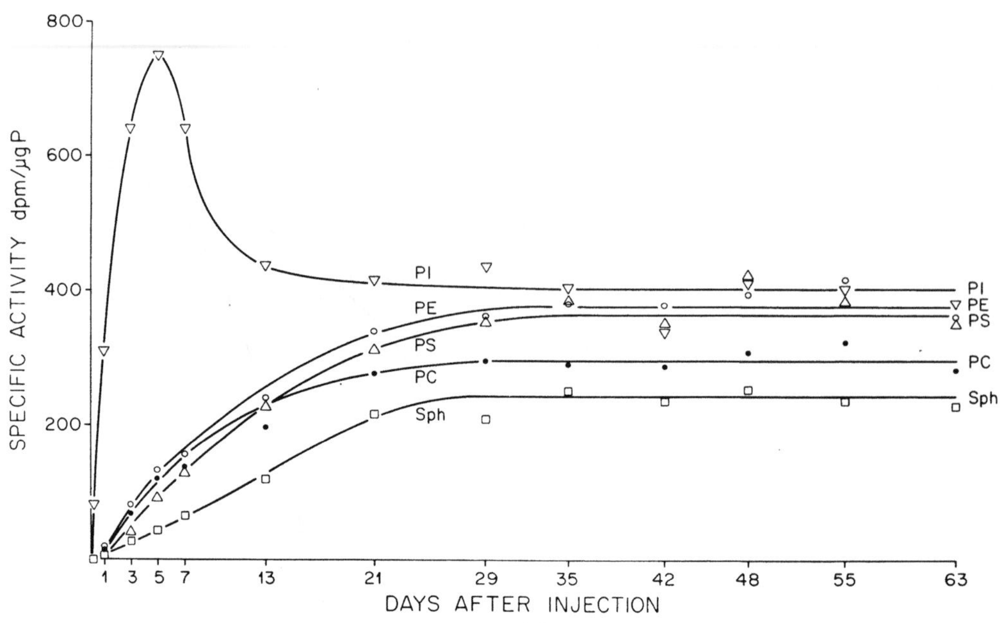

Fig. 2. Time course of incorporation of $H_3{}^{33}PO_4$ into rod outer segment phospholipids

---

so strong that they can only be broken by strongly ionic detergents (e.g. cetyltrimethylammonium bromide and sodium dodecyl sulphate) or denaturing solvents (e.g. chloroform-methanol).

We have studied the incorporation of phospholipids into ROS disc membranes, and their stability (or turnover) once they form a part of the membrane. We have also tried to answer the question of whether the incorporation of phospholipids into the disc membrane is dependent on the synthesis and incorporation of rhodopsin into the ROS disc.

Groups of frogs were injected intravenously in the light with either $H_3{}^{33}PO_4$, methyl-^{3}H-choline, or a mixture of ^{14}C-leucine and methyl-^{3}H-choline. At intervals after injection the frogs were sacrificed and the ROS were isolated. In double-label experiments with ^{3}H-choline and ^{14}C-leucine, a portion of the ROS was solubilized in 0.04 M cetyltrimethylammonium bromide (CTAB) in 67 mM phosphate buffer, pH 7.1, and the rhodopsin was purified by agarose-gel chromatography (12). In all experiments the phospholipids were extracted and separated by thin layer chromatography (13). The purified phospholipids were scraped off the plate and analyzed for phosphorus and radioactivity. The retinal residue remaining after separation of the ROS was treated in an identical manner.

When $H_3{}^{33}PO_4$ is used as a phospholipid precursor, the ROS phospholipids are only slowly labeled - with the notable expection of phosphatidyl inositol (PI) which undergoes a relatively rapid turnover (Fig. 2). The specific activity (SA) of the other ROS phospholipids increases slowly until 3 or 4 weeks after injection, then remains steady for the duration of the experiment. This indicates that when inorganic phosphate (Pi) is used as a tracer, the ROS phospholipids are stable once they are incorporated into the disc membrane structure. (We have previously shown that opsin undergoes no turnover once it is incorporated into the disc membrane structure (1).) An exception to this is PI, the SA of which increases rapidly for the

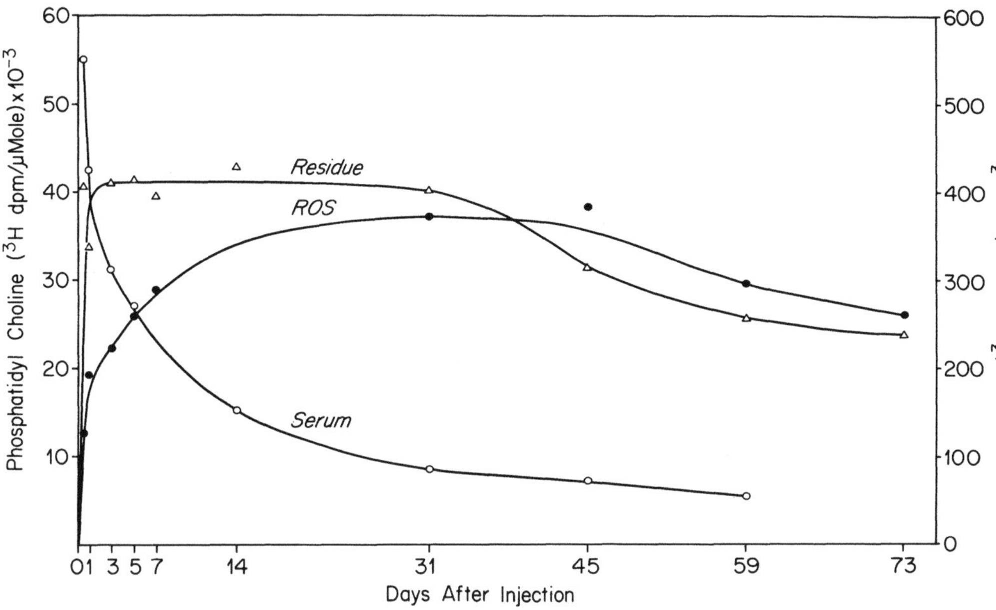

Fig. 3. Time course of incorporation of methyl-^{3}H-choline into retinal phosphatidyl choline

first 5 days after injection, then drops rapidly and eventually reaches a steady level at about 3 weeks after injection. The reason for this rapid turnover of PI is not known, but it is presumably related to some active role for this phospholipid in ROS function.

We have conducted a similar experiment using methyl-^{3}H-choline as a precursor, in which we studied the incorporation and turnover of phosphatidyl choline (PC) in ROS discs (Fig. 3). As with ^{33}Pi, the level of serum choline remains high for an extended period after injection, making a true pulse dose impossible. The SA of PC in both the retinal residue fraction, and in the ROS starts to drop between 4.5 and 6.5 weeks after injection, suggesting that some turnover of the base moiety of PC might be occuring. Choline is incorporated into outer segment PC much more rapidly than is Pi, and in fact approximates the rate of incorporation of leucine into rhodopsin during the first few hours after injection (Fig. 4). It thus seemed likely that rhodopsin and phospholipids might be co-incorporated into ROS discs. In order to study this, the following _in vitro_ experiment was conducted. Frog retinas were incubated for 30 and 60 minutes in the dark with 50 μg/ml of puromycin. At the end of the preincubation period, ^{14}C-leucine and ^{3}H-choline were added to each flask, and the incubation was continued for a total of 5 hours. The ROS were then isolated and the incorporation of ^{14}C-leucine into purified rhodopsin and ^{3}H-choline into purified PC was measured (Table II). While puromycin resulted in an 80 per cent inhibition of rhodopsin biosynthesis, the synthesis and incorporation of PC into ROS discs was inhibited by only 20-30 per cent. Since puromycin does not have any direct effect on phospholipid biosynthesis, the observed decrease in the incorporation of PC into the ROS disc must be related to the inhibition of rhodopsin biosynthesis. It thus appears that about 30 per cent of the PC is co-incorporated with rhodopsin into the disc membranes. When ROS are delipidated with non-bleaching organic solvents, more than half of the phospholipids are

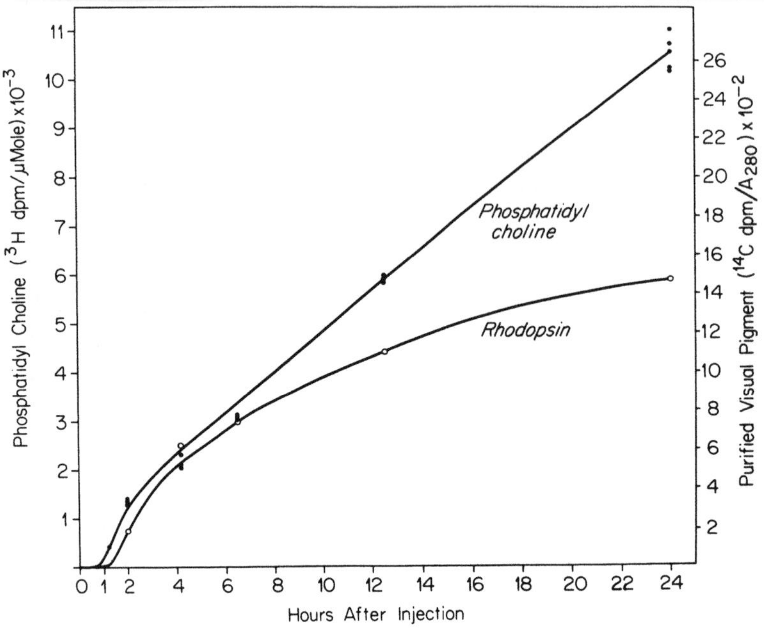

Fig. 4. Specific activity of outer segment rhodopsin and phosphatidyl choline as a function of time after injection of a mixture of ^{14}C-leucine and methyl-^{3}H-choline

very resistant to extraction (9). Strong detergents, such as CTAB, will completely delipidate rhodopsin (12,14), while others such as Emulphogene BC leave about 40 per cent of the phospholipid attached to rhodopsin (10). It is tempting to speculate that this tightly bound "core" of phospholipid might be that fraction whose incorporation into the disc membrane appears to be dependent on the incorporation of rhodopsin. This fraction might also contain those phospholipids which play a role in the regeneration of rhodopsin following bleaching (9,10). The final answer to some of these problems will only come when it is possible to isolate one or more rhodopsin "precursors" from the inner segment of the photoreceptor cell. This precursor can then be examined for the presence of retinal and for specific phospholipids.

Table II. Effect of puromycin on the incorporation of ^{14}C-leucine into rhodopsin and methyl-^{3}H-choline into phosphatidyl choline by the isolated retina

Puromycin	Pre-incubation (min)	^{3}H-choline into phosphatidyl choline (per cent)	^{14}C-leucine into rhodopsin (per cent)
None	30	100	100
50 µg/ml	30	79	23
50 µg/ml	60	67	15

Acknowledgement. The expert technical assistance of Rosemary Hoffman and Dennis Muehlenbach is gratefully acknowledged. This work was supported by United States Public Health Service Grants EY-00046, EY-00444, and EY-00331, and by a Fight for Sight Grant-In-Aid of the National Council to Combat Blindness, Inc., New York, New York.

References

1. HALL, M.O., D. BOK, A.D.E. BACHARACH: Biosynthesis and assembly of the rod outer segment membrane system. Formation and fate of visual pigment in the frog retina. J. Mol. Biol. 45, 397-406 (1969).
2. ROBINSON, W.E., A. GORDON-WALKER, D. BOWNDS: Molecular weight of frog rhodopsin. Nature New Biology 235, 112-114 (1972).
3. YOUNG, R.W.: The renewal of photoreceptor cell outer segments. J. Cell Biol. 33, 61-72 (1967).
4. YOUNG, R.W.: The renewal of rod and cone outer segments in the rhesus monkey. J. Cell Biol. 49, 303-318 (1971).
5. O'BRIEN, P.J., C.J. MUELLENBERG, J.J.B. de JONG: Incorporation of leucine into rhodopsin in isolated bovine retina. Biochemistry 11, 64-70 (1972).
6. YOUNG, R.W., B. DROZ: The renewal of protein in retinal rods and cones. J. Cell Biol. 39, 169-184 (1968).
7. O'BRIEN, P.J.: Incorporation of glucosamine into rhodopsin in vitro. Association for research in vision and ophthalmology. Sarasota, Florida (1972).
8. NEUTRA, M., C.P. LEBLOND: Radioautographic comparison of the uptake of galactose-^{3}H and glucose-^{3}H in the golgi region of various cells secreting glycoproteins or mucopolysaccharides. J. Cell Biol. 30, 137-150 (1966).
9. SHICHI, H.: Biochemistry of visual pigments. II. Phospholipid requirement and opsin conformation for regeneration of bovine rhodopsin. J. Biol. Chem. 246, 6178-6182 (1971).
10. ZORN, M., S. FUTTERMAN: Properties of rhodopsin dependent on associated phospholipid. J. Biol. Chem. 246, 881-886 (1971).
11. CONE, R.A.: Rotational diffusion of rhodopsin in the visual receptor membrane. Nature New Biology 236, 39-43 (1972).
12. HELLER, J.: Structure of visual pigments. I. Purification, molecular weight, and composition of bovine visual pigment$_{500}$. Biochem. 7, 2906-2913 (1968).
13. ANDERSON, R.E., M.B. MAUDE: Phospholipids of bovine rod outer segments. Biochemistry 9, 3624-3628 (1970).
14. HALL, M.O., A.D.E. BACHARACH: Linkage of retinal to opsin and absence of phospholipids in purified frog visual pigment$_{500}$. Nature 225, 637-638 (1970).

Discussion

M.W. Bitensky: It seems that incorporation of new components into membrane could occur at the junction of the inner and outer segments without obligatory migration through the connecting cilium. It seems just as plausible a route and would simply mean that membrane elongation occurs more closely to the inner segment. The connecting cilium would seem a necessary route for ATP and other metabolites.

M.O. Hall: Electron microscope evidence would argue against this idea - although would not disprove it. We interpret the most basal discs which are continuous with the plasma membrane as "growing" from an infolding of this membrane at some point distal to the connecting cilium. The exact site at which proteins (including rhodopsin) and lipids are added to the plasma membrane to form a ±100 Å thick membrane is of course unknown. However, the autoradiographic evidence suggests that proteins synthesized in the inner segment are transported through the connecting cilium before incorporation into disc structure.

S.L. Bonting: In the differential inhibitory effect of puromycin on phospholipid synthesis and rhodopsin synthesis, is it necessary to assume that the small inhibition of PL synthesis means association of PL with rhodopsin?

M.O. Hall: Since puromycin has no direct effect on phospholipid biosynthesis, it seems that the 30% decrease of PC incorporation into ROS discs must be coupled to the inhibition of rhodopsin synthesis - which in these experiments was inhibited about 80%.

G. Wald: Thomas Richardson in our laboratory has found the cytoplasmic space in the inner segments of mammalian rods just at the base of the outer segments to be rich in clumps of ribosomes (polysomes). Alexander Rich (MIT) has shown a close correlation between the sizes of polysomes and the molecular weight of protein synthesized. These rod polysomes contain about 12 ribosomes, agreeing well with the synthesis of a protein weighing about 40.000 g/mole - probably opsin.

The greater speed of incorporation of radioactivity from vitamin A in outer segments of light adapted eyes of frogs as compared with dark adapted ones may go with the much closer contact between the outer segments and the pigment epithelium processes in the light adapted eyes.

M.O. Hall: Some studies we have done on the uptake of injected vitamin A by the pigment epithelium indicate that the rate of uptake is the same in light and dark adapted eyes. However the release of vitamin A to the photoreceptor cell may be influenced by the state of light adaptation of the eye - or by other factors which we don't yet understand.

Enzyme and Phospholipid Patterns in the Compound Eye of Insects

Klaus M. Weber and Dietmar Zinkler
Institut für Tierphysiologie, Ruhr-Universität Bochum, Bochum-Querenburg/W. Germany

Insect visual pigments have been recently studied in detail (1,2). However, other compounds of the insect eye, such as enzymes and phospholipids, have received little attention. We have attempted to supplement this neglected aspect of insect visual physiology, with particular emphasis on comparison with visual systems in other animals.

Illumination of isolated eyes of Calliphora erythrocephala results in an increase of oxygen consumption and of monosaccharide content, while the glycogen content decreases (3-6). Also, an increased turnover of phosphate occurs, especially in the phosphatide fraction, along with an intensified degradation of ATP (7). These results have been interpreted as a consequence of elevated metabolic activity of the sensory cells when illuminated. However, this interpretation is uncertain because there is more than one cell type in the insect eye.

We have examined the metabolic activity in pigment and Semper cells by means of histochemical localization of certain key enzymes of biochemical pathways or of important metabolites. A first step was the demonstration of the presence of glycogen within the retina of light- and dark-adapted flies.

The white eyed mutant "chalky" of Calliphora was used in all experiments. This mutant is unusually well suited for histochemistry because the normal screening pigments are lacking.

Histochemical staining of glycogen using the PAS-reaction controlled by diastase treatment revealed that remarkable amounts of this compound are only to be found within pigment cells (Fig. 1a). The eyes of light-adapted animals show less glycogen than dark-adapted ones (Fig. 1b). The decrease of glycogen content of whole eyes under light conditions was interpreted formerly as the consequence of a functional load of the sensory cells which lose their glycogen. That glycogen is localizable only within the pigment cells indicates a metabolic connection between pigment and visual cells in the fly, involving nutritive supplementation by the pigment cells.

The distribution of some hydrolases within the retina of Calliphora also favours the interpretation of a nutritive function for the pigment cells. We found (8,9) that non-specific esterase (Enzyme Commission No. 3.1.1.1.) reacts within pigment cells and at tracheoles. Non-specific acid phosphatase (E.C. No. 3.1.3.2.) is similarly distributed, except at tracheoles. There is remarkable activity within Semper cells also, where some reaction product seems to be in a lysosomal form. Non-specific alkaline phosphatase (E.C. No. 3.1.3.1.) shows only weak activity within pigment cells.

Special attention was focused on specific alkaline phosphatases, mainly the ATPase (E.C. No. 3.6.1.3.). One reason for this interest arose from the biochemical findings of Langer (7), who reported a decrease in ATP in Calliphora eyes after illu-

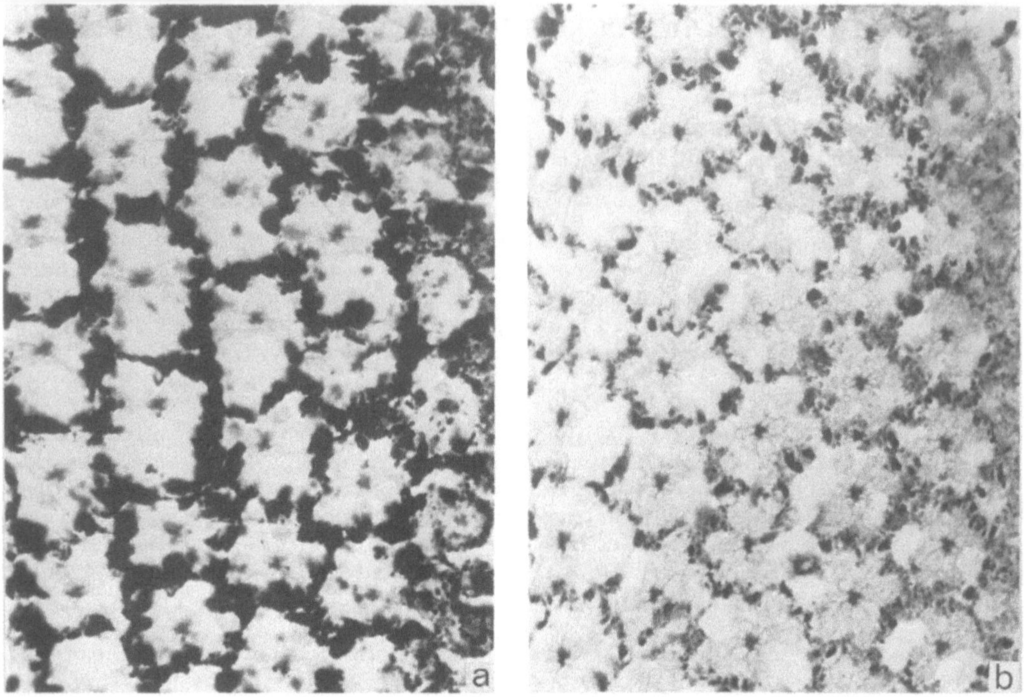

Fig. 1. Cross sections of eyes of <u>Calliphora</u> kept in the dark (a) and kept under daylight conditions (b) for 24 hrs. - a) Glycogen is located in pigment cells surrounding the sensory cells. b) Glycogen content is diminished. - The positive reaction within the interommatidial space is not due to glycogen. PAS-reaction, 7 µm, 550 x

mination, paralleled with an increase in ADP and inorganic phosphate. Part of the latter could be caused by the activity of non-specific acid phosphatase which shows a stronger reaction in the eyes of light-adapted animals than in dark adapted ones (8). The more important question with respect to visual excitation and its connection with ATP-consuming cation transport at photoreceptor membranes is, however, where in the retina hydrolysis of ATP takes place.

ATPase, as tested by the Wachstein and Meisel procedure, has been found only within pigment cells (8,9). This reaction, which occurs optimally above pH 8.0, seems to be insensitive to calcium but is stimulated by manganese. The result is in agreement with the fact that hydrolases are predominant within pigment cells. (Only within pigment cells could we find in addition aminopeptidase and α-glucosidase). A modification of the Wachstein and Meisel procedure led to the visualization of the reaction product also at rhabdomes. The medium employed successfully to demonstrate ATPase at rhabdomes is essentially equivalent to that used by Schulze and Wollenberger (10). These authors utilized Na-K-tartrate for the chelation of lead. With tartrate the medium remains generally clear during incubation, especially at higher pH. Thus artificial deposits of lead phosphate originating from spontaneous hydrolysis of ATP catalysed by free lead are diminished. We believe that the weak chelation of lead by tartrate reduces the spontaneous hydrolysis of ATP and also reduces the ATPase inhibiting effect of lead.

The histochemical reaction is not significantly stimulated by sodium and potassium and is only weakly inhibited by ouabain. With calcium there is a complete suppression and with manganese, a strong inhibition.

Thus, ATPases with different properties were demonstrable in different structures within the retina of the blow fly. However, quantifying these enzymatic reactions especially with respect to functional load on the organ by histochemical means is a very uncertain enterprise. Therefore Dr. Maria E. Rivera in our institute started to quantify and to specify the properties of the ATPases of Calliphora retinae by biochemical methods, by analogy to the histochemical procedure, using the method of Bonting and Caravaggio (11).

In homogenates of retinae consisting of only sensory, pigment, and Semper cells, Mg-activated ATPase was found in an amount of 0.70±0.09 µmoles P_i/mg protein/hr with a pH optimum of 8.5. Na-K-activated (transport) ATPase was present in an amount of 0.50±0.1 µmoles P_i/mg protein/hr with a pH optimum of 7.0.

The pH optimum of the Mg-ATPase corresponds well with that of the histochemically found cytoplasmic ATPase, whereas the pH optimum of Na-K-ATPase does not. The histochemically demonstrable rhabdome-ATPase shows a pH optimum at pH 7.6.

Lead in a concentration of 5 mM, in the same amount as used in the histochemical medium, suppresses ATPase activity by as much as 12%, but with 30 mM Na-K-tartrate this suppression is diminished and 29% of the former activity is regained. Tartrate per se has a stimulating effect, especially upon the Na-K-ATPase activity, as tested in the homogenate. These facts are in agreement with the accepted protective effect of tartrate by weak chelation of lead. Experiments with calcium also confirm the histochemical observations.

The other part of our investigations - phospholipid pattern in insect eyes - was prompted by the fact that phospholipids are believed to have not only a structural but also a functional role in biological membranes. For example, it has been shown recently that some structurally bound enzymes like transport-ATPase require phospholipids for their function (12). It also has been shown that phospholipids, though structural components, are metabolically active. Light-adapted eyes of blow flies have an increased turnover of phospholipids, as measured after incorporation of radioactive phosphorus (7).

A survey of the distribution of lipids in the eye of Calliphora was obtained histochemically by the Baker test. The rhabdomeres are especially intensively stained owing to the closely packed microvilli which form these structures (Fig. 2).

At the present state of our investigations, effective procedures for isolating insect rhabdomeres are lacking. Therefore, the analysis of whole organs constitutes the most attractive means for initial comparison of phopholipid content and composition.

Lipids were extracted (13) and separated by one or two-dimensional thin layer chromatography (TLC) on pre-coated silica gel 60 F_{254} plates (Merck). The quantitative analysis of phospholipids was based on chemical estimations of lipid phosphorus present in different spots (14). For comparative studies in these experiments we used not only the blow fly but also the moth Deilephila elpenor.

The phospholipid content of the blow fly was 7.31 µg lipid phosphorus /mg protein and that of the moth, 10.51 µg lipid phosphorus /mg protein. The eye tissues of blow

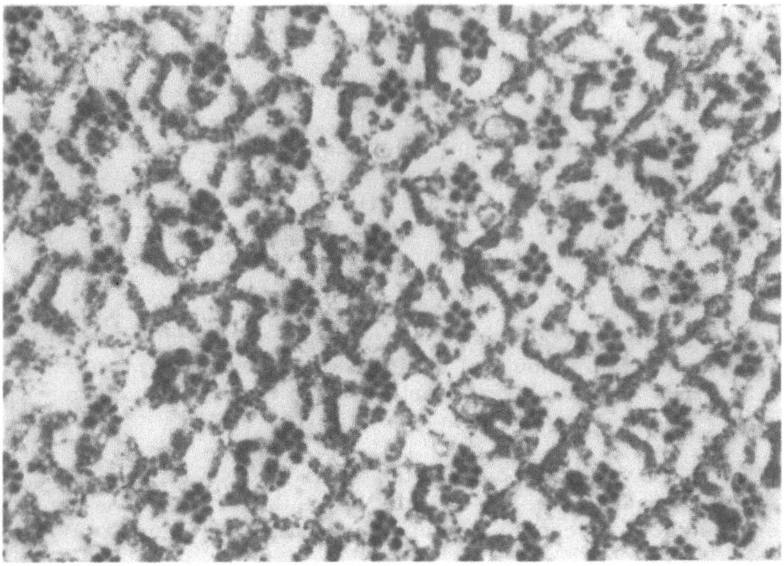

Fig. 2. Cross section of an eye of Calliphora. Black staining represents phospholipids. Baker test, 5 µm, 1500 x

fly and moth show similarity in their qualitative pattern but are quantitatively quite different (Table I).

The predominance of phosphatidyl ethanolamine (PE) is a characteristic of the Calliphora retina. The second major phospholipid is phosphatidyl choline (PC). Minor components are phosphatidyl inositol (PI), phosphatidyl serine (PS), lysophosphatidyl choline (LPC), lysophosphatidyl ethanolamine (LPE), diphosphatidyl glycerol (DPG), and an unidentified compound X. The phospholipid spectrum in Deilephila resembles that in mammals in that PC and PE are present in approximately equal proportions. Unusual phospholipid patterns in flies in comparison to other insects and to vertebrates have been shown in whole animal extracts as well (15,16). The difference therefore is correlated with species differences and not with eye function.

On the basis of these data we decided to examine the turnover of different phospholipids in Calliphora using radioactive precursors such as ^{32}P and ^{14}C-ethanolamine.

Table I. Phospholipid composition of insect eyes (per cent of total lipid-phosphorus)

	Blow fly	Moth
PE	57.8	38.9
PC	26.2	40.1
PS	6.1	6.9
PI	4.3	1.4
DPG	1.4	2.7
LPE	2.4	1.8
LPC	0.6	6.4
X	1.2	1.8

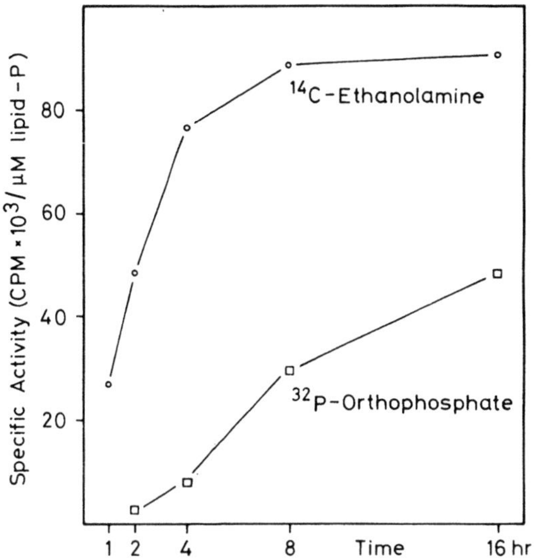

Fig. 3. Incorporation of ^{14}C-ethanolamine and ^{32}P-inorganic phosphate into phospholipids of Calliphora eyes

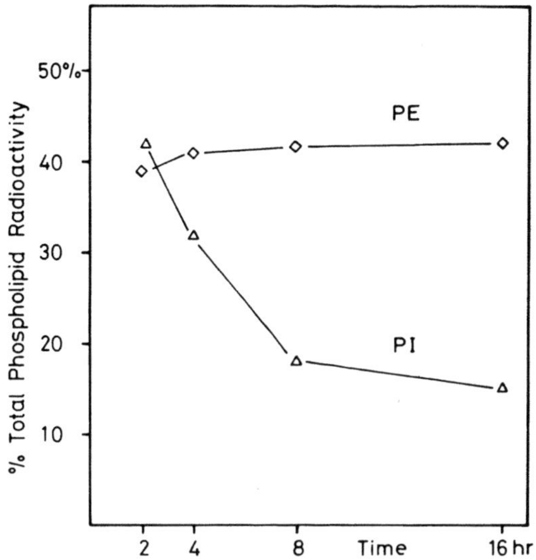

Fig. 4. Incorporation of ^{32}P-inorganic phosphate into phospholipid classes of Calliphora eyes

The results obtained are given in Fig. 3 and 4. The most active incorporation of ^{32}P was observed within phosphatidyl inositol. In comparison to other phospholipids this compound seems to undergo a relatively rapid turnover. In general, the incorporation of inorganic phosphate into phospholipids proceeded slowly, whereas the incorporation of ethanolamine label was much more rapid and extensive.

After only 4 hrs about 35% of the injected ^{14}C-ethanolamine dose was recovered in the lipids of the thorax and abdomen, and 1-2% in the lipids of the eyes.

A time dependent incorporation rate is also recognizable after autoradiography of eyes labeled with ^{14}C-ethanolamine and treated with a phospholipid stabilizing fixative. The morphological method, on the other hand, demonstrates a specific distribution within the retina (Fig. 5).

In general, a homogeneous labeling of the tissue takes place. Only over Semper cells are there more grains to be seen. We are certain that the label represents mainly PE; as ascertained by two-dimensional TLC, an average of 89% of the total recovered radioactivity was found to be associated with PE. As we cannot yet decide whether or not the rhabdomes are predominately labeled, experiments with ^{3}H-precursors are necessary.

At present, we have only a few results concerning the influence of adequate stimulation by light upon phospholipid turnover in Calliphora retinae. In agreement with Langer (7), we found a considerable increase of ^{32}P-incorporation into the overall phospholipid content of eyes of light-adapted flies. This increased incorporation took place in all phospholipid classes. Also with ^{14}C-ethanolamine we found an increased incorporation (28%) after illumination. This seems to be correlated specifically with eye function, because in light- and dark-adapted flies the incorporation into thorax and abdomen is the same.

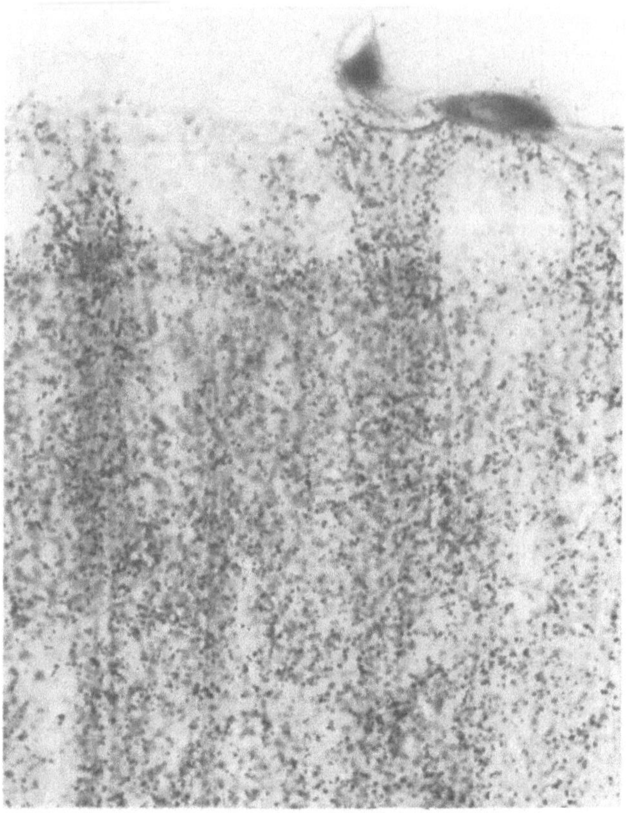

Fig. 5. Radioautograph of a longitudinal section of an eye of Calliphora injected with 1 µCi ^{14}C-ethanolamine and sacrificed after 16 hrs. Over Semper cells there are more grains than over rhabdomes or cytoplasm of sensory and pigment cells. Some other accumulations of grains are over Baker-positive globules of unknown significance. AR 10 stripping film exposed for 25 days, 3 µm, 700 x

References

1. HAMDORF, K., R. PAULSEN, J. SCHWEMER: Photoregeneration and sensitivity control of photoreceptors of invertebrates. This volume, pp. 155-166.
2. HÖGLUND, G., K. HAMDORF, H. LANGER, R. PAULSEN, J. SCHWEMER: The photopigments in an insect retina. This volume, pp. 167-174.
3. AUTRUM, H., K. HAMDORF: Der Sauerstoffverbrauch des Bienenauges in Abhängigkeit von der Temperatur, bei Belichtung und im Dunkeln. Z. vergl. Physiol. 48, 266-269 (1964).
4. HAMDORF, K., A.H. KASCHEF: Der Sauerstoffverbrauch des Facettenauges von Calliphora erythrocephala in Abhängigkeit von der Temperatur und dem Ionenmilieu. Z. vergl. Physiol. 48, 251-265 (1964).
5. LANGER, H.: Die Wirkung von Licht auf den chemischen Grundaufbau des Auges von Calliphora erythrocephala Meig. J. Insect Physiol. 4, 283-303 (1960).
6. LANGER, H.: Untersuchungen über die Größe des Stoffwechsels isolierter Augen von Calliphora erythrocephala Meigen. Biol. Zbl. 81, 691-720 (1962).
7. LANGER, H.: Der Phosphatstoffwechsel des Facettenauges im Dunkeln und im Licht. Helgol. Wiss. Meeresunters. 9, 251-260 (1964).
8. WEBER, K.M., G. SCHORRATH: Histochemische Untersuchungen am Auge der Schmeißfliege Calliphora erythrocephala Meig. I. ATP-ase, saure Phosphatase, alkalische Phosphatase und Glukose-6-Phosphatase. Histochemie 28, 137-144 (1971).

9. WEBER, K.M., G. SCHORRATH: Histochemische Untersuchungen am Auge der Schmeißfliege Calliphora erythrocephala Meig. II. Nukleosidphosphatasen, Thiaminpyrophosphatase, unspezifische Esterase und Cholinesterase. Histochemie 30, 131-149 (1972).
10. SCHULZE, W., A. WOLLENBERGER: Zytochemische Lokalisation und Charakterisierung von phosphatabspaltenden Fermenten im sarkotubulären System quergestreifter Muskeln. Histochemie 10, 140-153 (1967).
11. BONTING, S.L., L.L. CARAVAGGIO: Studies on sodium-potassium activated adenosinetriphosphatase. V. Correlations of enzyme activity with cation flux in six tissues. Arch.Biochem.Biophys. 101, 37-46 (1963).
12. BRUNI, A., A.R. CONTESSA, P. PALATINI: Functions of phospholipids in adenosintriphosphatases associated with membranes. Adv. exper. Med. Biol. 14, 195-207 (1971).
13. FOLCH, J., M. LEES, G.H. SLOANE-STANLEY: A simple method for the isolation and purification of total lipids from animal tissue. J. biol. Chem. 226, 497-509 (1957).
14. ROUSER, G., S. FLEISCHER, A. YAMAMOTO: Two-dimensional thin layer chromatographic separations of polar lipids and determination of phospholipids by phosphorus analysis of spots. Lipids 5, 494-496 (1970).
15. FAST, P.G.: A comparative study of the phospholipids and fatty acids of some insects. Lipids 1, 209-215 (1966).
16. BRIDGES, R.G., G.M. PRICE: The phospholipid composition of various organs from larvae of the housefly, Musca domestica, fed on normal diets and on diets containing 2-aminobutan-1-ol. Comp. Biochem. Physiol. 34, 47-60 (1970).
17. LANGER, H., CH. HOFFMANN: Elektro- und stoffwechselphysiologische Untersuchungen über den Einfluß von Ommochromen und Pteridinen auf die Funktion des Facettenauges von Calliphora erythrocephala. J. Insect Physiol. 42, 357-387 (1966).

Discussion

T.H. Goldsmith: What is the metabolic cost to the fly of being mutant vs. wild type?

D. Zinkler: Under darkness, there isn't any difference in the metabolic cost between the eyes of the wild type and the mutant (4, 17). Under light, the metabolic cost of the mutant is higher. To obtain the same increase in carbohydrate consumption, the intensity of illumination must be raised for wild type eyes by a factor of 100 (17). Under light, the eyes of mutant "chalky" have a significantly larger increase of free inorganic phosphate than wild type ones (7).

T.H. Goldsmith: Is there turnover of material in the axial "cavity" between the rhabdomeres as a function of retinal activity?

K.M. Weber: Till now, we know nothing about turnover of material in the interommatidial space. There is an indication that the substance(s) of the space participates in sensory cell function. The material, probably glycoproteins, stains more intensely with PAS in light-adapted eyes than in dark-adapted ones. Measured microphotometrically there is a difference in absorption of 10-12% (Weber, unpublished).

M.W. Bitensky: The light dark differences in glycogen content are very striking. I wonder if it would be possible to implicate cyclic AMP here by showing increased glycogenolysis following a light flash by using methyl xanthines or papaverine.

K. M. Weber: It would be interesting, but we are not yet able to give a comment to this aspect of the physiology of the retina.

W. T. Mason: I would like to comment that the results of the phospholipid composition of the blowfly rhabdomes strongly parallel our results on the lipid composition of the squid, Loligo pealii. We find also the reversal of the phosphatidyl ethanolamine/phosphatidyl choline proportions in pure rhabdomes, as well as similar distributions in other phospholipid classes. In spite of the fact that our results are from purified rhabdomes and yours from whole eyes, the nature of the two compositions is quite similar. Our analysis of the fatty acid distribution and the cholesterol content of the squid rhabdome both suggest that the microvillar photoreceptor membranes of the invertebrate squid are highly rigid structures and apparently in marked contrast to the fluid membrane of the vertebrate.

D. Zinkler: There is quite a difference between squid and blowfly eye phospholipid pattern. We couldn't find any reversal of the PE/PC proportions because PE is the most abundant phospholipid not only in blowfly eye tissue but also in extracts from whole animals.

S.L. Bonting: Since I would expect the microvillar membrane to be relatively stiff compared to that of the rod sacs, I think it would be interesting to know the cholesterol content and the percentage unsaturated fatty acids.

D. Zinkler: At present we have only a few results concerning the cholesterol content of Calliphora eye tissue which didn't suggest drastic differences between vertebrates and insects.

M.O. Hall: Could the differential rate of phospholipid synthesis in light and dark be due to the holdup of the precursor by some structure analogous to the pigment epithelium in the frog - rather than to a different rate of synthesis? - I might mention that Dr. D. Bok in our institute has shown that RNA synthesis in the frog photoreceptor is depressed both in darkness and intense light.

K. M. Weber: At present we can only speculate about the different incorporation rates of phospholipid precursors. It is possible that they are due to differences in ion transport in light and dark. We looked for RNA by means of histochemistry. The visual cells of flies (mutant chalky) kept in dark for 48 hrs show less cytoplasmic RNA than those of flies under light.

Cyclic AMP and Photoreceptor Function

M. W. Bitensky, J. J. Keirns, and R. C. Wagner
Yale University School of Medicine, New Haven, CT 06510/USA

Our recent work (1,2,3,4) has indicated that rod outer segment preparations from a number of different vertebrate retinas contain adenyl cyclase with a specific activity significantly higher than any previously recorded in any tissue. This photoreceptor cyclase is inactivated by light with the inactivation proportional to the bleaching of rhodopsin. There are also, in photoreceptor preparations, a phosphodiesterase (not regulated by light) and a cAMP activated protein kinase (with characteristics similar to its counterpart in brain). Other photoreceptors (ground squirrel cones, and the rhabdoms of gypsy moth, crayfish and lobster) exhibit a high specific activity of cyclase which does not however show sensitivity to illumination.

This data suggests either that cAMP can participate as an intermediate in visual excitation or that cyclase plays a role in other functions of the rod outer segment (e. g. light adaptation). Also, in view of the effects of homogenization and digitonin on rod cyclase (loss of light sensitivity) the integrity of disk membrane architecture appears necessary for coupling the photoinfluence of rhodopsin to photoreceptor cyclase activity.

There are a number of interesting studies which are related to the finding of a light sensitive cyclase in photoreceptors. These include description of a cyclic-GMP phosphodiesterase in photoreceptors (5) and subsequently a guanyl cyclase which does not however exhibit change in activity as a result of illumination (6).

Kühn and Dreyer (7) have found a phosphorylation of rhodopsin which is regulated by its photochemical state. Thus when native rhodopsin (in frozen and thawed disc membranes) is bleached, it then becomes a substrate for an intrinsic protein kinase. The unbleached material is not phosphorylated by this enzyme. Bownds et al. (8) have found a light dependent phosphorylation of rhodopsin which occured in isolated intact rod outer segments. These workers find that in the course of a one percent bleach of rhodopsin in these organelles, there is phosphorylation of fifty rhodopsins for every photon that is captured. This amplified phosphorylation thus includes rhodopsin molecules which are unbleached, a finding with contrasts with the results obtained by Kühn and Dreyer. The membrane integrity of these isolated rod outer segments is shown by their capacity to behave as osmometers. The ability of isolated rod outer segments to retain their osmometric integrity is somewhat surprising in view of their apparent permeability to γ-^{32}P ATP (a necessary component of the phosphorylation reaction). Bownds also reports that cyclic AMP in concentrations as low as 10^{-7}M is indispensible for complete recovery of the dark adapted sodium conductance found in isolated segments following a one percent bleach.

These findings are intriguing and suggest a meaningful role for cyclic nucleotides and / or phosphorylation reactions in the as yet unknown molecular events which underlie visual excitation. The many problems conjured up by apparent disagreements in the nature of these findings clearly will require further study of the biochemistry of rod outer segment membranes.

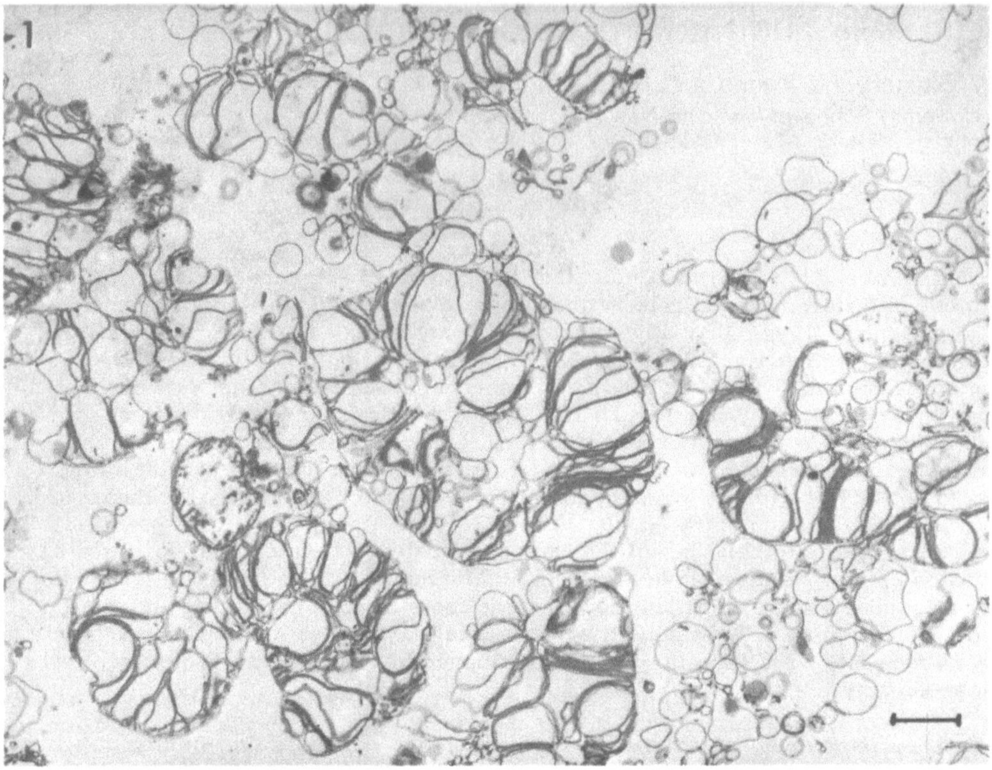

Fig. 1. Electron micrograph of retinal lamella fraction exhibiting primarily stacked lamellae and minimal amounts of membrane vesicles. The sample was incubated for 45 minutes in substrate free medium. Scale marker 1 μ m.9,200 x

---

Very relevant to these problems in photoreceptor biochemistry is the precise localization of the photoreceptor adenyl cyclase system. Previous work has indicated reasons for believing that cyclase is in the disc membranes, including the observation that disc membranes prepared with the continuous sucrose gradient technique (9) (following disruption with a glass on glass motor driven homogenizer) still exhibit as much as 40% of the cyclase activity found in unhomogenized, sucrose floated rod outer segments. Recent experiments with a specific cyclase substrate (5-adenylyl-imidodiphosphate; AMP-PNP (10,11)) and a specific cyclase inhibitor (alloxan (12)) using cyclase from isolated epididymal capillaries have permitted development of an apparently reliable and specific technique for the histochemical localization of adenyl cyclase in electron micrographs (13). Since AMP-PNP is not utilized by other membrane ATPases the appearance of a precipitate of lead and imidodiphosphate is suggestive of cyclase activity. Furthermore since alloxan inhibits cyclase but appears not to inhibit other membrane ATPase systems (both ouabain sensitive and insensitive) its inhibition of product formation (with AMP-PNP as substrate) further strengthens this technique as a procedure for localizing the cyclase system. This combination of reagents has been used with unfixed dark adapted rod outer segments prepared by flotation on sucrose. It has been found that there is an AMP-PNP utilizing, alloxan sensitive enzyme in disc membrane structures. This report is a work in progress description of the application of this technique to rod outer segments and the first findings which have been obtained with it.

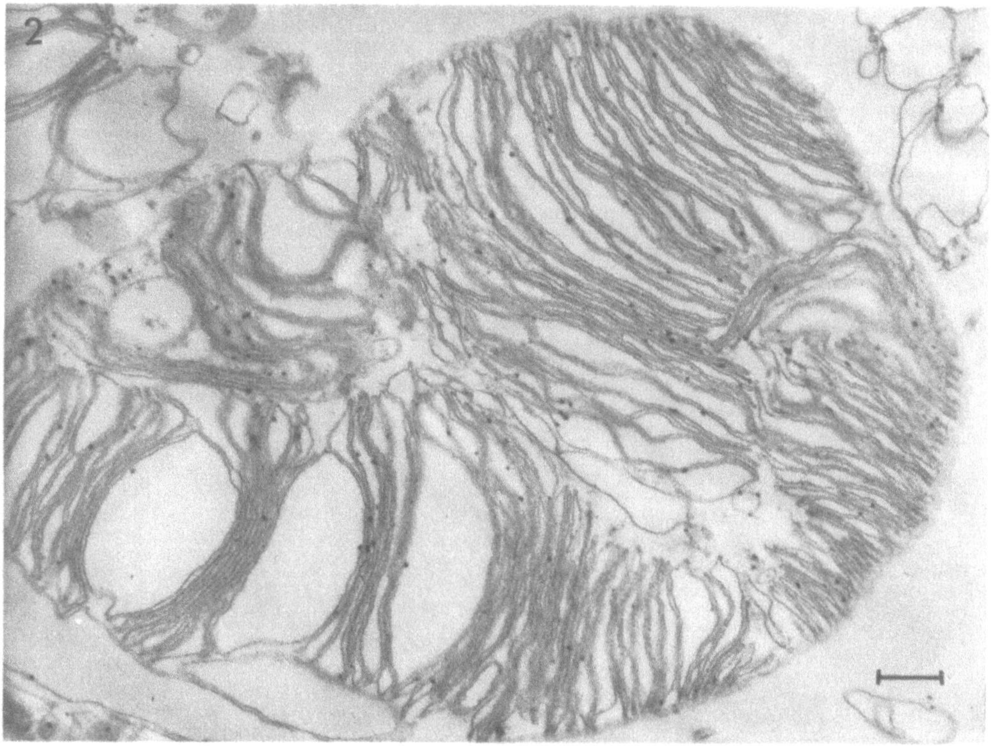

Fig. 2. Retinal lamellae treated for 45 minutes with 2 mM AMP-PNP. Reaction product (lead-PNP) is located almost exclusively within the lamellae. Scale marker 0.25 µm. 36,800 x

Materials and Methods

Rod outer segments were prepared as described previously (3). The incubation media for the cytochemical localization of cyclase consisted of 80 mM Tris-maleate buffer (pH. 7.2), 6% dextran M.W. 250,000, 2 mM $Pb(NO_3)_2$, 4 mM $MgSO_4$, 5 mM aminophylline with or without 0.5 mM adenylyl-imidodiphosphate (I.C.N. Ltd.) as substrate and with or without alloxan (5 mM). Unfixed bovine rod outer segments were incubated for 45 minutes with the incubation medium at room temperature, washed twice in 0.2 M Tris-maleate buffer (pH. 7.2) and fixed for 60 min. in 2% glutaraldehyde in Tyrodes-cacodylate buffer (pH. 7.4). After two washes in Tyrodes-cacodylate buffer, the tissue was refixed in 2% OsO_4 for one hour, dehydrated in ethanol, embeded in epon and sectioned with an LKB 1 ultramicrotome. Counterstained (uranyl acetate) and unstained sections were observed with an Hitachi 11B electron microscope.

Results and Discussion

Fig. 1 shows a control with AMP-PNP omitted. It illustrates the homogeneity of the outer segment preparations prepared by sucrose flotation of intact retinas which are dispersed only with a mechanical vibrator. The electron micrograph shows

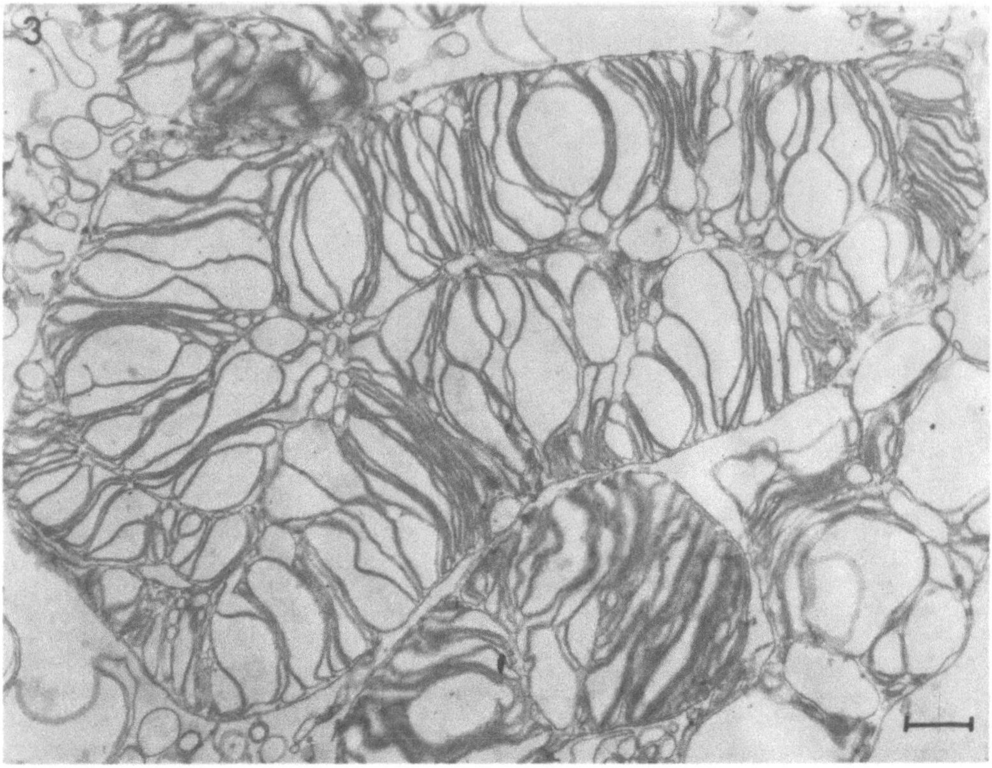

Fig. 3. Retinal lamellae treated with AMP-PNP and alloxan. Lamellae are for the most part free of reaction product. Scale marker 0.43 μ m. 20,200 x

excellent preservation of disc membrane architecture. There is no evidence of significant contamination by any other membrane species.

Fig. 2 shows the appearance of reaction product formed upon hydrolysis of AMP-PNP in the presence of lead nitrate. In Fig. 3 both alloxan and AMP-PNP have been included in the reaction mixture. In the presence of alloxan, there is a marked decrease in the amount of product formed in association with the disc membranes. These findings support the idea that an enzyme which can utilize AMP-PNP, and which is inhibited by alloxan, probably adenyl cyclase, is present in the disc membrane structures. Further studies with this system including the demonstration that illumination can markedly influence the accumulation of product will be needed to clarify and to define the question of where adenyl cyclase is located in photoreceptor cells.

Acknowledgement. This work was supported by USPHS grant 1-RO1-AM-15016, NEI grant 8-RO1-EY-00089, USPHS grants AM-03688, and TICA-05055. J.J. Keirns is a fellow of the Jane Coffin Childs Memorial Fund for Medical Research. This investigation has been aided by a grant from the Jane Coffin Childs Memorial Fund for Medical Research.

References

1. BITENSKY, M.W., R.E. GORMAN, W.H. MILLER: Adenyl cyclase as a link between photon capture and changes in membrane permeability of frog photoreceptors. Proc.Nat.Acad.Sci. USA 68, 561-562 (1971).
2. MILLER, W.H., R.E. GORMAN, M.W. BITENSKY: Cyclic adenosine monophosphate: Function in photoreceptors. Science 174, 295-297 (1971).
3. BITENSKY, M.W., R.E. GORMAN, W.H. MILLER: Digitonin effects on photoreceptor adenylate cyclase. Science 175, 1363-1364 (1972).
4. BITENSKY, M.W., W.H. MILLER, R.E. GORMAN, N.H. NEUFELD, R. ROBISON: Cyclic AMP in visual excitation. In: Advances in Cyclic Nucleotide Research I: International Conference on the physiology and pharmacology of Cyclic AMP, ed. P. GREENGARD, R. PAOLETTI, G.A. ROBISON. Raven Press. In press (1972).
5. PANNBACKER, R.G., D.E. FLEISCHMANN, D.W. REED: Cyclic nucleotide phosphodiesterase: High activity in a mammalian photoreceptor. Science 175, 757-758 (1972).
6. PANNBACKER, R.G.: personnel communication.
7. KÜHN, H., W.T. DREYER: Light dependent phosphorylation of rhodopsin by ATP. FEBS Letters 20, 1-6 (1972).
8. BOWNDS, D., J. DAWES, J. MILLER, M. STAHLMAN: Phosphorylation of frog photoreceptor membranes induced by light. Nature New Biology 237, 125-127 (1972).
9. MC CONNELL, D.G.: The isolation of retinal outer segment fragments. J.Cell Biol. 27, 459-473 (1965).
10. RODBELL, M., L. BIRNBAUMER, S.L. POHL, H.N.J KRANS: The glucagon-sensitive adenyl cyclase system in plasma membranes of rat liver. J.Biol. Chem. 246, 1877-1882 (1971).
11. YOUNT, R.G., D. BABCOCK, W. BALLANTYNE, D. OJALA: Adenylyl imidodiphosphate, an adenosine triphosphate An log containing a P-N-P linkage. Biochemistry 10, 2484-2489 (1971).
12. COHEN, K.L., M.W. BITENSKY: Inhibitory effects of alloxan on mammalian adenyl cyclase. J.Pharmacol.Exp.Ther. 169, 80-86 (1969).
13. WAGNER, R.C., P. KREINER, J.R. BARRNETT, M.W. BITENSKY: Biochemical characterization and cytochemical localization of a catecholamine-sensitive adenylate cyclase in isolated capillary endothelium. Proc.Nat.Acad. Sci. (Wash.) 69, 3175-3179 (1972).

Discussion

D. Bownds: Dr. Ann Gordon-Walker in our laboratory has found that the membrane fraction which contains light inactivated adenyl cyclase activity comes to an equilibrium position in ficoll gradient centrifugation different from that of the outer segments - it is more dense. Dr. Bonting tells me that a similar result has been obtained in his laboratory. Both of us find only 1% of the activity reported in your first paper, a 2-3 fold change in activity caused by light. I would suggest that you also analyze cyclase activity in a continuous gradient. Also, in the histochemical experiments, is it not necessary to demonstrate that light inactivate the cyclase activity you think you see in outer segments before it can be identified as the light inactivated cyclase you previously reported?

M.W. Bitensky: It should be emphasized that in one of our recent studies (3) on photoreceptor cyclase we included the observation that the fraction of photoreceptor membranes from the continuous gradient of McConnel which contained the bulk of the rhodopsin exhibited excellent adenyl cyclase activity although this activity appears to have lost light sensitivity as a result of homogenization. Secondly I am pleased that Dr. Bownds can report a 75% inactivation of adenyl cyclase activity associated with photoreceptor preparations. I have spoken with Dr. Bownds and with Dr. Daemen and Mr. Hendricks from Dr. Bonting's group and they have indicated to me that the gradient separation of cyclase from the disc membranes is not very complete and a residual fraction of the cyclase activity - as much as 20% - still exhibits, in their hands, identical migration with the rhodopsin containing membranes. I am certainly concerned about the differences in cyclase specific activity which we are finding and which you are finding. I think that this will require some comparison of methodology in order to find the cause of this disparity. I would agree with you that the histochemical data would be strengthened significantly by the incorporation of light inactivation as an additional element in identification of disc membrane cyclase.

S.L. Bonting: Mr. Th. Hendricks in our laboratory has repeated your cyclase assay on frog rods. Either when following your technique as closely as possible or when introducing various modifications in order to improve the results, we never find more than about 1% of your activity. Light at near-maximum bleaching does give a decrease in activity, but maximally 50-60%. In sucrose density gradient centrifugation we find about four times as large activity in the higher density white (non-rod outer segment) band as in the purple (ROS) band.

M.W. Bitensky: It appears that Dr. Bonting also is finding some cyclase in association with the purple band although he finds cyclase in other places in his sucrose gradient. This observation appears to support the idea of a light sensitive cyclase in photoreceptor materials. I have spoken again with Mr. Hendricks. There are some minor differences in methodology between our laboratories including the species of frogs that is being used and some of the conditions of membrane dispersion (sonication and sucrose concentration). Nevertheless, again we are particularly concerned about the differences in total activity and feel that it is important to reexamine our conditions and find the factors which are responsible for these differences. I hope that you will at least in part agree that the most significant aspects of your findings is the apparent confirmation of a light sensitive cyclase associated with photoreceptor membranes. The fact that there is cyclase in other parts of the gradient and the fact that the specific activities are not yet in full agreement are nevertheless disturbing and will require further clarification.

The Effects of Cyclic Nucleotide Phosphodiesterase Inhibitors on the Frog Rod Receptor Potential*

Thomas G. Ebrey and Donald C. Hood
Department of Biological Sciences and Department of Psychology, Columbia University, New York, NY 10027/USA

The events that link the absorption of light by rhodopsin to the initiation of visual excitation are not fully understood. Hagins and coworkers (1,2) and Tomita and coworkers (3,4) have shown that the electrical signal from the photoreceptors presumably responsible for visual excitation is a voltage transient produced when light turns off a current of sodium ions flowing in the dark. Thus it seems that light reduces the sodium permeability of the rod outer segment. The driving force for this current would be the difference in resting potential between the inner and outer segments, due to differences in ionic permeabilities (and perhaps ionic concentrations).

So far two explanations have been offered of what controls sodium permeability. Yoshikami and Hagins (5,6) have shown that calcium mimics the action of light; an increase in external calcium concentration reduced the dark current of both rods and cones. They have proposed that the absorption of light by rhodopsin triggers the release of calcium and, in turn, this increase in intracellular calcium in some way provides the link between light absorption and visual excitation. The second explanation derived originally from some biochemical studies. Bitensky, Gorman, and Miller (7,8) found a light inactivated adenyl cyclase in the frog rod outer segment. Adenyl cyclase is an enzyme that catalyzes the conversion of adenosine triphosphate (ATP) to adenosine-3':5'-cyclic monophosphate (cAMP). Previously, cAMP has been implicated in mediating a large number of physiological functions including membrane permeability (9). On the basis of their findings, these investigators proposed a role for cAMP in photoreceptor function. They suggested that the cAMP mediates the effect of light in decreasing the sodium permeability of vertebrate photoreceptors. The sodium permeability of the outer membrane of the rod would be kept high in the dark by a constant supply of cAMP made by the adenyl cyclase; the action of light would be to directly or indirectly inactivate the adenyl cyclase; the cAMP level after inactivation would be lowered by the enzyme cyclic nucleotide phosphodiesterase; and this decrease in cAMP would decrease the sodium permeability causing a voltage transient - the receptor potential. The enzyme which inactivates cAMP, phosphodiesterase, is known to be present in large amounts in the rod outer segment (8,10). Curiously this enzyme seems to work better on guanosine-3' : 5'-monophosphate (cGMP) than on cAMP (10).

The proposed roles of calcium and cAMP do not necessarily represent alternate explanations of how sodium permeability is regulated. In a number of systems a close relationship between calcium and cAMP has been proposed (11). For instance, it is possible that light absorption triggers calcium release which in turn inhibits adenyl cyclase. It is also possible to devise hypotheses giving calcium or cyclic nucleotides a role in controlling the sensitivity of the receptor rather than a role in visual excitation. Bownds and coworkers (12) reported that when the sodium

*Supported by P.H.S. grants EY-00433 and MM-19322

permeability of isolated rod outer segments was reduced by bleaching rhodopsin (0.1 to 30%), the sodium permeability, but not the rhodopsin, could be restored by the addition of dibutryl cAMP to the medium.

In this study we have examined the role of cyclic nucleotides in visual excitation by recording receptor potentials from frog rods under conditions designed to increase intracellular cAMP. Two methods are commonly used to demonstrate the role of cAMP in mediating a physiological event (13). One is to add exogenous cAMP or dibutryl cAMP, a derivative of cAMP with presumably a higher permeability to membranes, and to look for an effect on the physiological event. A more common procedure is to add an inhibitor of the enzyme which destroys the cAMP, phosphodiesterase. The inhibition of this enzyme would allow the intracellular levels of cAMP made by the adenyl cyclase to rise and thus would be equivalent to adding intracellular cAMP. In most studies of the role of cAMP in mediating physiological responses, this procedure has been found to be more successful than the first.

Methods

The procedures we used allow the recording of gross electrical activity from the frog's red rods. Sillman, Ito, and Tomita (14) reported that replacing part of the chloride in frogs Ringer's solution with aspartate (Asp) isolates the electrical activity from the receptors of the frog retina. Thus, it is possible to study the receptor potential of the isolated frog retina using electroretinographic recording techniques. Witkovsky, Nelson, and Ripps (15) have shown that although a long latency, slow potential of nonreceptor origin remains after aspartate treatment, the early portion of the evoked response, including the initial amplitude, is generated in the receptor layer.

All of our experiments were done under conditions where previous work indicated the receptor potential is dominated by the 502 nm rod of the frog retina (16; Hood and Hock, in prep.).

Completely dark adapted frogs (Rana pipiens) were used for these experiments. A dark adapted frog was decapitated and then an eye enucleated and hemisected. The retina was carefully teased away from the pigment epithelium and dropped into a dish of NaAsp Ringer's solution. This was made according to Furakawa and Hanawa's (17) normal Ringer with 100 mM NaAsp replacing 100 mM NaCl. The retina was then floated, receptors up or down, onto a cotton pad soaked in the NaAsp Ringer's solution. The entire procedure was performed under red light (Wratten filter no. 29).

The cotton pad with the isolated retina was placed in a plastic chamber. This chamber, similar in principle to that of MacNichol and Svaetichin (18), had a moat filled with Ringer's solution surrounding, but not touching the retina. A mixture of 95% O_2 and 5% CO_2 flowed slowly into the chamber, creating a moist, oxygenated atmosphere.

Two techniques were used to apply the various test substances to the retina. One involved placing the retina, receptors down, on a cotton pad soaked in NaAsp Ringer. After initially testing the retina to determine the size of the receptor potential, it was removed from the recording set-up and immersed in 10 ml of the test solution and then placed receptors down on a new thoroughly soaked cotton pad. This method should give the solutions maximum access to the rod cell. In order to follow the receptor potential changes at short times after the solution was applied, a different method was used. In this case, 0.2 ml of the solution to be tested was squirted on the retina, receptors up, as it sat in the recording apparatus.

The Ringer soaked cotton wick electrodes in contact with Ag-AgCl wire were connected to the differential input leads of a low level a. c. preamplifier (Tektronix type 122), or a d. c. preamplifier (made at Rockefeller University). The output of the preamplifier was connected to an oscilloscope (Tektronix type 502A). The low frequency cutoff of the a. c. amplifier was set at 0.2 Hz and the high frequency cutoff at 250 Hz.

The isolated retina was stimulated with a projection system that entirely flooded it with light. The wavelength of stimulation was controlled by a Baird-Atomic interference filter with a peak transmission at 487 nm and a half-band width of 6 nm. The light stimulus in all of these experiments was a 50 msec flash with an intensity of $3.5 \cdot 10^4$ photons/sec / μm^2.

Results

Effects of cyclic nucleotides. Our initial experiments tried to determine if cyclic nucleotides had any effect on the receptor potential of the dark adapted photoreceptor. We found that neither cAMP, dibutryl cAMP, nor cGMP, in concentrations up to 10 mM, had any discernible effect on the receptor potential.

A second set of experiments was performed to see if cAMP could effect the restoration of the receptor potential's sensitivity when part of the rhodopsin had been bleached. In order to do this the sensitivity of the receptor potential was measured by determining the intensity necessary to evoke a 10 μV response: then about 20% of the rhodopsin in the retina was bleached. After bleaching, the sensitivity was monitored for about 30 minutes until it had reached a new dark-adapted steady-state value, a decrease of 0.3 log unit. Dibutyryl cAMP (0.2 mls of 10 mM) was then squirted on the retina. It had no effect on the sensitivity of the preparation, that is, the light intensity associated with the dark adapted sensitivity still evoked a 10 μV response. Moreover, if dibutyryl cAMP was squirted on the retina immediately after bleaching, it did not affect the time which it took the retina to reach its dark-adapted steady-state threshold value. It should be recognized that for us to see a larger receptor potential not only would the suspected increase in cAMP have to restore the sodium permeability of the cell for a substantial length of time, but also light would have to be able to turn off this increased permeability.

Effects of phosphodiesterase inhibitors. Although there was no effect on the receptor potential in any of the above experiments, this does not rule out cyclic nucleotide involvement in receptor electrical activity. The receptor could either have a low permeability to these nucleotides or the phosphodiesterase of the receptor could act so fast that it would be impossible for us to see the effects of the exogenous cyclic nucleotides.

Since photodiesterase breaks down the cyclic nucleotides, it is possible to increase the internal cyclic nucleotide concentration by applying phosphodiesterase inhibitors. Therefore we decided to test whether any phosphodiesterase inhibitors could affect the receptor potential.

We used four different phosphodiesterase inhibitors, papaverine, theophylline, caffeine, and a new inhibitor developed by Squibb, SQ20006. The most potent inhib-

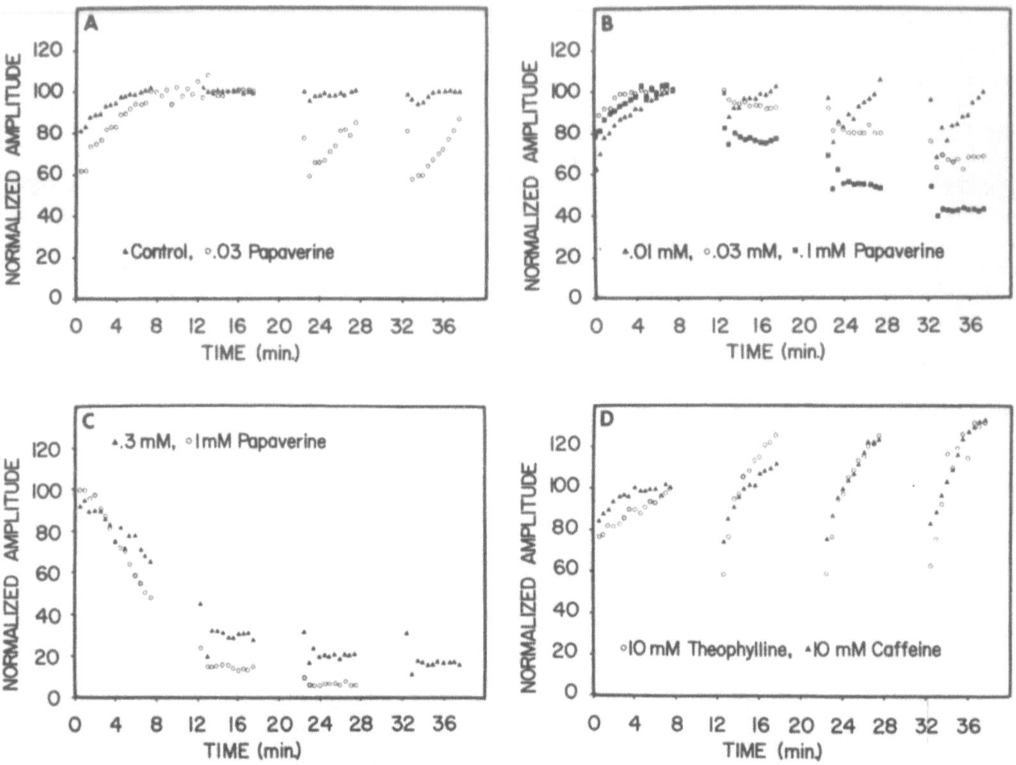

Fig. 1. The effect of papaverine, theophylline, and caffeine on the frog receptor potential. A 50 msec 487 nm flash was presented every 30 sec for $7\frac{1}{2}$ min and the amplitude of the receptor potential measured. Except for the retina in A treated with .03 mM papaverine, all retinas were dark-adapted for 5 min and then tested with flashes every 30 sec for 5 min. This procedure was repeated 2 more times. The amplitude of the responses are normalized by the largest amplitude they obtained during the initial test period.

itor as shown by tests on rod outer segment phosphodiesterase, was papaverine (10; Pannbacker and Reed, private communication). With respect to the phosphodiesterase of other systems, papaverine has also been found to be a more effective inhibitor than theophylline or caffeine (19).

Papaverine did not appear to alter either the waveform or the latency of the receptor potential. However, even at very low concentrations, it has a very dramatic effect on response amplitudes. If a retina which has attained a constant response to test flashes every 30 sec is allowed to remain in the dark for several minutes and then presented with a test flash, the receptor potential evoked after this wait in the dark will be smaller. The longer the wait in the dark, the smaller will be the receptor potential. However, if after this dark period the program of flashes every 30 sec is begun again, the amplitude does not continue to decrease and even occasionally increases (Fig. 1B). The difference between the effects of light and dark adaptation on the receptor potential can easily be seen by comparing the 0.03 mM papaverine records in Fig. 1A and 1B. This decrease in amplitude seen after dark adaptation is in direct opposition to the effects usually associated with dark adapta-

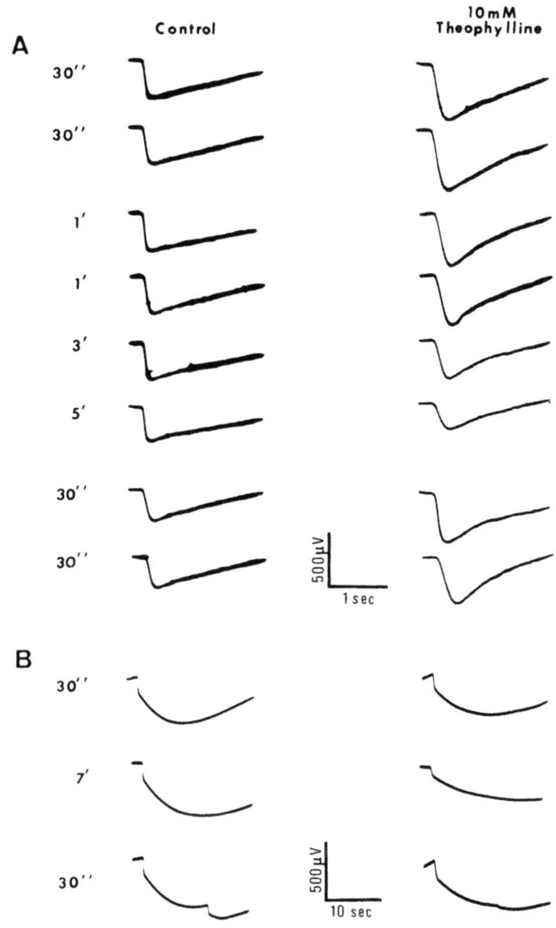

Fig.2. The effect of theophylline on the isolated receptor potential. A) A.C. Recording. The records in the Control column are from a retina soaked in NaAsp Ringers. The times are the intervals between 50 msec flashes of 487 nm light. The first two records were recorded after a constant size response was obtained to flashes given every 30 sec. The next four records were recorded in response to flashes presented 1, 2, 5, and 10 min after the flash evoking the second record. Then flashes were presented every 30 sec until the amplitude of the response was constant. The last two records shown were taken after the amplitude had reached a stable level. This retina was then immersed in 10 mM theophylline and the records in the right hand column obtained using the same procedure. Note that with theophylline the response becomes smaller as the interval between the flashes increases. B) D.C. Recording. Here the control and the theophylline records are from different retinas subjected to identical procedures. Flashes were presented every 30 sec until a constant size response was obtained. The steady state responses are shown as the first records in each column. Then, after 7 min of dark adaptation, a test flash was presented and the second record obtained. Flashes were again presented every 30 sec until a steady state amplitude was obtained. The response to the last of this series of flashes is shown along with a response to a flash presented only 12 sec after this test flash

tion. At the higher papaverine concentrations, from 0.1 to 1 mM, a similar decrease after dark adaptation is observed, except that even with repetitive stimulation (light adaptation) there is a substantial decrease in receptor potential amplitude (Fig. 1C).

The methyl xanthines, caffeine (10 mM), and theophylline (10 mM), and SQ20006 (1 mM) share with papaverine the ability to drastically decrease the size of the receptor potential upon dark adaptation (Fig. 1D and Fig. 2). In addition these three drugs also affect other properties of the rod receptor potential. First, the waveform of the receptor potential with either a.c. or d.c. recording is broadened and the latency to peak amplitude is increased (Fig. 2). The amplitude of the isolated receptor potential always increases, sometimes as much as twofold, after bathing in, for instance, 10 mM theophylline. Moreover, the size of the receptor potential evoked by a second flash 12 sec after an initial flash is much smaller in a retina treated with 10 mM theophylline than in a control retina. All of these effects as well as the reduction of the receptor potential during dark adaptation are shown in Fig. 2.

Of the various effects on the receptor potential noted above, the only one common to all of the phosphodiesterase inhibitors tested and the only one seen with papaverine was the decrease in the amplitude of the receptor potential as the retina was dark adapted. We investigated whether this reduction could be due to a reduction of the sodium gradient across the rod outer segment membrane. A retina was dissected out in 10 mM NaAsp Ringers and was stimulated every 30 sec until the amplitude of the receptor potential had stabilized. To test for the effect of papaverine without the possible collapse of the sodium gradient, we created an artificial gradient across the membrane after treating the retina with papaverine. To do this we immersed the retina in Ringer's solution containing 1 mM papaverine but with choline (chloride) replacing all but 15 mM of the sodium. This procedure should allow papaverine to presumably increase the internal cAMP by inhibiting the phosphodiesterase while at the same time equilibrating the intracellular and extracellular spaces with low molarity sodium. The retina was then dark-adapted for 15 min. Under these conditions the receptor potential was now about 10 per cent of its initial amplitude. At the end of the 15 min dark period in low sodium Ringer, normal Ringer (100 mM Na (chloride), 10 mM NaAsp) also with 1 mM papaverine was then squirted onto the retina and test flashes presented every 30 sec. A receptor potential of almost normal magnitude was evoked which then decreased.

This experiment suggests that the reduction in amplitude of the receptor potential seen with papaverine is not due to any direct blocking of the sodium permeability control process but rather is due to the collapse of the sodium ion concentration gradient across the rod membrane.

Discussion

We have attempted to test the hypothesis of Bitensky, Gorman, and Miller (7) and of Bownds *et al.* (12) concerning the role of cAMP in mediating permeability changes in the rod outer segment. We have found that neither cAMP, dibutyryl cAMP nor cGMP affected the receptor potential. Likewise, when we bleached some of the rhodopsin and as a consequence increased the threshold by 0.3 log unit, the application of up to 10 mM dibutyryl cAMP did not affect the increased dark-adapted threshold nor the rate of attaining this threshold. These results do not, however, rule out the hypothesized role of cAMP in increasing sodium permeability.

At least two problems exist with the above procedures. First, the applied solutions may not reach the site of action. A second problem involves the time course of action of the phosphodiesterase. It could be the case that the enzyme that destroys cAMP phosphodiesterase, acts extremely fast. If this were true, then it might be possible that none of our procedures would enable us to see the effects of the exogenously applied cyclic nucleotides. The inability to see any effects of exogenously added cAMP is the usual case in physiological systems where cAMP acts (13).

To circumvent these difficulties we used a more common way (13) of showing the suspected involvement of cAMP in mediating a physiological response, which is to use an inhibitor of the cyclic nucleotide phosphodiesterase which should increase the intracellular concentration of cAMP. We asked whether phosphodiesterase inhibition produces changes in the receptor potential that could be attributed to an increase in cAMP. The only common feature of all of the phosphodiesterase inhibitors examined in this study was that the receptor potential decreased if the preparation was dark-adapted. It was also shown that the receptor potential could be restored if a sodium gradient was reestablished across the rod outer segment membrane.

These results are most easily explained if the effect of these inhibitors is to partially fill up the rod outer segment with sodium ions; this would decrease the sodium gradient across the rod outer segment membrane and would cause a reduction in the size of the receptor potential.

If the phosphodiesterase inhibitors cause the rod cell to fill up with sodium ions, this would also explain the difference in amplitude of the receptor potential we find when flashes are repeated every 30 sec for 5 min compared with just five minutes of dark adaptation. The light adaptation program, according to the effect of light on the sodium permeability of the rod outer segment discussed in the introduction, would be expected to decrease the sodium permeability and thus slow up any influx of sodium ions tending to abolish the gradient. On the other hand, dark-adapting for five minutes would be expected to restore a high sodium permeability and allow a much larger sodium influx. Thus dark adaptation would accelerate any tendency to abolish the sodium gradient.

There are at least two possible ways the phosphodiesterase inhibitors could allow the rod outer segment to become filled with sodium ions. First, all of the drugs could be acting, in some unknown manner, to prevent the normal influx of sodium ions from being pumped out. This might happen if either the energy supply for the pump, presumably ATP, were halted or if the sodium pump were poisoned - that is, if all of these drugs were acting as ouabain acts. In fact, 10^{-4} mM ouabain gives results quite similar to 0.03 mM papaverine. Such an explanation could also explain the results of Miller, Gorman, and Bitensky (8) that the phosphodiesterase inhibitors theophylline and aminophylline depolarize the Limulus photoreceptor.

However, there is no strong evidence for the phosphodiesterase inhibitors acting in the manner suggested above. First, there is no evidence that these inhibitors affect the ATP supply. Moreover, it is unlikely the sodium pump itself is inhibited since enzyme assays show that neither papaverine nor theophylline affects the activity of the Na-K activated, ouabain sensitive ATPase usually associated with the sodium pump (D. Reed, private communication).

A second explanation, one which is consistent in part with the cAMP hypothesis, is that cAMP can control the sodium permeability of the outer rod membrane. In the dark when the sodium permeability is already high, the increase in cAMP produced by the phosphodiesterase inhibitors is sufficient to increase the sodium permeability to the point where the ionic gradient cannot be maintained. Consequently the outer segment becomes partially filled with sodium ions, and the receptor potential is reduced.* It is useful to devide this hypothesis into two subhypotheses. The first is that changes in cAMP concentrations caused by light change the sodium permeability and produce the receptor potential. This subhypothesis would appear to be excluded if our artificial ion gradient experiment is considered in more detail. Here we obtained a normal receptor potential after the intracellular cAMP concentration was increased and, when according to the hypothesis, the sodium permeability should have been increased. However, also according to this hypothesis, light should inactivate the adenyl cyclase and hence turn off the supply of cAMP. But if we have in fact inhibited the phosphodiesterase, then the intracellular cAMP concentration will remain high even after the stimulus flash, the sodium permeability should not decrease, and thus there should be little or no receptor potential. Thus, it seems that the simplest role for cAMP in producing the receptor potential cannot be correct for it is possible to obtain a receptor potential presumably without greatly changing the concentration of cAMP.

The second alternative of the cAMP hypothesis circumvents this difficulty. Here it is proposed that cAMP keeps the membrane permeability high in the dark, but it is not lack of cAMP which reduces the sodium permeability as a consequence of light. Rather, some other light initiated event (e.g. calcium release) turns off the sodium permeability. Here, the inactivation by light of the adenyl cyclase would only affect the permeability of the receptor over time periods much longer than the millisecond time range - the time course of the receptor potential.

We believe that these results indicate that cAMP does not have a role in visual excitation but may, perhaps, be involved with longer time course physiological events. This remains to be demonstrated, however.

Acknowledgement. We thank R.H. Pannbacker and D. Reed for showing us their unpublished results, and F. Ratliff for the use of his D.C. Amplifier, E. Galanter for making available some of his facilities for this work, and B. Grover for his technical assistance.

References

1. PENN, R.D., W.A. HAGINS: Signal transmission along retinal rods and the origin of the electroretinographic a-wave. Nature (Lond.) 223, 201-205 (1969).
2. HAGINS, W.A., PENN, D.R., S. YOSHIKAMI: Dark current and photocurrent in retinal rods. Biophys.J. 10, 380-412 (1970).

---

*It is possible that the changes of waveform and latency seen with caffeine and theophylline are attributable to increased cAMP. However, if this were the case then some concentration of papaverine should also cause the same effects. These effects were never seen at any concentration of papaverine. On the other hand, caffeine, and probably the structurally similar theophylline, are known to have a number of effects on calcium binding and transport in such tissues as sarcoplasmic reticulum (20). It may be that these effects of caffeine and theophylline are responsible for those receptor potential changes not seen with papaverine.

3. TOYODA, J., M. NOSAKI, T. TOMITA: Light induced resistance changes in single photoreceptors of Necturus and Gekko. Vision Res. 9, 453-463 (1969).
4. TOMITA, T.: Electrical activity of vertebrate photoreceptors. Quart. Rev. of Biophys. 3, 179-222 (1970).
5. YOSHIKAMI, S., W.A. HAGINS: Ionic basis of dark current and photocurrent of retinal rods. Biophys. Soc., Abstr. 10, 61a (1970).
6. YOSHIKAMI, S., W.A. HAGINS: Light, calcium, and the photocurrent of rods and cones. Biophys. Soc., Abstr. 11, 47a (1971).
7. BITENSKY, M.W., R.E. GORMAN, W.H. MILLER: Adenyl cyclase as a link between photon capture and changes in membrane permeability of frog photoreceptors. Proc. Nat. Acad. Sci. U.S. 68, 561-562 (1971).
8. MILLER, W.H., R.E. GORMAN, M.W. BITENSKY: Cyclic adenosine monophosphate: function in photoreceptors. Science 174, 295-297 (1971).
9. JOST, J-P, H.V. RICKENBERG: Cyclic AMP. Ann. Rev. of Biochem. 40, 741 (1971).
10. PANNBACKER, R.G., D.E. FLEISCHMAN, D.W. REED: Cyclic GMP phosphodiesterase in rod outer segments. Fed. Proc., Abstr. 30, 1267 (1971).
11. RASMUSSEN, H.: Cell communication, calcium ion, and cyclic adenosine monophosphate. Science (Wash.) 170, 404-412 (1970).
12. BOWNDS, D., A. GAIDE-HUGUENIN, A. BALL, W.R. ROBINSON, J. DAWES: Frog photoreceptor membranes: Analysis and in vitro excitation. Presented at the meeting of the Assoc. for Res. in Vision and Ophthalmology, Sarasota, Florida (1971).
13. RALL, T.W., A.G. GILMAN: The role of cyclic AMP in the nervous system. Neurosciences Research Program Bulletin 8, 221-323 (1970).
14. SILLMAN, A.J., H. ITO, T. TOMITA: Studies on the mass receptor potential of the isolated frog retina. -I. General properties of the response. Vision Res. 9, 1435-1442 (1969).
15. WITKOVSKY, P., J. NELSON, H. RIPPS: Spectral properties of the isolated receptor potential of the carp retina. Presented at the meeting of Assoc. for Res. in Vision and Ophthalmology, Sarasota, Florida (1971).
16. HOOD, D.C., A.F. MANSFIELD: The isolated receptor potential of the frog isolated retina: Action spectra before and after extensive bleaching. Vision Res. 12, in press (1972).
17. FURAKAWA, T., I. HANAWA: Effects of some common cations on electroretinogram of the toad. Japan J. Physiol. 5, 289-300 (1955).
18. MAC NICHOL, E.F., G. SVAETICHIN: Electric responses from the isolated retinas of fishes. Am. J. Ophth. 46, 26-36 (1958).
19. TRINER, L., Y. VULLIEMOZ, I. SCHWARZ, G.G. NAHAS: Cyclic phosphodiesterase activity and the action of papaverine. Biochem. Biophys. Res. Comm. 40, 64-69 (1970).
20. WEBER, A., R. HERZ: The relationship between caffeine contracture of intact muscle and the effects of caffeine on reticulum. J. Gen. Physiol. 52, 750-759 (1968).
21. MILLER, R., J. DOWLING: Intracellular responses of the Müller (glial) cells of mudpuppy retina: Their relation to b-wave of the electroretinogram. J. of Neurophysiol. 33, 323-340 (1970).
22. GRABOWSKI, S.R., L.H. PINTO, W.L. PAK: Adaptation in retinal rods of axolotl: Intracellular recordings. Science (Wash.) 176, 1240-1243 (1972).
23. BAUMANN, CH.: Rezeptorpotentiale der Wirbeltiernetzhaut. Pflüger's Arch. 282, 92-101 (1965).
24. TOYODA, J., H. HASHIMOTO, H. ANNO, T. TOMITA: The rod response in the frog as studied by intracellular recordings. Vision Res. 10, 1093-1100 (1970).

25. SICKEL, W.: Energy in vertebrate photoreceptor function. This volume, pp. 195-203.
26. CURTIS, D.R.: Amino acid transmitters in mammalian central nervous systems. Proc. 4th Internat. Congr. Pharmacol., Vol. 1, 9-31 (1969).

Discussion

M.W. Bitensky: The use of phosphodiesterase inhibitors is complicated by their nonspecific effects. The papaverine however, at 10^{-5} molar is perhaps the most specific of all since at least at these concentrations it is not inhibiting cyclase while the other inhibitors do inhibit cyclase at the concentrations (7 millimolar) in which they are used.

T.G. Ebrey: Yes, papaverine seemed to have fewer effects on the receptor potential than the other phosphodiesterase inhibitors.

Ch. Baumann: We have now reasonably good evidence about the waveform of frog receptor potentials (23,24,25). Yours look rather different as they contain a slow component which probably arises from the Müller system.

W. Ernst: Unfortunately even after aspartate treatment the trans-retinal rod P III may consist of two components: a fast one probably generated by the rods and a slow one which may be produced by the Müller cells. These two components are clearly seen in rat retina, and preliminary evidence suggests that they are both present in the trans-retinal P III of frog. It is quite possible that the trans-retinal P III which you measured reflects receptor electrical activity, but I think you should perhaps interpret your electrophysiological correlations with some caution.

T.G. Ebrey: At low intensities our recordings look like the intracellular recordings from frog (24) and axolotl (22). At higher intensities (see Fig. 2B) the aspartate isolated potentials have a slow component. Witkovsky et al. (15) have shown that in the carp this long latency, slow potential is generated proximal to the receptors. However, the early portion of the aspartate-isolated response, including the initial amplitude, is generated in the receptor layer. It is likely that the slow potential is generated by the Müller cells much like P II is thought to be (21). The amplitudes of the potentials we measured could be slight underestimates of receptor potentials due to the low frequency cut-off of the AC preamplifier or slight overestimates due to the slow potential. However, this would hardly change the results of this study.

A.I. Cohen: It seems probable that the effects seen with aspartate and glutamate would be seen with all acidic amino acids as suggested by the iontophoretic experiments of Curtis (26).

T.G. Ebrey: Probably so. In our experiments we have used only aspartate.

General Discussion

Moderator and Reporter: S. L. Bonting
Department of Biochemistry, University of Nijmegen, Nijmegen/The Netherlands

The Symposium was concluded by a 2-hr discussion of several topics, selected for their timeliness from the subjects of the Symposium papers by the moderator. He invited the participants to leave aside slides and data, and to comment, speculate and ruminate on these topics.

Chromophore Conformation and Red Shift

S.L. Bonting: We have heard about a 12 s-cis conformation in the 11-cis retinaldehyde and about twisted bonds in the chromophore. We have also learnt about a 345 nm pigment in Ascalaphus. What does all this mean for our understanding of the absorption characteristics of the visual pigments?

E.W. Abrahamson: We have made calculations of the λ_{max} by means of the Pariser-Parr-Pople method, assuming that a single negative charge is placed $4\frac{1}{2}$ Å above the π-electron system of the retinaldehyde chain and is shifted from the β-ionone ring towards the aldimine bond. We find that when the negative charge is moved above the C atom adjacent to the aldimine bond, the maximum would be at 376 nm, while the unperturbed position is 438 nm. Thus, a rather striking blue shift is possible. Another point shown by our calculations is a rather large increase in λ_{max} (26 nm) for the 3-dehydro form in a pigment absorbing around 550 nm, while this increase is only 4 nm around 475 nm, and for absorption still further in the blue region the difference increases again. It would, therefore, be interesting if Dr. Hamdorf could prepare a 3-dehydro derivative of the Ascalaphus pigment to determine the shift in that case.

S.L. Bonting: But your calculated shift of the maximum to 376 nm is not yet sufficient to explain the 345 nm pigment.

G. Wald: We should keep in mind that the possibilities of adapting the model for these calculations are almost infinite. Dr. Abrahamson assumes that the Schiff base is protonated, but if you take the proton away, you are already down to 365 nm. Then you need only a slight twist in the retinylidene chain, due to some interaction with opsin, to shift another 20 nm down to 345 nm. Where is the maximum of the first photoproduct?

K. Hamdorf: The first transition by light from the 11-cis to the all-trans form shifts the maximum to 365-370 nm, and the extinction is increased by a factor 1.75, near that for free all-trans and 11-cis retinaldehyde.

G. Wald: That's great! That would support my suggestion of an unprotonated Schiff base in this pigment. The slight twist in the native 345 nm pigment would be relieved by the light-induced isomerization and attending conformational change in the opsin, bringing λ_{max} up to near that for free retinal.

E.W. Abrahamson: The absence of protonation should be proven. This now appears to be possible by means of the resonance-Raman spectrum, as has been shown by Rimai et al. (Biochem. Biophys. Res. Commun. 41, 492; 1970) for cattle rhodopsin. One doesn't need a large quantity of the pigments for this.

M. Karplus: I want to add a word of caution on these theoretical calculations. It is not permissible to say that a particular model is correct, merely because it gives the right values of the wavelength of maximal absorption. It will be useful to incorporate extinction values and other data into the calculations.

J. Heller: To complicate matters, I should like to call to your attention the pigment, which D. Oesterhelt and W. Stoeckenius (Nature New Biology 233, 149-152; 1971) have recently isolated from halophylic bacteria and which has retinaldehyde as a chromophore. The linkage seems to be similar to that in rhodopsin, and it is also reduced by $NaBH_4$ only after some kind of denaturation. However, this pigment upon illumination does not show the spectral shift associated with rhodopsin. I feel, therefore, that it is not simply the chromophore as such which is responsible for the interaction with light, but that the protein is involved too.
(Note: The so-called bacteriorhodopsin does show a reversible spectral shift (570-558 nm) upon illumination by blue or red light, respectively. This situation is, therefore, comparable to that in invertebrate visual pigments.)

P.A. Liebman: Going back oncemore to Abrahamson's model and calculations, I believe that it does not predict the right cross-over point for the difference in λ_{max} of the retinaldehyde and 3-dehydroretinaldehyde pigments in the 435 region.

At this point a discussion about the occurrence of porphyropsin and rhodopsin in tadpoles developed.

C.D. Bridges: Is Dr. Liebman sure that the green rod pigment in tadpoles, which he assigns a λ_{max} very close to the rod pigment in adult frogs, is in fact a porphyropsin? I have found that when tadpoles (R. clamitans and R. catesbeiana) arrive from the commercial suppliers, they may have as much as 30% rhodopsin (see also: Wilt, Devel. Biol. 1, 199; 1959). This may be increased even more, if they are kept in darkness or dim illumination (Bridges, Nature 227, 956; 1970).

P.A. Liebman: The tadpole green rod pigment has a higher λ_{max} and broader absorption band than the frog green rods and the tadpole absorption peak matches the 3-dehydroretinaldehyde pigment nomogram. I found no evidence for retinaldehyde-based pigments in Rana pipiens tadpoles by the usual criteria of bandwidth and product λ_{max} nor for 3-dehydroretinaldehyde pigments in adult Rana pipiens. My specimens were kept in a well-illuminated environment until the time of experimentation.

G. Wald: Ernest Baldwin once explained an embarrassing discrepancy by saying, "Evolution has not ceased". When I first analyzed the visual pigments of bullfrog tadpoles, I found by straight extraction and $SbCl_3$ tests almost exclusively porphyropsin and vitamin A_2; and only a trace of vitamin A_1, but just the opposite in adults. It was, therefore, a tremendous surprise to me when Tom Reuter found considerable amounts of porphyropsin in the adult bullfrog (Rana catesbeiana). There seemed to be a seasonal variation. I still don't see how we missed that.

T. Reuter: If you keep tadpoles in light for a few weeks, as we did, you find almost pure porphyropsin.

G. Wald: My tadpoles of many years ago were taken right out of a New Hampshire lake and analyzed at once. They were on natural day and night shifts in the middle of summer.

Returning to the matter of the 345 nm pigment,

A. Knowles: Could Dr. Abrahamson suggest an explanation for the fact that in this case the metapigment absorbs at longer wavelength than the native pigment, while parent pigments of higher λ_{max} have metapigments at shorter wavelengths?

E.W. Abrahamson: The only suggestion I can make is that in the 345 nm pigment the opsin configuration relative to the π electron system is radically different.

S.L. Bonting: Unless in this case the native pigment is unprotonated, as Dr. Wald suggested, and the metapigment would be protonated. Why not?

E.W. Abrahamson: The unprotonated Schiff base isn't nearly as easily polarized as the protonated form. Thus, it would be much easier to perturb the protonated Schiff base than the unprotonated one.

S.L. Bonting: Yes, but we need only a very little twist or pertubation, if the 345 nm pigment is unprotonated, in order to shift the maximum from 365 to 345 nm.-Well, if there are no more comments on this topic, I would like to turn to the next topic:

Photolytic Intermediates

We now have some new ones: hypsorhodopsin and bathorhodopsin, proposed by Dr. Yoshizawa. The latter term is a new name for prelumirhodopsin, I understand. Would Dr. Yoshizawa like to say something more about this?

T. Yoshizawa: In my paper I spoke mainly about the circular dichroism of the intermediates, and so I did not have enough time to speak about hypsorhodopsin. I showed that rhodopsin, isorhodopsin, hypsorhodopsin, and bathorhodopsin are all perfectly interconvertible. I believe that the first photoproduct is either hypsorhodopsin or bathorhodopsin, or both. This poses the interesting question in which structural aspect hypso- and bathorhodopsin differ from each other.

S.L. Bonting: Do you think that hypso- and bathorhodopsin really occur in the physiological state, or are they artefacts of the very low temperatures used in your in vitro experiments?

T. Yoshizawa: Cone has shown by means of laser flash photolysis (5 psec flash, 540 nm) that bathorhodopsin is a true intermediate at physiological temperature. The same may be true for hypsorhodopsin, but we have no evidence for this at the moment.

G. Wald: Rentzepis and co-workers have shown with a laser pulse and a time resolution of 6 picoseconds that bathorhodopsin is the immediate photoproduct (Busch, G.E., M.L. Applebury, A.A. Lamola, and P.M. Rentzepis, Proc. Nat. Acad. Sci. U.S. 69, 2802; 1972). I wonder whether isorhodopsin might be the precursor of hypsorhodopsin. If this is so, then the latter substance would not be in the direct line of photolytic products of rhodopsin.

W. Sickel: Does isorhodopsin produce any electrical or psychophysical effects upon bleaching?

T.P. Williams: We have started experiments to test this. Our first attempt was to see whether we could produce isorhodopsin in the eye of the living albino rat, as we can produce it by photoisomerization in solution. We could indeed do this. Then we investigated whether the animal is able to get rid of it in the dark. Over a 24-hr period in the dark there was no loss of isorhodopsin. Apparently it stays, either until it is bleached by light or is removed by the sloughing off of discs to the pigment epithelium. Next we intend to produce a high concentration of isorhodopsin in the rat eye with a laser, and test the sensitivity of that eye compared to a control eye.

C.D. Bridges: In this connection, I like to mention that Lewin and Thompson (Biochem.J. 103, 36 P; 1967) have shown in rats, fed on a vitamin A-free diet supplemented with 5.6-mono-epoxyretinaldehyde, formation of a small amount of visual pigment (λ_{max} 467 nm). They observed that the visual threshold was lower than that in the control animals receiving the unsupplemented diet. In other words, this unnatural 5.6 epoxypigment did have a visual effect.

S.L. Bonting: May I now turn our attention to metarhodopsin III or pararhodopsin? I have the feeling that its position as a photolytic intermediate has been weakened. It seems not to be in the mainline of photolysis from Dr. Baumann's experiments, and it doesn't always appear during photolysis as Dr. Reuter showed for crucian carp porphyropsin.

T.G. Ebrey: I can add another example. Cattle rhodopsin solubilized in Ammonyx-LO does not yield metarhodopsin III, while it does when solubilized in digitonin.

G. Wald: But this is not in the eye!

T.G. Ebrey: That is right, and it may say more about detergents than about metarhodopsin III.

Ch. Baumann: I do not agree with the suggestion that metarhodopsin III is not in the mainline of photolysis. In frog we find at 20°C that about $\frac{2}{3}$ of metarhodopsin II decays via meta III, at 10°C as much as 80%. In freshly excised human eyes we find at 36°C more than 50% of the meta II decay going through meta III. This is in agreement with the results of Ripps and Weale (Nature 222, 775; 1969) for the intact eye.

K.O. Donner: On the other hand, in the crucian carp retina we find no meta III formation upon bleaching of porphyropsin, while in detergent extracts of this pigment the substance is observed.

J.J.H.H.M. de Pont*: Is in the isolated retina sufficient NADPH present to permit the normal transformation to retinol by means of retinoldehydrogenase? In other words, could a lack of NADPH, such as observed earlier by Bridges in washed rods, explain the occurrence of metarhodopsin III?

Ch. Baumann: The retinoldehydrogenase system appears to work equally well in the excised eye as in the in situ situation, hence I presume that a lack of NADPH cannot explain the occurrence of metarhodopsin III.

C.M. Kemp: Experiments with isolated rat retina, in which I have deliberately starved the retina, indicate no effect on the decay route of metarhodopsin II.

---

* Department of Biochemistry, University of Nijmegen, Nijmegen, The Netherlands; as visitor.

G. Wald: Actually we cannot exclude the occurrence of metarhodopsin III (pararhodopsin), since when you place a frog retina in sunlight the color goes to orange, a much deeper color than that of retinaldehyde. Yet there is something strange in the appearance of this intermediate. In going from metarhodopsin I to metarhodopsin II, we must assume that we lose the proton from the aldimine-N; and then, just before the retinaldehyde comes off and is reduced to free retinol, we obtain an intermediate in which the aldimine-N appears to be protonated again, as in pararhodopsin. It doesn't make good sense.

W. Sickel: I should like to point out another seemingly contradictory phenomenon. For the reduction to retinol a high NADPH/NADP ratio is required, and we find that a high ratio coincides with oxygen uptake, i. e. is energy-linked. On the other hand, the experiments of Futterman and of Rotmans indicate that the presence of oxygen hinders isomerization and regeneration. So these conditions are contradictory, and presumably in the physiological situation a compromise between them must be attained.

S.L. Bonting: Now oxygen has been mentioned, I think we are in a good position to turn to our next topic:

Regeneration

K. Hamdorf: You posed the question of the physiological relevance of photoisomerization, and I should like to comment. When you have rhodopsin (R) and metarhodopsin (M) which are capable of convertible photoisomerization and where a thermal regeneration (t) from metarhodopsin to rhodopsin exists, the system is described by the following equation:

$$[R](\lambda) = \frac{\alpha_M(\lambda)\cdot\gamma_M(\lambda)\cdot I + t}{\alpha_R(\lambda)\cdot\gamma_R(\lambda) + \alpha_M(\lambda)\cdot\gamma_M(\lambda)\cdot I + t}$$

$$R \underset{\alpha_M(\lambda)\cdot\gamma_M(\lambda)\cdot I \;;\; t}{\overset{\alpha_R(\lambda)\cdot\gamma_R(\lambda)\cdot I}{\rightleftharpoons}} M$$

where at the wavelength of illumination (λ):

$R(\lambda)$	= equilibrium concentration of R
$\alpha_R(\lambda)$, $\alpha_M(\lambda)$	= probabilities of quantum absorption of R and M
$\gamma_R(\lambda)$, $\gamma_M(\lambda)$	= absolute quantum efficiencies of R and M
I	= number of incident light quanta per R-molecule and time unit
t	= rate of the thermal reaction

It contains a temperature-dependent term for the thermal regeneration reaction. At low light intensity this reaction will return all metarhodopsin to rhodopsin, so that the rhodopsin content is virtually 100%. At high light intensities the thermal reaction component becomes relatively less important, and the photoisomerization equilibrium system dominates.

S.L. Bonting: Your comment seems to imply two points: 1) thermal regeneration is possible in the insect eye, 2) it is a matter of light intensity, which determines when photoregeneration takes over. My question is, do you feel that these light intensities are within the physiological range for these insects?

K. Hamdorf: Yes, the intensities of the sunlight, into which the insect eye has to look, are three log units higher than that required to reach photo-equilibrium in the receptor cell.

T.P. Williams: I do not see how - as Dr. Hamdorf seems to suggest - in a closed system the ratio of metarhodopsin to rhodopsin would depend on the light intensity.

It would seem to me that if one allows a longer time period at low intensity, one would still obtain the equilibrium mixture.

H. Langer: Your remark is correct, but you must consider one additional factor. The ratio of metarhodopsin to rhodopsin at equilibrium depends on the wavelength. The chemical reaction may bring back all metarhodopsin to rhodopsin at any wavelength, if the intensity is low. Photoregeneration will not do this at certain wavelengths, because the wavelength determines the equilibrium composition.

G. Wald: Hamdorf's system is similar to what we have observed for iodopsin. Illumination of iodopsin converts it to bathoiodopsin, and this can at -196°C be photoisomerized back to iodopsin. When the temperature of bathoiodopsin is raised to about -170°C, it goes back to iodopsin in the dark, i.e. by thermal regeneration.

S.L. Bonting: Having thus considered the situation for the insect eye, let us ask the question of the physiological relevance of photoregeneration in the vertebrate eye.

H. Langer: Would Dr. Reuter comment on this for the photoregeneration, which he and Donner observed in the frog eye?

T. Reuter: There is no reason to believe that the observed photoregeneration in the frog eye has any importance for vision. Here, thermal regeneration is the only physiologically relevant process.

S. Futterman: The term "thermal regeneration" is misleading. You can incubate all-trans retinaldehyde as long as you like at 37°C and it does not isomerize. What we are speaking about is a catalytic regeneration as opposed to a photic regeneration.

M. Akhtar: Let us call it simply "dark regeneration".

S.L. Bonting: More important than the name, whether "thermal", "catalytic" or "dark" regeneration, is how this process goes about. We have had some spirited discussion earlier at this Symposium between Dr. Akhtar and Dr. Rotmans, whether the retinal "isomerase" activity studied by each of them is really an isomerase or whether it is an isomerizing factor produced by bacteria, and also whether the product is rhodopsin or isorhodopsin.

W. Sickel: I want to raise a point with regard to the energy involved. In my experiments I do find extra energy uptake during regeneration in the dark, but I do not find extra energy uptake during prolonged illumination that might indicate thermal regeneration under those conditions.

S.L. Bonting: Your failure to find evidence for "dark regeneration" during illumination from your energy balance may be due to the smallness of the effect. There should be some. At any rate, it occurs in darkness. I feel there is all the more reason for the biochemists among us to search for an isomerase, which might isomerize all-trans to 11-cis retinaldehyde.

G. Wald: It is perfectly certain that in rats and frogs there is a way to convert all-trans to 11-cis retinaldehyde in the dark. The mechanism is not clear, and we should try to elucidate it. Another as yet unsolved phenomenon occurs in the lobster eye. Several years ago, we observed that all the vitamin A in the body of the lobster is located in the eye, and that it is within the limits of error completely in the 11-cis form. How do these animals keep their vitamin A store in the 11-cis form, although energetically the all-trans form is favored?

H. Stieve: With regard to the bacterial isomerization, did I understand correctly that the bacteria supply a factor, which perhaps is normally present in the retina but in limiting amount?

J.P. Rotmans: There may be several factors, which can initiate isomerization of all-trans retinaldehyde. Since we have not yet identified the bacterial factor, we cannot say at this moment whether this factor is the same as the one acting in the eye. It could still be dihydroriboflavin, which according to Futterman's experiments may be such an isomerizing factor. The partly purified preparation of our factor does not show the absorption peak of this substance, but we must keep in mind that the factor may be active in very minute quantity and thus may not show up in the absorption spectrum of the partly purified preparation. On the other hand, this factor forms 9-cis rather than 11-cis retinaldehyde, and so it is also possible that our factor may not have any relevance for the physiological isomerization process.

H. Stieve: Do all bacteria, which you tried, produce this factor?

J.P. Rotmans: Nearly every one.

S.L. Bonting: Would Dr. Akhtar like to add anything?

M. Akhtar: No. I want to go back to the laboratory and do some more experiments.

C.D. Bridges: Shouldn't we look for the isomerase in the pigment epithelium? The experiments with frog retina, which I reported, clearly show the importance of the pigment epithelium for regeneration, at least in the frog. I should add however, that cones seem to regenerate quite well in the isolated frog retina without the pigment epithelium.

S.L. Bonting: May I turn our attention to the problem of the

Electrophysiology of Rods

S. Yoshikami: asked to comment on Zuckerman's findings:

There is agreement between us on the reduction in sodium permeability of the outer segment membrane by light. The difference is that Zuckerman (Nature 234, 29; 1971) concludes from his experiments the existence of two dark current loops: one from the inner to the outer segment, as we find, and one from the inner segment to the synaptic end of the cell. The former would be driven by an electroneutral pump located anywhere along the cell membrane, the latter by an electrogenic sodium pump, located at the inner segment. His experimental evidence for this is a transient increase in dark voltage upon addition of ouabain. Ouabain inhibits both types of pump, but the effect on the current would be immediate only for the current generated by the electrogenic pump. We have repeated his experiments, using frogs and both his as well as our Ringer solutions and combinations of these. We could not duplicate his finding of a transient increase in dark voltage caused by ouabain. We observed a steady decrease in dark voltage. There was only one way in which we could obtain a voltage profile as described by Zuckerman, and that is when electrodes appeared to impale and damage a cell, causing a region of negativity to appear for 2-3 min in the cell body layer. Perhaps the leakage of potassium from the damaged cells caused local depolarization of neighboring cells and hence altered the voltage profile. However, even in this case we did not observe the transient increase in the dark voltage he describes. We cannot explain this discrepancy at present.

S.L. Bonting: Must I assume then, from what you said just now, that Zuckerman's contention that you used the wrong buffer in your earlier experiments, has been disproven by your recent experiments applying the same solutions as he used?

S. Yoshikami: Yes, and the aspartate treatment used by him has no effect either.

G. Arden: I should like to come back to the differences in results between Yoshikami and myself. I have followed his technique very closely, using the same solutions, except that I use the entire retina. However, I do not see the prominent dark voltage, which he reports, although I obtain a very prominent photovoltage. Its distribution for the outer and inner segment, determined against the position of the microdrive for the electrods, is the same as he finds. Why do I obtain different results?

S. Yoshikami: I don't think either our experimental conditions or our solutions were the same. You sometimes use plasma, and at other times you add EDTA. There is nothing that you have presented, which necessarily contradicts our findings. At the very low calcium concentrations, which you used in your experiments, the ionic conductances of inner and outer segments are very different from those in 1 mM calcium, which we usually use. Now if in bright light the ratio of potassium to sodium ion conductance (G_K/G_{Na}) of the outer segment plasma membrane is higher than for the inner segment, perhaps owing to a high G_{Na} in the inner segment at low Ca^{++}concentration, the external current will flow from outer to inner segment at the peak of the light response, as you find. We also see this in some experiments. Such results are quite consistent with the dark current-photocurrent model previously suggested by Tomita and ourselves.

G. Arden: I copied your conditions as accurately as I could! Of course, I also varied my conditions. However, even when my conditions are the same as yours, I cannot repeat your findings that the dark voltage is larger than the photovoltage.

H. Rüppel: I think I can offer a possible explanation for Dr. Arden's finding that the photovoltage exceeds the dark voltage. We must consider the space constant. In the normal dark-adapted system I calculate a value of about 20 μm, which is nearly the length of the outer segment (H. Rüppel and W.A. Hagins, this book, p. 260 (Table I)). If you lower the Ca^{++} concentration, the space constant may go down to 6 μm. As the source of the dark current is found in the inner segment, this would lower the dark voltage over the outer segment, and would under certain conditions make the photovoltage higher than the total dark voltage. This could very well happen in Dr. Arden's low Ca^{++} system. In Dr. Yoshikami's experiments the photovoltage/ dark voltage ratio decreases, when he increases his Ca^{++} concentration.

G. Arden: I appreciate your point about the change in the space constant, but I wish it would explain my results. The discrepancy is enormous, and not only under conditions where one would expect a lower space constant, but also under conditions where one would expect a high space constant. More serious yet are my observations in the presence of only 4 mM Na^+, where the dark voltage gradient should be unobservable in Yoshikami's experiments, but where I still observe large photovoltages. There seem only two ways to explain this: either the model of Hagins is wrong, or the responses do not come from the photoreceptors at all. My experiments indicate that the responses do come from the photoreceptors and that they appear to do so in the total absence of a dark current along the length of the photoreceptors.

S.L. Bonting: It appears that we cannot resolve this problem here and now, but I suggest that the three of you try to find some time to go over the matter. (Later announcement by Dr. Yoshikami: Dr. Arden and I are working things out.) Meanwhile we might turn our attention to the pump: is it electrogenic or neutral and where is it located?

W. Sickel: My experiments suggest the pump is a neutral one. I measured oxygen uptake and recorded the electrical on-response and off-response. When I lowered the temperature, the oxygen uptake went down and also the off-response, but not the on-response. An electrogenic pump is characterized by a high temperature coefficient, but the on-response has a low temperature coefficient. This means that if the on-response depends on the dark current, the latter is apparently not produced by an electrogenic pump.

S.L. Bonting: I am not sure that you may draw this conclusion, since the on-response is recorded lower down the chain. The dark current, which reflects the pump activity more directly, is inhibited very rapidly by ouabain according to Yoshikami and Hagins. This seems to me to be a strong argument in favour of an electrogenic pump. More important to me, however, is the question: where is the pump located? From the papers of Hagins and Yoshikami, I gather that they place it at the junction of the inner segment and the outer segment. When I was looking this afternoon at Dr. Cohen's slides, seeing the narrow cleft there, I was worrying all the time about this poor pump sitting there in that cleft and having to generate all this dark current from that awkward spot.

S. Yoshikami: There is nothing in our previously published findings, from which we can deduce the location of the pumps. Perhaps they are located along the inner segment, but it will take further work to be sure. The finding that ouabain abolishes the dark current, as measured with external electrodes, cannot be used to distinguish whether the ionic pumps are electroneutral or electrogenic.

S.L. Bonting: I am relieved to hear that we do not have to assume the pump to be located at the junction. May I add another question? The pump enzyme in a cell would normally be distributed over the entire membrane. Dr. Arden placed the pump in his model also along the outer segment membrane. Do you not do that?

S. Yoshikami: Against the argument that it is unusual to find that the pump is located only at one end of the cell, I could point out that the visual pigment is also located at one end of the cell! Any pump in the outer segment membrane would short-circuit the dark current, and this would be an inefficient use of energy.

S.L. Bonting: I quite agree, but the contention that it is more efficient not to have a pump there is not sufficient proof that it really is not there. The question is: does anybody know?

D. Bownds: Over the last two years we have tried to find Na-K ATPase activity in the frog rod outer segment. With the best possible fractionation method, trying to exclude microsomal and mitochondrial fractionation, we do not find a significant Na-K ATPase activity.

S.L. Bonting: According to recent results in our laboratory, I would be inclined to agree with you in part. In sucrose gradient isolation of cattle rod outer segments we find a much higher Na-K ATPase activity in the white band of higher density than in the purple band containing the outer segments. We are investigating this in more detail at the moment.

T.G. Ebrey: If there is a pump in the outer segment, and you add ouabain to an outer segment suspension, you would expect them to swell. Does Dr. Bownds see this?

D. Bownds: No, we don't see swelling upon addition of ouabain.

S.L. Bonting: That is not a conclusive argument, since cells will not swell in the presence of ouabain if there is a 1:1 exchange of Na^+ and K^+. We have observed this in several tissues. On the other hand, we and also Ostrovski and co-workers (V.F. Antonov, A.L. Afanasev, M.A. Ostrovski, I.B. Fedorovich, Biofizika 16, 78, 1971) have clearly observed a loss of K^+ and a gain of Na^+ in outer segments exposed to ouabain in the dark. However, these experiments do not distinguish between a pump located in the outer membrane and the membranes of the sacs.

H. Langer: Is there any histochemical evidence for the presence of the Na-K ATPase system in vertebrate outer segments? As Dr. Weber showed this afternoon, ATPase activity has been found in insect rhabdomeres and its existence in the eye has been confirmed by chemical assay.

J.J.H.H.M. de Pont: By means of the chemical assay we have found a high Na-K ATPase activity in Sepia rhabdomes (Comp.Biochem.Physiol. 39B, 1005; 1971). Matschinsky (in: Biochemistry of simple neuronal models, Adv.Biochem.Psychopharmacol. 2, 217, 1970) finds little or no Na-K ATPase activity in outer and inner segment strips, micro dissected from frozen-dried rabbit retina sections. However, our experiments suggest that he may have lost a significant part of the activity during the dissection procedure. We still feel that there is a significant Na-K ATPase activity in vertebrate outer segments.

S.L. Bonting: Unfortunately, the histochemical staining techniques for Na-K ATPase have so far been unreliable, since the lead ions needed to precipitate the inorganic phosphate released by the enzyme strongly inhibit its activity. Although e.g. Scarpelli and Craig (J.Cell Biol. 17, 279, 1963) have demonstrated histochemically ATPase activity in vertebrate rod sac membranes, there is no proof that this represents Na-K ATPase. Dr. Weber's results in insect rhabdomes were obtained with an improved technique, where the lead ions are complexed with Na-tartrate. Hence, this activity may really represent Na-K ATPase activity. This method should also be tested on vertebrate rods.

J.J.H.H.M. de Pont: Dr. Yoshikami was talking a little earlier about a 4-min period required for complete exchange of the Na^+ in the outer segment, and this morning he mentioned a 1-min period. What is the true figure?

S. Yoshikami: Both are correct. The larger frog rod requires 4 min, whereas the smaller rat rod needs only 1 min.

Role of Calcium

S.L. Bonting: There is now some indirect evidence that Ca^{++} ions would play a role in controling the Na^{+}-permeability of the receptor membrane. In rods Ca^{++} ions would be released by the rod sac upon illumination and go to the outer membrane, where they would close the Na^{+}-channels. More or less the reverse would occur in rhabdomes, where light appears to increase the Na^{+}-permeability of the microvillar membrane. Who would like to comment on this?

M.W. Bitensky: We have observed an ATP-dependent binding of Ca^{++} by suspensions of isolated, intact rod outer segment disc membranes (Biochem. Biophys. Acta 266, 67; 1972). We are not absolutely sure that we have eliminated all mitochondrial participation, but it is our opinion that the major part of the effect is due to the photoreceptor disc membranes. The Ca^{++} binding is completely reversed upon exposure to sulfhydryl compounds (DTE or DTT). I doubt, whether we can at this time distinguish between binding to the surface of the disc membranes and sequestration of Ca^{++} within the rod sacs.

D. Bownds: We have done similar experiments. We find that outer segment suspensions will take up only about 0.03 mole Ca^{++} per mole rhodopsin in the presence of ATP. We have 30-50% mitochondrial contamination, which explains the lowered uptake in the presence of oligomycin and antimycin. The problem in trying to do meaningful experiments is that a tiny amount of Ca^{++} bound or sequestered by the rod sacs could be physiologically significant, but would be undetectably chemically.

S.L. Bonting: I am not sure that one should say that. I could see that for the physiological effect of only a few quanta absorbed the amount of Ca^{++} bound or released would be far too small to be measured, but if one would fully bleach the amount of Ca^{++} involved would be large enough to be detected.

D. Bownds: We have tried very hard to observe a light effect, but we have been unable to do so. This, however, does not in any way rule out the Ca^{++}-hypothesis, since a very small release of Ca^{++} would be sufficient.

H. Stieve: It seems fair to say that, although the Ca^{++}-hypothesis is a plausible explanation for the elusive transmitter between rod sac and outer membrane, this hypothesis remains unproven at this moment.

S.L. Bonting: Unfortunately, technically it is rather difficult to obtain definite proof. May I ask your views on the role of Ca^{++} in the excitation of the invertebrate photoreceptor? There the Ca^{++} ions would have to function just the other way around. In the rod light would make Ca^{++} ions go to the outer membrane, where they close the Na^{+}-channels, whereas in the invertebrate receptor light would have to remove Ca^{++} from the Na^{+}-channels.

H. Stieve: I do not see a problem here. The assumption is that in vertebrates light causes a hole in the rod sac membrane through which Ca^{++} flows out. In the invertebrate microvillar membrane light also causes an opening by releasing Ca^{++} from the outer surface of the membrane. Through this hole Na^{+} and Ca^{++} could flow in.

S.L. Bonting: Yes, but we have the problem that in cones, where the Na^{+}-permeability behaves as in rods but where the outer membrane is the photosensitive membrane as in rhabdomes, the Ca^{++} ions would have to act in an opposite way as in the invertebrate receptor. Another point, which puzzles me, is that Dr. Yoshikami seems to suggest that both extracellular Ca^{++} and Ca^{++} released from the rod sacs would participate in closing the Na^{+}-channels in the outer membrane.

S. Yoshikami: One of the reasons for assuming that extracellular Ca^{++} may participate in rod excitation is that there is rhodopsin present in the outer membrane, so that this membrane may act like the cone membrane.

S.L. Bonting: Why have then all these rod sacs? They cannot be there without a purpose?

D. Bownds: The sacs constitute a large light-sensitive membrane surface, which is not in the path of the dark current but still can trigger excitation. In this way rods are much more efficient than cones, because they sequester a large area of "triggering" membrane without making that membrane a part of the dark current circuit which is driven by metabolic energy. In cone cells, on the other hand, the visual pigment containing membrane is also the membrane across which dark current is moving.

S.L. Bonting: I quite agree, but if both the extracellular Ca^{++} and the Ca^{++} released from the rod sacs can close off the Na^{+}-channels, then it seems you are losing most of the efficiency of the rod sac system. There are always plenty of free Ca^{++} ions present on the outside, while the rod sacs would keep the intracellular, extrasaccular space relatively free of Ca^{++} during the dark period by their binding or sequestering of Ca^{++} ions. If you assign the extracellular Ca^{++} ions an equally significant role as those released by the rod sacs during illumination, then the rod sacs would contribute very little in triggering excitation and the rod would behave virtually as a cone. It seems to me that we could say that in principle extracellular Ca^{++} ions could act on the Na^{+}-channels of the outer membrane, but that this action is not necessary and will not happen at low light intensities.
Would Dr. Yoshikami agree to this?

S. Yoshikami: Yes, it is not necessary to have extracellular Ca^{++} ions participate in rods.

D. Bownds: When we assume the release of a transmitter from the rod sac upon illumination, we must keep open a number of options for the mechanism. E.g. the enzyme hypothesis is still a real possibility. A fast enzyme system has a turnover of 1 to 100 molecules per millisecond. We need to generate only some 50-100 transmitter molecules per quantum absorbed. Thus, the activation of a single enzyme molecule might suffice to produce the required number of transmitter molecules.

S.L. Bonting: And the conceptual difference between an enzyme in the normal sense of the word and a rhodopsin-containing membrane system releasing Ca^{++} is rather small.

T.G. Ebrey: Has anybody ever tried to determine, whether rhodopsin itself can bind Ca^{++}?

D. Bownds: Those of us who have done Ca^{++} uptake experiments have in a sense been looking at Ca^{++} binding. The amount of binding is very small; we found a maximum of about 0.03 mole Ca^{++} per mole of rhodopsin, as I said before. Compared to other membranes or proteins this is very little. There is certainly nothing like a stoichiometric binding of Ca^{++} to rhodopsin under any of the conditions we have tried. The figure 0.03 mole Ca^{++} per mole rhodopsin works out to an average 1 mM Ca^{++} concentration inside the rod cell.

G. Wald: Is that free ionic or total calcium present, i.e. if you would ash the rod, would you find this amount of calcium or more?

D. Bownds: We don't know yet. The Ca^{++} turnover is very slow, as Cone has found with tracer experiments.

S.L. Bonting: With tracer experiments and also with atomic absorption spectrophotometry one would not be able to distinguish between free and bound Ca^{++}. To do that, one would have to use a Ca^{++} ion-sensitive electrode.

At this point the moderator had to close the discussion on account of the time. He expressed the thanks of all participants to Dr. Langer and his coworkers for making possible an enjoyable and stimulating symposium and voiced the hope that all participants would have a good return journey and that in the years to come they would see each other again to continue the discussion of the fascinating problem of vision.

Index of Contributors

Information Processing in the Visual Systems of Arthropods

Editor: R. Wehner

With 263 figures.
XI, 334 pages. 1972
Soft cover DM 36,–
US $ 13.40

Contents:

Anatomy of the Visual System. – Optics of the Compound Eye. – Biochemistry of Visual Pigments. – Intensity – Dependent Reactions. – Wavelength – Dependent Reactions. – Pattern Recognition. – Visual Control of Orientation Patterns. – Storage of Visual Information. – Methods of Quantifying Behavioral Data.

Fields of interest:

Sensory Physiology, Neurophysiology, Neuroanatomy, Behavioral Physiology, Biological Cybernetics.

It is now generally accepted that the visual systems of insects provide a suitable model for the study of information processing in neuronal networks. A European conference on compound eye systems, held in Zurich in March 1972, had as its main theme systems analysis.

However, in the case of neuronal networks, systems analysis cannot be a "black-box" maneuver, so that the speakers necessarily went over some ground already covered in anatomical, neurophysiological and behavioral studies. The first three sessions were thus devoted to light and electron microscope studies on the structure of the visual system, the optics of the compound eye and the biochemistry of visual pigments; further sessions dealt with reactions specific to various intensities and wavelengths, pattern recognition, visual control of orientation patterns and storage of visual information. All the authors attempted to coordinate neuroanatomical and neurophysiological results with behavioral studies.

The conference proceedings will give readers interested in sensory systems a greater appreciation of the advantages of using arthropods rather than vertebrates to study the uptake, transmission, and processing of optical data.

Springer-Verlag
Berlin
Heidelberg
New York
München London Paris
Sydney Tokyo Wien

Prices are subject to change without notice.

Journal of Comparative Physiology

Founded in 1924 as Zeitschrift für vergleichende Physiologie
by K. von Frisch and A. Kühn

Editorial Board

H. Autrum, München
K. v. Frisch, München
G. A. Horridge, Canberra City
D. Kennedy, Stanford
A. W. Martin, Seattle
C. L. Prosser, Urbana
H. H. Weber, Heidelberg

The increasing emphasis on the comparative aspects in many branches of biology plus the impetus derived from new findings at the cellular and subcellular level have led to a major growth of comparative physiology. Research results in molecular biology often have a bearing for comparative physiology studies dealing with more complex organisms and even for exploring ecological problems such as temperature control or the physiological control of behavior. As its broad coverage embraces new areas of investigation and the still important classical ones, this journal mirrors the growing diversification of comparative physiology.

Sample copies available upon request

Springer-Verlag
Berlin · Heidelberg · New York
München · London · Paris · Sidney · Tokyo · Wien

Prices are subject to change without notice.

MIX
Papier aus verantwortungsvollen Quellen
Paper from responsible sources
FSC® C105338

FSC
www.fsc.org

If you have any concerns about our products,
you can contact us on
ProductSafety@springernature.com

In case Publisher is established outside the EU,
the EU authorized representative is:
Springer Nature Customer Service Center GmbH
Europaplatz 3, 69115 Heidelberg, Germany

Printed by Libri Plureos GmbH
in Hamburg, Germany